Foundations
of Cyclopean
Perception

If this figure is illuminated from behind, the random-dot array on it and the one printed on the other side will form a combined image, and a text becomes visible. This presentation is not cyclopean and is included merely to imitate, for the sake of the reader who has no stereoscopic depth perception, the kind of phenomena random-dot stereograms convey.

If this figure is illuminated from behind, the random-dot array on it and the one printed on the other side will form a combined image, and a text becomes visible. This presentation is not cyclopean and is included merely to imitate, for the sake of the reader who has no stereoscopic depth perception, the kind of phenomena random-dot stereograms convey.

Foundations of Cyclopean Perception

Bela Julesz

The University of Chicago Press

Chicago and London

BELA JULESZ
is Head, Sensory and Perceptual
Processes Department,
Bell Telephone Laboratories,
Incorporated

Holders of copyright on materials
reprinted in this volume are listed in
the Acknowledgments (p. 369)

International Standard Book Number:
0–226–41527–9
Library of Congress Catalog Card
Number: 70–149594
The University of Chicago Press,
Chicago 60637
The University of Chicago Press, Ltd.,
London

PRE

To Margit

Anaglyph form of figure A

The mythical cyclops looked out on the world through a single eye in the middle of his forehead. We too, in a sense, perceive the world with a single eye in the middle of the head. But our cyclopean eye sits not in the forehead, but rather some distance behind it in the areas of the brain that are devoted to visual perception. One can even specify a certain site in the visual system as being the location of the cyclopean eye. For instance, we can locate the cyclopean eye at a place where the views of the two external eyes are combined.[1] In this case, normally we think of the information registered by the cyclopean eye as very similar to that presented to our external eyes. Accordingly, we assume that this cyclopean eye receives little more information than each of the external eyes alone obtains. All that is added, it would seem, is a somewhat richer impression of depth.

During the last decade, however, the cyclops within us has begun to collaborate with the psychologist and the digital computer. The result of this collaboration has been a series of unexpected insights into the nature of visual perception—enough, finally, to warrant being collected into a book.

What is cyclopean perception? How did we succeed in awakening the dormant cyclops? Rather than answer these questions—the whole book is devoted to that task! —let me offer an example. The square on the left in figure A looks like a random as-

Preface

A   Random-dot stereogram which when monocularly viewed appears as an aggregate of random dots. However, when stereoscopically fused (by crossing your eyes or by viewing the same display in anaglyph format), a diamond is perceived hovering over the random background. The anaglyph version of this figure is on the facing page; it should be viewed with the red and green viewer provided with the book. The red filter should be placed over the right eye in order to get the same effects as described in the text. The reversing of the filters reverses the depth effect. This is the only anaglyph version of a stereogram that is printed within the text. All other stereograms that are marked with an asterisk after the figure number have their anaglyphs collected at the end of the book following the appendix.

1. According to Hering (1879; 1964, p. 232) the term "cyclopean eye" was originally coined by Helmholtz to denote a hypothetical eye that was introduced by Hering in order to explain identical binocular directions. This hypothetical eye incorporates the two real eyes into a single entity (with two overlapping retinae) and lies midway between the two real eyes. It is literally the *center* eye of the cyclops, and the mythological allusion is very fitting since such an eye does not exist. I took the liberty of using "cyclopean" in a more abstract and yet more concrete sense than is customary. While Hering's cyclopean eye is still an external eye and merely a geometrical concept, I use the same term to denote a *central* processing stage inside the brain having a concrete neuroanatomical existence. Although I will always use cyclopean in this new way, in order to keep terminology straight and stress physiological connotations, after the Preface the term "cyclopean retina" will be adopted in place of the "cyclopean eye." For further details see chapter 1.

semblage of dots, and it is that. The one on the right is equally devoid of form. If you view one pattern at a time with one eye at a time, that is all that you can see. But something quite amazing happens when your cyclops views the figures, when each eye looks at its own square simultaneously (this you do by crossing your eyes or, equivalently, by looking at the anaglyph version of figure A with the red-green goggles that are supplied with the book).

Where your external eyes were presented with randomness, your cyclopean eye receives something new and definite. A clear-cut diamond is perceived hovering over the random-dotted background. Here there is a sharp, determinate, distinctive perception of form that must originate somewhere in the visual nervous system beyond the retinae of the eyes. As psychologists, then, we have to a certain extent managed to look into the "black-box" of the visual system: we have found out something about where a particular perceptual process takes place. This is a procedure that is seldom employed in the traditional experimental psychology of perception; there the workings of the black box typically are not localized, beyond hypothesizing that something is going on somewhere inside to produce an output from certain inputs.

Suppose that when your cyclops looked at a similar figure it saw not a diamond, but the familiar Müller-Lyer arrowheads (as shown in fig. 2.6-2*). And suppose that (as is indeed the case) it saw one line as illusorily longer than the other. Then we would know that the processes responsible for that illusion can occur farther along in the visual system than is often thought. We would again have added to our knowledge of the internal structure of the black box: the illusion occurs beyond the point where inputs from the two eyes are combined. We can even state from recent neuroanatomical evidence that this site must be at least six synaptic levels past the retinal receptors. Such probings of structure and order in the visual system are the counterpart, using purely psychological means, of the neuroanatomist's quest. At times I have thought of calling this book "Foundations of Visual Psychoanatomy!"

Although random-dot stereograms were the starting point for my research and stereoscopic vision was the source of the term "cyclopean," this book is by no means limited to discussions of a single technique or a single sort of perception. Both the technique and the concept that I label "cyclopean" have been generalized. Random-dot "cinematograms" display information to a different cyclopean eye, at the site where movement is perceived. Computer-generated complex patterns can be created that selectively stimulate some other cyclopean eyes. With these developments cyclopean perception has broadened and ramified, to such an extent that virtually all of the traditional topics in perception—and a number of the most exciting new ones—come in for mention at one point or another.

The main subject matter of this book is visual perception, although some auditory implications are given. Some findings have relevance to cognition (e.g., localization of eidetic memory, cooperative phenomena and models, perceptual learning, etc.). Others bear on purely "sensory" processes (e.g., localization of simultaneous contrast and some color phenomena). The primary subject, however, is perception.

Chapter 1 introduces the reader briefly to what the book is about. Chapters 2 and 3 review the main findings of cyclopean perception together with the necessary background material of neuroanatomy and neurophysiology. Chapters 4, 5, and 6 summarize the main advances in monocular and binocular perception obtained by the use of random textures. Finally chapters 7, 8, and 10 are devoted in great detail to experimental findings and an overview of what has gone before. All of these chapters are based on experimental evidence, while chapter 9 is more speculative, being devoted to problems of invariance and form. The only section that contains psychological evidence not directly checked by me is 7.4. It reviews a recent development, in which cyclopean techniques are used on a remarkable subject with eidetic imagery—a "photographic" memory.[2]

Most of the reported experiments are demonstrated by the aid of stereo-viewers supplied with the book. Since the necessary background material is included in the text, the book can be used to some extent by undergraduate students of psychology. However, the book is primarily aimed at mature graduate students and workers in visual perception. Mathematics is confined to a minimum and is presented without proofs, yet at an advanced level. The essence of these few theoretical sections is explained in nonmathematical terms, and the mathematically sophisticated reader is referred to the literature.

The book may also serve the clinician. The published plates can be used for unambiguous and unfakable tests for stereopsis. Early detection of stereopsis deficiency is of great importance and is similar to the accepted procedure of testing for color deficiency. There is evidence (reviewed in § 8.2) that some of the plates can reveal neurological abnormalities. Similarly, some of the techniques and stereograms might be used as powerful stimuli during the microelectrode probings of the visual cortex of the monkey.

The techniques reported here can serve the mathematician or the designer in the visualization of complex surfaces; and last, but not least, some of the plates may delight the layman or the student in the visual arts.

The study of perception has an aspect that is unique among the various branches of scientific inquiry. This aspect is that some of the main phenomena of perception can be directly experienced, and that often the implications of these phenomena are self-evident. Perhaps it is this feature (rather than the style of presentation that grew out of my training in the natural sciences) that has attracted the interest of engineers, physicists, and mathematicians to several of the topics in this book. I would be immensely pleased if this book were to bring problems of psychology closer to the natural scientist.

Finally, let me emphasize the monographic character of this book. Although I

2. The reader who is interested only in cyclopean perception (and is knowledgable in visual perception) can take the following shortcut: read only chapters 1–3, 7, 8, and 10. The reader who is interested primarily in binocular depth perception can take another shortcut: read only chapters 1, 3, 5, 6, § § 7.3, 7.5–7.8, 8.1–8.5, 9.2, and 10.3. The reader who is interested in this book mainly as a monograph may skip the following reviews: §§ 3.3, 3.4, 4.3, 4.6, 4.9, 5.2, 5.6, 5.9, 7.4, 8.5, and chapter 9. Finally, the reader who is annoyed by conjectures, speculations, suggestions, and personal opinions should skip the following material: §§ 3.5, 4.1, 4.2, 4.3, 4.9, 6.6, 6.7, 7.4, 8.5, 10.2, and chapter 9. (On the other hand, some of this material is recommended for the graduate student in search for thesis topics.)

wanted to refer to all published works that used cyclopean techniques, the rapid development of this field made even this attempt questionable, as discussed in chapter 1. With respect to the other references no attempt was made for completeness, and the selections reflect entirely my taste and limitations. They served primarily as background material, as illustrations, or as related evidence for cyclopean findings. I found nevertheless that the published findings (whether obtained in my laboratory, or elsewhere) were insufficient for presenting a coherent picture of cyclopean perception. In order to fill large gaps, I had to undertake many new experiments that are reported here for the first time.

This book could not have been written without the help of many colleagues and friends. My warmest thanks go to Dr. John R. Pierce, Executive Director of Communication Research at Bell Telephone Laboratories, whose understanding and interest for more than a decade made this research possible. Thanks also to my many co-workers, co-authors, and students whose names are included in the reference list.

Let me explicitly thank Mr. Richard A. Payne for his most skillful and broad technical assistance, ranging from instrument design to the running of experiments.

Beyond acknowledging those colleagues whose help and encouragement led to the findings reported in this book, additional thanks are due those who made the actual writing possible: Miss Rosanne Hesse and Mr. Enrico Chiarucci for programming and displaying a large number of stimuli for this book and for developing several sophisticated input-output routines, and Mr. Walter J. Kropfl for designing a computer-controlled facsimile system that greatly facilitated the transformation of outlined drawings into cyclopean stimuli. I am indebted to Dr. Charles S. Harris, Mr. Leon D. Harmon, Dr. Davida Y. Teller, Dr. Colin Blakemore, and Anthony D. J. Robertson for their valuable comments on the manuscript. Particular thanks go to Bell Laboratories' management and to Professor Hans-Lukas Teuber, Chairman of the Psychology Department of Massachusetts Institute of Technology, who made it possible for me to spend the spring semester of 1969 as a visiting professor of experimental psychology. Most of this book was written during that period and delivered to a graduate seminar frequented by many students and faculty members. As a result of their active participation, many parts of the book were rewritten, omitting most conjectures and speculations. This seminar served as a catalyst in starting a cyclopean study of eidetic memory. I am much obliged to Dr. Charles F. Stromeyer III for his first-hand reports on some ingenious experiments he undertook on an eidetiker (although I only refer to his published findings). My particular thanks go to Dr. Harry L. Frisch for helping me to develop an operator theory of symmetry perception that proved to be more than just an elegant notation. Space does not permit me to thank everyone who helped me with this venture, and I cannot sufficiently express my gratitude to my wife for her patience and encouragement. In particular, I thank her for drawing my attention to an interesting feature of the Hungarian language: a normal person is considered to be a cyclops, in that his two eyes are referred to in the singular, while a one-eyed person is called "half-eyed."

These rules, the sign language and grammar of the Game, constitute a kind of highly developed secret language drawing upon several sciences and arts, but especially mathematics and music (and/or musicology), and capable of expressing and establishing interrelationships between the content and conclusions of nearly all scholarly disciplines. The Glass Bead Game is thus a mode of playing with the total contents and values of our culture; it plays with them as, say, in the great age of the arts a painter might have played with the colors on his palette. . . . [The Game] was an exercise in memory and improvisation quite similar to the sort of thing probably in vogue among ardent pupils of counterpoint in the days of Schütz, Pachelbel, and Bach.

Hermann Hesse

The invention of counterpoint in the sixteenth century had a revolutionary impact on musical sensibilities. This polyphonic music was based on two (or more) different musical textures that produced a new melody in the listener's mind. However, although most of us would agree that a melody conveyed by counterpoint is more than merely the sum of its parts, it is certainly not independent of its constituent melodies. Thus it is understandable why, in music, no one has tried to produce from two *random* melodies a third, completely new, nonrandom melody. It is precisely this that was done with the first random-dot stereograms in the fall of 1959. With the technique of "random counterpoint" (which is free from the strict rules of musical harmony) it was possible to portray images or events binocularly or binaurally that do not exist even physically on the left and right retinae or cochleae. Such a random-dot stereogram is shown in figure 1.0-1*. Throughout the book all the stereograms will be printed as black and white displays with their left and right fields side by side. In addition, the majority of the stereograms will be also printed as red and green anaglyphs. From now

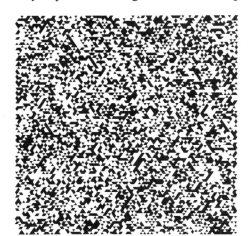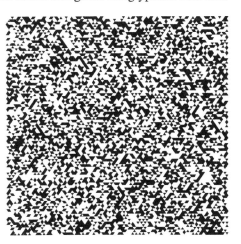

1.0-1* Random-dot stereogram composed of two arrays of black and white randomly selected cells of triangular shape. When monocularly viewed, the arrays appear as formless random textures. However, when stereoscopically fused, a large triangle is perceived above a background in vivid depth.

on in each figure reference an asterisk will be used after the figure number to indicate that the stereogram in question is printed both as an ordinary stereogram and as an anaglyph. The ordinary stereograms will be found at appropriate places in the text, whereas the anaglyphs are collected at the back of the book. For comments on viewing the ordinary stereograms and the anaglyphs see the appendix.

This complete separation of binocular percepts from monocular ones has great potentialities. Therefore, I was surprised to find, after producing the first *random-dot stereograms* on a computer (Julesz 1960a,b) that no similar attempts have been tried before in psychological studies.[1] There had been a few experiments years earlier in audition using binaural white noise as stimulus, but the purpose of these studies never was to produce a predetermined new pattern. The basic idea of the counterpoint is not simply to produce a new sensation (such as stereoscopic depth or stereophonic localization or the percept of a chord), but to use this new sensation in order to portray a new pattern that varies in the *same dimensions* as the constituent patterns. Thus if the basic melodies are developing in time, the counterpoint melody should vary in time as well. Similarly, if the monocularly presented textures are portrayed on a two-dimensional spatial array, then the binocularly perceived new pattern should occupy the same spatial coordinates.

The binocularly portrayed patterns have the same random texture as their surroundings; instead of the classical way of portraying images by brightness and color changes, here the information is portrayed by depth changes. Similar to early baroque music which avoids loudness changes (the crescendos and diminuendos) and portrays structure by pitch changes, random-dot stereograms avoid global brightness and hue changes and portray structure by depth gradients alone.

The principles inherent in such "visual counterpoint" are not restricted to binocular vision. Random-dot stereograms are a special case of a more general class of stimuli, and it is this class which I refer to as "cyclopean." Inspired by the mythical one-eyed cyclops, Hering and Helmholtz used the term "cyclopean" to refer to a hypothetical single "eye," the "mind's eye," that "sees" a single stereoscopic image

---

1. The fact that random-dot stereograms were never used before by psychologists is the more curious since the idea of such stimuli is as ancient as the notion of animal camouflage. There were many stereograms published of camouflaged scenes in which the hidden object was very difficult to trace monocularly but jumped out in the stereoscopic view. Even handmade random-dot stereograms were generated, as I learned after the publication of my first computer-generated random-dot stereograms. It is very difficult to trace the history of the first random-dot stereograms, since they might not have been published or were published in some obscure journals. As far as I know, I could trace only two such attempts. One was a photogrammetry study by Aschenbrenner (1954), who scattered an equal number of black and white paper disks over an area until it was completely covered. This textured surface served as the background. From a similarly textured surface letters were cut out and placed above the background. This display was then photographed from two views. In a study of stereo-radar systems Cutler (1956) produced a similar stereogram painted by hand. In these displays, the letters can be seen when binocularly fused. However, since these displays were not computer-generated, the minute imperfections, gaps, shadows, and texture elements that are cut in half by overlapping straight boundaries give rise to monocular cues. Thus, as in the case of animal camouflage, the monocular forms are difficult to see, but with scrutiny can be seen. Nevertheless, these stereograms enclosing animal and artificial camouflage could have been used in perceptual studies, yet were only presented as curiosities.

given appropriate stimuli in the two eyes. With random-dot stereograms it is possible to portray information on the "mind's retina"—that is, at a place where the left and right visual pathways combine in the visual cortex.

But there are other related techniques. For instance, in the technique of *random-dot cinematograms,* two (or more) random arrays presented in rapid temporal succession portray a desired image made up of elements that move in a similar way. As an example, one can portray a letter "A" (global information) by means of thousands of tiny, randomly oriented line segments that rotate clockwise surrounded by a similar array of line segments that rotate at twice the speed (Julesz and Hesse 1970). Such cinematograms share the following property with random-dot stereograms: retinal mechanisms (at least in the cat, monkey, and man) cannot process the movement information; this information is first extracted and thus portrayed at some more central level. Just as there is some locus where information from the two eyes is first combined, the cyclopean retina, so there is a locus where motion is first dealt with, and this I also call a cyclopean retina. Thus, my use of "cyclopean" is in one way broader than Hering's: it extends beyond binocular vision. On the other hand, even for binocular vision, I do not call a binocular view cyclopean unless it differs markedly from the monocular views.

The essence of cyclopean stimulation is this formation of a percept at some central location in the visual system by using stimuli that could not possibly produce that percept at an earlier location. It is as though we were operationally "skipping" the peripheral processes and stimulating some central location. Thus, cyclopean stimuli can be a powerful tool for investigating the structure and information flow in the visual system. Consider again the cinematogram portraying an "A" by rotating line segments. Since this global information does not exist prior to movement perception, we can say that with cinematograms the input is at a stage where movement extraction takes place. If we perceive the vertical bilateral symmetry of "A" portrayed by local movement gradients, we conclude that symmetry perception occurs *after* movement perception. In this way we have established a new input stage, a cyclopean retina for movement perception.

We can portray the same global information, the same "A," for example, by utilizing different perceptual phenomena to serve as local elements. Depending on whether global information is portrayed by local differences in orientation, movement, depth, vernier acuity, or whatever, we assume that this information is cast on different cyclopean retinae. A given global percept may or may not vary depending on the percept used for its local portrayal. Thus, we are able to trace the information flow. For instance, if we portray with movement differences a horizontal bar and a vertical bar of the same length and find that this display yields the same classical optical illusion portrayed by brightness gradients, we conclude that the process responsible for this optical illusion occurs after movement perception. Of course, "before" and "after" do not mean physical localizations. We are merely establishing an information flow chart. When process B utilizes the output of process A, we contend that A is before B.

We can ask about almost any perceptual phenomenon whether it occurs peripherally or centrally. In other words we can profitably "repeat" most of the experiments of classical visual perception using cyclopean stimulation. If the cyclopean phenomena are identical to the classical ones, their central origin has been established. If, on the other hand, a classical phenomenon cannot be obtained under cyclopean stimulation, one can conjecture its peripheral origin until it is proven to be central. Finally, if the cyclopean stimulation produces entirely new, hitherto unobserved phenomena, one can conclude that central processes are at work that were either not adequately stimulated by the oversimplified classical targets or had been overshadowed by the powerful peripheral processes.

For those readers who like the counterpoint analogy for random-dot stereograms, the same analogy can be extended to random-dot cinematograms. As we know, musical counterpoint is not restricted to the "beating" of two melodies into a third one, but also covers techniques that modulate certain rhythmic patterns in order to produce a new one, a technique called syncopation. Nevertheless, the way successive random arrays produce a desired form by velocity gradients is only faintly reminiscent of syncopation. On the other hand, random-dot stereograms are literally the visual analogues of the musical counterpoint.

The study of both binocular depth perception and visual perception in general have benefited from the methodology of cyclopean stimulation. In addition to simplifying and clarifying many classical problems of binocular depth perception, this paradigm has led to new questions and has revived the interest of many researchers in stereopsis. Part of this book is devoted to findings in stereopsis by White, Kaufman, Marlowe, Lawson, and Gulick, to mention only a few researchers outside my laboratory who have used random-dot stereograms.

The techniques of random-dot stereograms and cinematograms provide us with novel insights into stereopsis and movement perception. Particularly important is the realization that contrary to previous belief, binocular vision is simpler than monocular vision inasmuch as the formidable problems of semantics can be avoided to some extent. The demonstration that binocular localization is a much earlier process than form recognition persuaded several neurophysiologists to study binocular problems first. Because of the rapid growth of research using these techniques and their diverse applications, ranging from testing for brain lesions to automatic contour-map plotting, there is a genuine need for collecting, reviewing, and evaluating all these results in a single book.

Were it just such a summary, I would never have been able to persuade myself to undertake the tedious task of writing a book. However, there were two other challenging tasks that I wanted to pursue for some time. I decided to combine these efforts with book writing, and what is more, regard them as the two main themes of the book. One of these objectives was merely a hunch. I felt that as a result of random-dot stereograms and a century of research in stereopsis, ample evidence had been collected to formulate a model for stereopsis. I hoped that prolonged and undisturbed concentration might reveal some common structure underlying the many seemingly unrelated known phenomena. These efforts resulted in the spring-coupled dipole model of stereopsis which is published here for the first time. The model is simple enough to be grasped without mathematical background, yet is sophisticated enough to exhibit multiple stable states that characterize stereoscopic phenomena and help to explain many other observed results. Chapters 5 and 6 summarize the main findings of stereopsis and build up the model. The model has its weaknesses but can be further improved, and testing some of its predictions might lead to new experiments.

My other motivation, and by far the more important, was to use random-dot correlograms—much beyond the study of stereopsis and movement perception—for the investigation of visual perception in general. The first step in this direction was the proposal by Papert (1961—just a year after the technique of random-dot stereograms was published) to use this technique to study optical illusions and to differentiate between peripheral and cortical processes. Hochberg (1963) studied a few optical illusions portrayed by random-dot stereograms and also reported that a Necker-cube similarly portrayed can be reversed in depth, but these findings were never published. Since no one pursued these studies further (perhaps because of the expensive technology) I became interested in exploring visual perception by this method in a systematic way. Particularly, I wanted to "repeat" many perceptual phenomena by portraying the necessary stimulus on a "cyclopean retina" much beyond the anatomical retina. Furthermore, I became interested in the question of whether one could go beyond the separation of retinal and cerebral processes. In several cases it became possible to take two perceptual phenomena, choose one to be the local elements of the second phenomenon and be able to evoke this second phenomenon by the use of the first phenomenon. The available computer-controlled displays permitted the generation of such complex stimuli. We have already discussed how such stimuli can help us to trace the information flow in the visual system. During 1969 and 1971 many such experiments were tried, and the more interesting ones are demonstrated here for the first time. If this monograph could convey to the reader that random-dot stereograms and cinematograms are not restricted to the study of stereopsis and movement perception but can aid in the localization of perceptual processes in general my efforts will have been well rewarded.

Some of the most interesting of the cyclopean phenomena are those that are entirely new, that simply cannot be observed with the usual sorts of stimuli (Fender and Julesz [1967], Julesz and Spivack [1967], Julesz and Payne [1968]). Shipley (1965) demonstrated a new cyclopean closure effect. Julesz and Johnson (1968a,b) produced ambiguous stereograms portraying two (or more) general surfaces out of which only one surface can be perceived at a given time. Blakemore and Julesz (1971) obtained a cyclopean aftereffect that is genuinely three-dimensional.

Such phenomena could be evoked only during the "eclipse" of the powerful monocular cues. The astronomer often waits for the eclipse of the sun to observe a weak celestial object otherwise overshadowed by it. He is usually not interested simply in taking measurements on a certain weak star or planet. Rather, he is trying to check on some basic rules (such as the predictions by general relativity theory of the motion of the perihelion of Mercury) that cannot be studied during everyday conditions. Similarly, cyclopean research may contribute not only some second-order perceptual effects that have been overshadowed until now by peripheral processes, but also a deeper understanding of more familiar perceptual phenomena as well.

The basic theme of the book is the tracing of the information flow in the visual system, particularly the separation of peripheral (retinal) processes from central (cerebral) ones. This inquiry is as old as psychology itself. From binocular color mixing to binocular transfer of aftereffects, researchers classically have used binocular phenomena to tell retinal processes apart from cerebral ones.

However, *dichoptic* methods, so often used by workers in vision in order to separate cerebral processes from peripheral ones, are usually inadequate for this purpose. Dichoptic methods consist of presenting two different stimuli to each eye, respectively, in order to achieve the percept of a combined binocular stimulus. In those few instances when the outcome is positive, (i.e., the two part-stimuli combine into a whole-stimulus and yield the desired perceptual phenomenon), the conclusion that a cerebral process is acting is well-founded. On the other hand, the literature abounds with negative dichoptic results which are interpreted as evidence for peripheral processes at work. Such conclusions, however, are most questionable. Since the essence of dichoptic stimulation is the difference between the left and right views, binocular rivalry is always present. Because of binocular rivalry the two eyes' views do not combine binocularly, and the failure to obtain a certain perceptual phenomenon is more likely to be the result of this fact than of the possibility of a retinal process operating. Only random-dot stereograms can portray a binocular combination of monocular stimuli without rivalry. Therefore, all reports that are based on negative dichoptic findings should be retested.

In addition to binocular techniques several psychological and semipsychological methods have been utilized for process localization. For instance, perceptual phenomena that transfer from one retinal locale to another (such as eidetic images that do not change position during scanning movement of the eyes) are certainly not retinal. Another powerful technique is the pressure blinding of the retina during the observing of an aftereffect. If this manipulation does not diminish the aftereffect, its central origin is fairly well-established. Electroretinograms and cortical-evoked potentials correlated with perceptual phenomena show great promise for process localization.

Cyclopean techniques are a useful addition to these already-existing methods. The perceptual phenomena studied in this book are primarily those for which the above listed localization attempts failed, but cyclopean techniques were successful. In telling retinal and cerebral processes apart, cyclopean techniques represent one class within several other procedures. However, while the other methods stimulate both retinal and cerebral processes simultaneously, cyclopean techniques permit us to, as it were, skip the early processing levels and portray the desired information only at a central level. This contributes to a much finer localization than is possible by classical techniques.

The separation of retinal and central processes may be regarded as of rather little importance by some researchers. After all, the decorticate monkey exhibits very limited "perception" if any. But who knows what serious defects result from the ablation of large brain areas. Furthermore, with increased sophistication it is not at all immaterial whether the site of certain inhibitory effects, such as contrast enhancement, is retinal or not. Even if retinal processes were shown to be of limited interest for form perception, cyclopean techniques permit us to determine whether a certain perceptual process occurs after global stereopsis (or global movement perception) or before. With increased evidence of global stereopsis occurring not prior to Area 18 in the monkey (Hubel and Wiesel 1970), many perceptual processes that occur in simple and complex units in Area 17 (and even perhaps higher) can probably be bypassed (operationally) by cyclopean stimulation, and their contribution to perception can be assessed!

This, then, raises the question whether cyclopean perception is really based on purely psychological foundations. Of course, the stimulation of a certain cyclopean retina is achieved by strictly psychological means. However, the site of a selected cyclopean retina cannot be guessed by psychological methods. As a matter of fact, lack of knowledge about neuroanatomy or neurophysiology may lead to trivial results. For example, in the rabbit movement extraction is a retinal process as shown by Barlow and Hill (1963), and the cyclopean retina for movement coincides with the anatomical retina. On the other hand, I already mentioned that for the cat and monkey movement extraction occurs in the visual cortex as revealed by Hubel and Wiesel (1962, 1968). That for apes and humans movement is also cortically extracted is almost a certainty; thus by using random-dot cinematograms we are able to skip several synaptic levels by purely psychological means and display the information at a cyclopean retina located in the cortex.

Let us note that several classes of cyclopean stimuli contain the global image as conveyed physically by the local features (orientation, movement, etc.) in the stimulus; only their extraction is beyond the capabilities of the peripheral mechanisms. However, random-dot stereograms do not contain, *even physically,* the global information in the left and right retinal projections. It is only the *relation* between the left and right patterns that produces a pattern of the desired kind.

Some readers, who have backgrounds in audition or cognition, are accustomed to diffuse interconnected systems and might be irritated by my frequent usage of words "before" and "after" in denoting relations between two perceptual processes. I think these readers miss two important points. First, the neopathway (geniculate-cortical pathway) of the visual system (believed to be the primary site for perceptual processes) is a highly directional system with well-specified connections. Second, and more importantly, "before" (or "after") are merely definitions, for one process is being utilized by another one (or vice versa). They have analogous meaning to neuroanatomical usage in which certain neurons are said to synapse on others. When histological studies of neural degeneration trace the connection of fibers from one site to another, or electron-micrographs reveal the presynaptic or postsynaptic sites of a junction, then the direction of the information flow is established between two neurons or two aggregates of neurons. Whether additional complex feedback loops exist between these structures is another question and can be separately traced. We will discuss these problems in detail.

The experimental methodology of cyclopean perception is purely psychological, yet its background and quest are neurophysiological. This is exactly what is traditionally regarded as physiological psychology. But whereas physiological psychology is usually a passive discipline that tries to explain psychological findings by physiological evidence, cyclopean perception is an active discipline that can tell the neurophysiologist where to search (or not to search) for a certain perceptual process.

In spite of encouraging results, cyclopean perception is a very new branch of psychology; and this book, I hope, reports only its beginnings. The stimuli necessary for this research could not have been generated twenty years ago, since only a fast digital computer and high-resolution displaying device can "play" our counterpoint melodies. Yet to reach virtuosity in real-time we must wait for the arrival of an order-of-magnitude faster computer with larger random-access memory. Those who do not believe in this dependency on technology and would cite the works of pointillists, particularly the paintings of Seurat, overlook a practical matter. Whether the execution of an image requires years of effort or can be programmed and displayed in a few hours is more than a mere quantitative difference. In addition, disregarding the artistic beauty, Seurat's density-modulated color textures often appear simple compared to the millions of dots with complex topological and statistical properties that are used in our random-dot stereograms and cinematograms.

The pointillist paintings of Seurat are not genuine cyclopean stimuli since the global information can be retinally processed. Nevertheless, the global information is not directly portrayed by brightness gradients but by thousands of local brightness variations whose average value is sensed. Such a separation between local and global information is a necessary (but not sufficient) characteristic of cyclopean methodology.

In spite of the fact that some of the artistic creations like musical counterpoint and pointillist paintings are only superficially analogous to cyclopean techniques, they nevertheless inspired the founding of cyclopean perception. Since time immemorial artists have used complex local percepts to portray global messages, particularly composers and poets, and this is the basic methodology of cyclopean perception, too. If the artist fails and is unable to portray his global message by the chosen constituents, his style degrades to empty numerology—a fad in art of using imperceivable mathematical relations. Similarly, if the scientist fails in portraying a perceptual phenomenon by the use of some other phenomenon, he cannot determine which of the processes responsible for the phenomenon occurred before the other. Thus both artist and scientist require a certain virtuosity in finding perceptual or cognitive phenomena that evoke more complex phenomena. Recently painters returned for inspiration to visual perception and used many visual phenomena from moiré patterns to vernier acuity perception. Perhaps some of the reported phenomena in this book might also be used by the artist.

I expected to end the introduction right here. Lately, however, I became aware of some misconceptions about cyclopean notions, which must be clarified as early as possible.

One such erroneous view is that "cyclopean" is just another name for "central."

Although it is true that a cyclopean retina refers to a central site,[2] the converse is false. A central site is regarded as cyclopean, if—and only if—it can be selectively stimulated. For the neurophysiologist, who can place his electrodes in every central location, central would be equivalent to cyclopean. However, for us psychologists, there are only a few cyclopean retinae that we can selectively stimulate by purely psychological means. As neurophysiologists continue to discover an increasing number of hierarchical feature extractors, the number of cyclopean retinae may also increase. For instance, one can exploit the neurophysiological finding that neural units exist in the cortex that detect perpendicular edges. Thus, a display can be generated that contains tiny needles in all random orientations, except for an A shaped area that is covered by needles which are perpendicular to their nearest neighbors. Of course, this display is cyclopean only if the letter "A" can be perceived as a distinct entity in the background of nonperpendicular needles. In other words, a cyclopean retina must have both perceptual and neurophysiological significance.

Another misconception is that the cyclops is really the old homunculus in disguise. Usually, the homunculus is construed as an active perceiving agent (the mysterious "little man" in the head) that "sees" the same things as we do. On the other hand, a cyclops of a given kind is simply a physiological mechanism that responds to certain stimulus features. In other words, a cyclops is an aggregate of neural feature extractors of the same type. Until the existence of these feature extractors is suggested by perceptual evidence alone, the cyclops is merely a hypothetical entity. However, when these feature extractors are actually found and localized by neurophysiologists and neuroanatomists, the cyclops becomes a reality. Thus, if we select proper stimulus features for the portrayal of some perceptual phenomenon, we can prevent our cyclops from metamorphosing into the enigmatic homunculus.

This brief introduction to cyclopean thinking is followed by a chapter that is devoted to the same task but elaborates it in greater detail. It is only after reading through chapter 2 that the essence of cyclopean methodology becomes clear. The reader who is interested only in methodology might well stop there. But those who are also interested in the obtained results or want to be experts in cyclopean techniques may regard even chapter 2 as introductory.

Let me conclude this introduction with a personal note: While writing the book it happened that in my immediate environment several discoveries were made using cyclopean techniques that forced me to rewrite some sections. Had I regarded this book as merely a summary of known material, such late modifications would have only amplified the existing opinion in the futility of writing a scientific book in the second half of the twentieth century. However, since I regarded the writing as a

2. Workers in peripheral processes might question this restriction of cyclopean to central. They might argue, for instance, that contrast detection by bipolar cells in the retina uses the output of several receptors and with respect to these receptors, the bipolar layer is a cyclopean retina. This comment is valid and I would be very pleased if cyclopean thinking would be of use to "peripheralists." However, this book is on perception, written by a "centralist," and cyclopean will refer to central sites.

scientific venture, the more unexpected and significant these discoveries were, the less I minded modifying my thinking and permitting these results to have an impact on the final form of the book. Who knows what its content might be were it written a year from now!

Some of these new developments were expected and were communicated to me just in time to be incorporated in the book in a coherent way. For instance, Edward Bough (1970) was able to train monkeys to respond to changing depth cues in random-dot stereograms, thus providing the missing link between animal neurophysiology and human psychophysics. The recent finding by Hubel and Wiesel (1970) that in Area 18 of the macaque monkey complex and hypercomplex binocular disparity sensitive units exist that are absent in Area 17 is important. This means that the cyclopean retina for stereopsis must be at least on the sixth synaptic level or even higher. Some other findings were less expected, even surprising, but not contrary to certain assumptions. Such was the report by Thomas Bower (1968), who observed that newborn infants would stop their scanning eye movements and stare at an elevated center square in a random-dot stereogram.

There was, however, an extremely intriguing discovery which if substantiated might be regarded as a most interesting application of random-dot stereograms. This development took place during my sabbatical stay at the psychology department at M.I.T., where I gave a weekly seminar to graduate students and faculty members on the first version of this book. My audience received the ideas and experiments of cyclopean psychology with active interest, though regarded one aspect of random-dot stereograms—the unfakable test for stereopsis—mainly as a theoretical point. (I may add that all of my students possessed good stereopsis.) The importance of such an unfakable test, however, was most dramatically demonstrated one day, when two participants in my seminar, Dr. Charles Stromeyer and Mr. Joseph Psotka (from Harvard) surprised all of us with the following report. They said they showed one of the random-dot stereograms of tens of thousands of picture elements to one eye of a twenty-three-year-old eidetiker, who presumably cortically stored it and days later could recall this image and stereoscopically fuse it with the corresponding array which was then presented to her other eye. This eidetiker could report the cyclopean forms correctly and trace their outlines with great accuracy. The experiments were carefully executed, and often the investigators themselves were unaware of the hidden cyclopean forms contained in the stereograms (which were freshly generated for this purpose). some of these first results are now published (Stromeyer and Psotka 1970; Stromeyer 1970).

The importance of these findings, particularly that a long-term, detailed texture memory can exist and that the site of eidetic storage occurs before stereopsis, made it imperative to include them in this book. I have devoted § 7.4 entirely to these results. The contrast between this section and the rest of the book is remarkable. While in my own research I tried to share my findings with everyone by demonstrating most of my results, here I am referring to some incredible feats that were obtained on a single and extremely rare individual. These findings merely confirm the many earlier reports and anecdotal stories of earlier investigators. The only new factor is the objectivity of the test and the possibility of quantifying the capacity of this memory for the first time and locating its site. For the reader who remains skeptical I can only advise him to regard § 7.4 merely as a description of an objective method that would permit him to test eidetikers, were he ever to encounter one!

The material reviewed in § 7.4 has been carefully selected. Only eidetic feats under strongly cyclopean stimulation were included. On the other hand, there are a great number of remarkable feats which this eidetiker reported under noncyclopean conditions and which are dependent on the naiveté and introspective abilities of the subject. Since in case there were some technical oversights in conducting these experiments, these incredible reports could harm physiological psychology; and so I refrain from mentioning them. These results should await publication until verified on several other eidetikers. The only exception I take on this self-imposed ban is to report very briefly a series of introspective findings in which eidetic imagery is used as a new cyclopean technique. As such, they can be regarded as merely hypothetical examples for didactic purposes in order to help in mastering cyclopean thinking. But more importantly, none of these eidetic-cyclopean findings were ever in conflict with the cyclopean results obtained by using random-dot correlograms.

These findings have such fundamental implications to perception that they could not be omitted from this book, in spite of the fact that, as of now, no other eidetikers with similar skills have been found. The potential significance of these findings does not mean, however, that I have more reason to believe these reports than the reader who has access now to the same information that I had obtained from firsthand communications. That the technique of random-dot stereograms played an important role in these findings and the thrill of learning about these results as they unfolded is irrelevant to the fact, namely, that I did not verify these findings myself! Even if these reports were further substantiated, until a few other eidetikers of similar talents are found, the existence of a detailed eidetic memory is still questionable.

Besides stereopsis and psychoanatomical studies, there are several sections on random textures. In all these investigations, including eidetic memory, problems of semantics could be skipped, and as a result of this simplification certain insights could be gained. However, I felt that a book on perception which did not mention semantic problems might be too limited in its scope. Because of this, I included § 4.9 and chapter 9 which are less coherent than the rest. Probably the reader can find better review articles on these complex problems of attention, invariances, multidimensional scaling of form, and so on, in the literature. I was seriously tempted to take out these questionable sections which belong rather to a textbook than a monograph. Yet, it is these complex, often ad hoc processes of form recognition which characterize the field of perception. They give some introduction to the even more complex and enigmatic problems of cognition. The sudden break in the quality of the book in these sections primarily reflects another level of hierarchical complexity. I had another impetus to retain these sections. Originally, I outlined this monograph to cover visual processes that do not require memory. I was convinced that memory storage is a highly central process; and, by excluding memory, all the semantic problems of form recognition will be outside the scope of my interest. Then came the report that detailed memory can exist before stereopsis! With this possibility, memory suddenly became a subject matter of my book. After this realization, I decided to retain the sections on higher perceptual processes in spite of all their limitations.

During the time which elapsed between submission of the manuscript to the publisher and receipt of the galley proofs, another important insight was gained that was incorporated in the book but not to the extent it deserves. For two years I was aware of mounting evidence of spectrum analysis occurring in the visual system mainly as a result of the work of Campbell and others, particularly Blakemore and Sutton (1969) and Pantle and Sekuler (1968c). However, all these phenomena were demonstrated at threshold, and the effects obtained were relatively small. Because of this, I was somewhat skeptical; the more so since the decomposition of visual scenes into periodic gratings appeared rather remote from our everyday experience. Yet, recently, we were able to verify these findings at levels 30 dB above threshold or even higher, using one-dimensional masking noise (Julesz and Stromeyer 1970). In the light of these results there is indeed a "critical band" in vision; and, if the masking noise does not fall into the critical band around the frequency of a sinusoidal grating, the grating remains visible in spite of any amount of masking noise energy. Using filtered noise, we were also able to mask the fundamental of a square-wave grating such that its third harmonic became visible. These findings are of greatest importance not only to visual perception but to the basic theme of this book as well. The selective masking of certain spectral components by filtered noise gives rise to a new cyclopean technique. One can present a stimulus that is perturbed beyond recognition by some gratings. However, it is possible to apply properly filtered noise that will mask the perturbation gratings yet have much less effect on the original stimulus. Thus, at the site of this neural spectrum analysis one can make an image reappear that prior to this stage was completely hidden. For details, see § 3.4 and particularly § 10.2.

Throughout the book I refer to neurophysiological evidence, but rather cautiously. I am well aware that in spite of the startling findings in neurophysiology (obtained in recent years by single-neuron recordings) their usefulness to perception is greatly handicapped by the fact that most of these findings can be related only to local perceptual processes. Some of the simplest percepts, such as that of a dot, edge, slit, or spatial periodicity and phase, are mediated by unexplored global processes and cannot be explained by the neural-feature extractors as described by Kuffler, or Hubel and Wiesel.[3]

However, the technique of random-dot correlograms brings stereopsis and movement perception closer to neurophysiological treatment by separating local and global perceptual phenomena. Local stereopsis can now be related to the recently discovered binocular-disparity-sensitive cortical units. Furthermore, cyclopean experiments have led to a global model of stereopsis and have thus predicted the structure

---

3. For instance, these "bar detectors" cannot discriminate between two physical bars that have different widths when the narrower bar has greater contrast. A *real* bar detector is a global unit that evaluates the outputs of several overlapping simple units. If the number of these simple units is small, then the number of discriminable brightness-transient shapes is severely limited. An inability to discriminate between such transients has been demonstrated in humans by

of global stereo units. The correctness of this prediction remains to be verified by the neurophysiologist, but without this help from the psychologist it is unlikely that the neurophysiologist will be able to find global perceptual units with increasingly complex properties.

This personal note was intended to serve as an epilogue, since it could not have been written before the completion of the book. However, I decided to place it right at the beginning, since these comments will help the reader to understand the structure of the book.

After the essential ideas of cyclopean perception were formulated, my greatest problem was to find the most effective way in which the many new notions and results could be presented. Often the best didactic order is the historic one: trying to preserve the sequence of ideas as they evolve. I was seriously tempted to follow this order and start the book with the physiological and psychological facts about binocular depth perception, then introduce the technique of random-dot stereograms and review its impact on the study of binocular vision, and, after this, to formulate a model of binocular depth perception which can be easily generalized to encompass movement perception—this would naturally have led to random-dot cinematograms and would have gradually broadened the horizon until the reader would have been drawn into cyclopean thinking.

Instead, I started with cyclopean perception and tried to lay a broad foundation by using analogies from many disciplines, including music. I wanted to impress upon the reader that cyclopean methodology is far more general and useful than just one technique for studying stereopsis. Furthermore, cyclopean perception is founded on such a solid basis that at times one can afford to relax and play the "bead game," when deeper insight is gained into one's own specialty by finding analogies between various scientific disciplines and arts, than by quoting concrete examples from this very specialty.

Julesz (1959), who reprocessed familiar photographs by a computer such that all the fast brightness transients (regardless of their shapes) were substituted with straight lines without affecting perception. Similarly, a simple unit will fire optimally for a spatial grating whose half-cycle matches the width of its excitatory region, but this cannot explain the perception of spatial periodicity and phase. Here, again, several adjacent simple units might form a global unit that could extract periodicity and phase information. That a spatio-temporal-coherence principle might cause summation or interference in complex units depending on the spatial position (phase) of two bars is suggested by D. A. Pollen, J. R. Lee, and J. H. Taylor (*Science,* in press). They observed that complex units in the cat's cortex changed their latency of firing as a bar changed its position across their receptive fields. If it turns out that these authors are right and the cortex uses, in addition to a most intricate spatial mapping, phase-coherent temporal pulse trains, then the complexities of its structure would increase by orders of magnitude.

2 Foundations of Cyclopean Perception

2.1 Human Visual Perception and the "Cyclopean Mind"

Cyclopes allegedly were not very smart creatures but were possessed of a keen vision mediated by an eye in the center of their foreheads. When we restrict our attention to human visual perception alone and ignore the higher cognitive processes together with the peripheral sensory processes, this part of the mind is essentially cyclopean. Indeed, this cyclopean mind is a giant since the great majority of all the neural input of our nervous system enters into it. It is also a simpleton, incapable of the symbolic manipulations so essential in languages, logic, and mathematics; and it lacks the ability of abstraction. Furthermore, it possesses a concrete, vivid imagery unaffected by abstract thoughts. This book is restricted to human visual perception by the cyclopean mind. One might hope that such a simplified mind is easier to study, and some of the enigmatic perceptual processes can be clarified.

Minds with limited cognitive abilities are represented by the speechless hemisphere of split-brain patients of Sperry (1964) and by some idiots or idiot savants and the famous mnemonist described by Luria (1968) whose perceptual capabilities much surpass our own in concreteness, accuracy, and speed and are often accompanied by eidetic memory. Furthermore, the cyclopean eye can be directly stimulated at some cortical level, bypassing the many peripheral processes that transform and modify visual information. Such a project has been recently undertaken by Brindley and Lewin (1968), who implanted a few dozen gross electrodes in the visual cortex of a blind woman and were able to evoke spatial patterns of visual phosphenes. Other examples of visual phenomena that occur without retinal stimulation are the mental images in dreams, hallucinations, and hypnotic states.

At this point I must quickly assure the reader that this book will not deal with bizarre phenomena of hypnotic suggestions nor will it survey findings on hypothetical split-brain patients with cortically implanted electrodes. On the contrary, it reports findings that were obtained by classical methods of experimental psychology on normal human subjects. "Classical methods" refers to the presentation of images under usual binocular or monocular retinal stimulation. However, the stimuli themselves belong to a very special class which will be called "random-dot correlograms," a stimulus technique I devised in 1959; and one subclass, called "random-dot stereograms" (Julesz, 1960a,b), has been used vigorously since then in perception by me and by others.

These stimuli have two essential characteristics. First, the random textures used deprive the subject of all familiarity cues; therefore, the cognitive processes can be made inoperative. Second, these stimuli individually (i.e., when monocularly presented) do not contain any global information on the retinae. It is only at some central level, on the cyclopean retina where two or more retinal images are combined to portray a cyclopean image. This cyclopean information is conveyed by some specified correlations such that the central perceptual processes can extract them yet stay hidden from the peripheral processes. Thus the revelation of a cyclopean mind is not the result of drastic surgical intervention, but of strictly psychological techniques. Before we discuss in detail how such random-dot stereograms, cinematograms, and other cyclopean stimuli are generated, let us investigate their usefulness for perceptual studies.

2.2 Separation of Peripheral and Central Processes

We still lack a physiological psychology of human visual perception. One of the greatest obstacles to its attainment is our inability even to guess the neural levels where certain perceptual phenomena might occur. Any hint that might elucidate the peripheral or cortical origin of a psychological phenomenon has great heuristic value. After all, there are at least four synaptic junctions in the visual pathways from the receptors to the first stage of the visual cortex, and at each synaptic level each neuron can be connected to thousands of others. If it were possible to bypass three, four, six, or even more synaptic levels and portray the desired information at some cortical level, many neurophysiological and psychological problems could be simplified.

For instance, it is fairly well-established that the left and right visual pathways remain separate at the lateral geniculate nucleus levels and are first combined at some level of the visual cortex. The work of Hubel and Wiesel (1962, 1968) on the cat and the monkey indicate that the majority of cortical cells respond to binocular stimulation in Area 17 of the cat and in Area 18 of the monkey. Recent work on the cat by Barlow, Blakemore, and Pettigrew (1967) and by Pettigrew, Nikara, and Bishop (1968) indicates that some of these cortical units respond to various binocular disparities and, as such, can be regarded as the first stage of stereoscopic depth perception. Hubel and Wiesel (1970) showed similar binocular disparity sensitive units in Area 18 in the monkey cortex. Bough (1970) demonstrated that monkeys actually can perceive depth in random-dot stereograms. Fox and Blake (1970) reported behavioral evidence for stereopsis in the cat and are planning to use random-dot stereograms to be sure that no other cues will be used inadvertently.

It is almost certain that for humans random-dot stereograms portray information (in the form of changes in depth) at this cortical level, if not higher. Since in psychology we cannot open the "black-box," the exact location of some of this processing must wait for the knife and the probing of the neurophysiologists. However, it is possible to "repeat" all the experiments of textbook psychology which do not study brightness and color for their own sake, by using random-dot stereograms or other cyclopean stimuli. As I discuss in the following sections in great detail, the outcome of an experiment under cyclopean-stimulation may or may not be identical with the classical results. If it is identical, we have established that the perceptual phenomenon in question is the result of central processes. If the cyclopean phenomenon quantitatively differs from the classical counterpart, one may assume that also peripheral processes are at work. The more our neurophysiological knowledge improves, the more sharply we can make this distinction between peripheral and central perceptual phenomena. In return, the psychological screening for central processes and their rough localization may help the neurophysiologist to decide what to look for.

Cyclopean stimulation is not restricted to stereograms and binocular vision. It can be extended to any stimulus pair, even of different modality, as long as the central nervous system has the ability to combine them and extract the inherent relations which were purposefully introduced. For instance, a person with strong synaesthesia (the ability to experience cross-modality couplings) might perceive certain colors for certain sounds of various pitch values. He might even experience an enhanced color sensation if simultaneously with color the "right" pitch is sounded (Luria 1968). Provided we had found such a subject, or perhaps produced one by hypnotic suggestion (Blum 1960), it then would be possible to present a random sequence of colors to his eyes and a random sequence of tones to his ears which in unison would convey a predetermined stimulus pattern in space and time. This information would be the correlated pattern between color and tone, existing only for the "cyclopean eye-ear" at the cortical level where the two modalities combine. Thus the cyclopean stimulus can be regarded as a generalized counterpoint, the two stimulus patterns being independent "melodies" and the combined pattern ("melody") completely independent of the individual ones. It is the hidden relationship between the two information channels that conveys the third message.

The previous example was meant only for illustration. We will stay strictly in the domain of vision. But after this fundamental idea of counterpoint is grasped, it is possible to create a whole set of correlograms. The basic rule is to use two or more patterns in space and time containing correlations which *can* be extracted by the perceptual processes. In the following section I will discuss a class of correlograms that produce thousands of dots in random movements; however, certain areas portraying, say, a letter "A," are made up of random-dots that move more slowly than their surroundings. In that example two or more frames of a movie strip are random textures when statically viewed, yet when presented in rapid temporal succession certain areas within the frames are perceived to move at different rates of apparent motion.

We have thus seen that cyclopean stimulation is not restricted to random-dot stereograms and binocular fusion. Instead of presenting a stimulus pair simultaneously to both eyes, the pair can be presented in temporal succession (or alternation) to one eye alone. However, these different presentations will stimulate different cyclopean retinae. Nevertheless, if these cyclopean retinae are close to each other, then for most purposes we can use either one of these techniques. For example, if the reader has no stereopsis, he can still verify most of the demonstrations by presenting the random-dot stereograms in rapid temporal alternation to one eye. Of course, instead of seeing the cyclopean figures hovering over a random background, the cyclopean figures will be seen as oscillating from left to right as though they were cut out and moving as solid sheets.

We are now in a position to clarify the difference between cyclopean perception and classical perception. Both disciplines study the same system and adhere to the methodological tabu that the enclosure of the system cannot be opened. In systems engineering this is called the black-box approach, and usually the best one can do is to establish a relationship between the input and output variables. Traditionally, many experimental psychologists regard their discipline as basically the black-box kind, contenting themselves with determining stimulus-response relations.

If the central nervous system had only one input and output, pure psychology would be unable to extend its scope beyond stimulus-response studies. However, there are many inputs and outputs in the central nervous system (CNS) even within the same modality. The multiplicity of outputs is obvious since the visual perception of depth is as different from that of velocity or color as it is from auditory pitch or gustatory experience. The multiplicity of inputs within the same modality is not restricted to our having two retinae and two cochleae whose transmitted information is combined at some relatively high level after several processing stages. We have seen that it is also possible to skip operationally the peripheral input and to portray global information on a higher level. It is this "operationally bypassing peripheral stages" that is the key notion of cyclopean perception and must be grasped with all its possibilities and limitations.

At this point the reader might ask an important question about cyclopean stimulation. In the examples mentioned above, no single stimulus physically contained the global information. This information was provided only by the correlation between two (or more) stimuli. On the other hand, one might regard a single stimulus that physically contains the desired information, but in a form that is beyond the extraction capabilities of the peripheral processes, as a cyclopean stimulus. What, then, is our criterion for a cyclopean stimulus? Is it enough to define it by saying that the peripheral system cannot cope with the desired information, or do we insist that this information should not even physically exist at peripheral levels? These questions will be discussed next.

2.3 Two Classes of Cyclopean Stimulation

Classical stimuli convey information by brightness and color changes. An example of a commonly used stimulus is shown in figure 2.3-1; it portrays a T-shaped cluster that consists of black dots in a white surround. There is nothing simple about this type of stimulus except that it is simple to draw it. Objects of real life are covered by textured surfaces, and it would be most unusual if the neighboring areas had uniform colors.

2.3-1

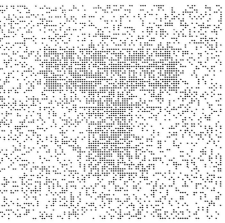

2.3-2

2.3-1 Classical stimulus, portraying the information (of a letter T) by brightness (or color) gradients.

2.3-2 Stimulus in which the global information (T-shaped area) is portrayed by texture-density gradients.

Nevertheless, such black and white outlined drawings have been the test targets in psychological experiments for over a century. Not only is the information presented on the retinae, but it seems that the retinae possess the ability to begin to process this information. Figure 2.3-2 portrays the same global information (of a T-shaped area) by different local means. The T-shaped area is composed of random dots having slightly different densities from the surround (the density ratio is 0.6/0.4). Probably some of this information is already extracted at the ganglion cell levels of the retina, and therefore it is not a cyclopean stimulus.

2.3-3 Weakly cyclopean stimulus, portraying the global information by line segments of various orientations.

2.3-4 Weakly cyclopean stimulus, portraying the global information by minute breaks in lines that are in the vernier acuity range.

2.3-3   2.3-4

A more sophisticated presentation is shown in figure 2.3-3. Here again the same global information is portrayed by small line segments of various orientations. The T-shaped area is clearly visible; and although this global information is contained in the retinal image, its extraction is beyond the capabilities of the retina and visual pathways. This we know from the work of Hubel and Wiesel (1965a, 1968), who found that almost all neural units in the visual cortex of the cat and the monkey responded to slits or edges of certain orientations when presented on the retinae. However, as Kuffler (1953) discovered in the optic nerve of the cat, and Hubel and Wiesel (1960) in the optic nerve and lateral geniculate nucleus of the monkey, there are neural units having receptive fields of concentric shapes such that a center disk might be excitatory and be surrounded by an inhibitory annulus or vice versa. The diameters of these receptive fields vary; and in addition to units which are sensitive to brightness changes, there are similar units for hue changes (Wiesel and Hubel 1966) in the monkey LGN. I will dwell on some of these physiological findings in detail; here I mention them to emphasize that stimuli like the one shown in figure 2.3-3 might be extensively processed on peripheral levels. Nevertheless, no edge detectors were found in the retina.and lateral geniculate nucleus of the monkey. Thus, figure 2.3-3 can be regarded as cyclopean, since the global information that is conveyed by two sets of line segments scattered around the horizontal and vertical orientations, respectively, is evaluated by central processes.

Stimuli that physically contain the information on the retina, but require central processes for its extraction, will be called "cyclopean stimuli in the weak sense." These central processes that are required for extracting the global information (such as the orientation of a line segment) should not be confused with the even higher processes required to recognize letters. Thus we have here a hierarchy of stimulus information. The lowest level of stimulus information consists of brightness (or color) values or gradients in a small neighborhood of a stimulus point. The next level of stimulus information is some function defined on the brightness (color) distribution in a sufficiently large neighborhood of the stimulus point. Finally, the highest level of stimulus information is contained in the entire stimulus. These hierarchies are not strict, and whether the information is local or global depends on the context in which it is used. For instance, the notion of an edge of a given orientation and velocity is a global concept when compared to the local parameters of brightness or extent of a dot. However, such edges are local parameters in figure 2.3-3 where a set of edges defines a global entity of a letter in a different surround. In this section I simply regard global information as complex enough to be beyond the processing capabilities of the first peripheral levels.

Stimuli that even physically do *not* contain the global information—but the relations between two or more such stimuli—will be called "cyclopean stimuli in the strong sense."

The next example is intermediate between the cyclopean stimuli in the weak sense and those in the strong sense. When figure 2.3-4 is viewed from a distance of 25 cm or more, the only cues that convey the global information (a V-shaped area) are the minute breaks in the diagonal lines which are not more than a few sec of arc in extent. We know from anatomical findings as well as from psychological experiments that visual resolution of the receptor mosaic in the retina is at best 30 sec of arc, and the much finer resolution of breaks in a line (of a few sec of arc) is due to vernier acuity, a central process. Thus the inability of the peripheral processes to obtain superresolution is not based on neurophysiological findings, but on the geometry of minimal receptor distance. Nevertheless, the retinal image contains the breaks in the lines, and therefore cannot be regarded as cyclopean stimulation in the strong sense.[1]

The image pair of figure 2.3-5 shows cyclopean stimulation in the strong sense. The two images are composed of uncorrelated, random black and white cells of a 100×100 Cartesian array. However, a T-shaped area in the center is identical in the two images. Since the uncorrelated and the correlated areas have the same random distribution of black and white cells, when viewed alone, they appear uniformly random; indeed they do not contain the global information by themselves. Yet when presented in temporal succession and alternating at an 8–30 Hz frame repetition rate, the impression is of a dynamic noise (as seen on TV screens in the absence of transmission) in which a static random texture is seen portraying a T-shape. Such stimuli and their reverse (when the T-shaped area is formed by dynamic noise and its surround is static) were first used by me in 1959 and reported in several contexts (Julesz

1. In agreement with mathematical custom, any cyclopean stimulus that is not strong is regarded as weak. Thus figure 2.3-4 represents a weakly cyclopean stimulus.

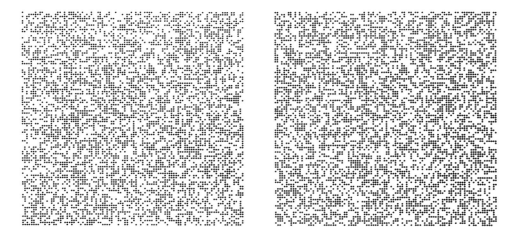

2.3-5 Strongly cyclopean stimulus, portraying the global information as a random-dot cinematogram. When the two images are shown in rapid temporal succession, a T-shaped area of dynamic noise appears in a static noise surround. An anaglyph version of another random-dot cinematogram is portrayed by fig. 7.2-15*.

1965a, Julesz and Payne 1968). This stimulus class is strongly cyclopean for cats, monkeys, apes, and humans; but as Barlow and Hill (1963) have shown, the rabbit retina possesses movement detectors, thus for the rabbit, these stimuli are not cyclopean. The cyclopean nature of these stimuli thus depends on the inertia (or memory) of the peripheral system. Only for a retinal process with inertia is the global information physically present.[2] For a peripheral process that has no memory, the two successive stimuli are independent and therefore do not individually contain any global information. These stimuli will be called "random-dot cinematograms." We have good reason to believe that their perception occurs at a somewhat higher level than that of the random-dot stereograms, to be discussed next.

2.4 Monocular, Binocular, and Cyclopean Stimulation

The prototype of cyclopean stimulation in the strong sense is represented by random-dot stereograms (Julesz 1960a,b) as illustrated by figure 2.4-1*. Here the two images do not contain any global information until the left and right images are actually combined at some neural level. Here we do not have to know anything about the memory span of the processes in question, only the anatomy. It is an anatomical fact that the left and right visual pathways do not meet prior to the lateral geniculate nucleus (LGN) as shown in figure 2.4-2. Furthermore, as we have already discussed, we know from microelectrode recordings that at the LGN level the left and right channels are still separated and synapse only at cortical levels.

The way figure 2.4-1* was generated is shown schematically in figure 2.4-3 using a small array. The left and right images are identical random-dot textures except for certain areas that are also identical but shifted relative to each other in the horizontal direction as though they were solid sheets. The shifted areas (denoted by A and B cells) cover certain areas of the background (denoted by 1 and 0 cells); and owing to the shift, certain areas become uncovered (denoted by X and Y cells). If the hori-

2. The inertia (memory) that is incorporated in cortical movement detectors is different from that inertia due to retinal afterimages.

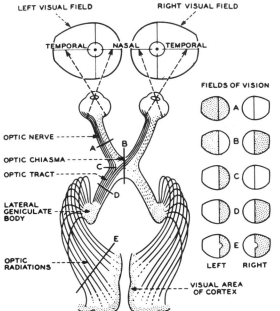

LEFT VISUAL FIELD   RIGHT VISUAL FIELD

TEMPORAL   NASAL   TEMPORAL

FIELDS OF VISION

OPTIC NERVE

OPTIC CHIASMA

OPTIC TRACT

LATERAL GENICULATE BODY

OPTIC RADIATIONS

VISUAL AREA OF CORTEX

LEFT   RIGHT

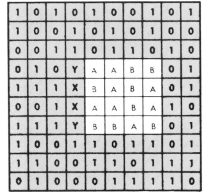

2.4-1* Strongly cyclopean stimulus, portraying the global information as a random-dot stereogram. When the images are monocularly inspected they appear uniformly random, yet when stereoscopically fused a center square is seen above the surround in vivid depth.

When the anaglyph of this stereogram is viewed with the red-green viewer, a center square is seen in front of the surround (or behind—depending whether the right or left eye sees through the red filter).

2.4-2 Anatomy of the visual pathways. Fibers of the same hemiretinae meet for the first time at the corresponding cortical hemispheres. (Modified drawing after Homans 1945.)

2.4-3 Illustration by a small array of the way figure 2.4-1* has been generated.

Left array:

1	0	1	0	1	0	0	1	0	1
1	0	0	1	0	1	0	1	0	0
0	0	1	1	0	1	1	0	1	0
0	1	0	Y	A	A	B	B	0	1
1	1	1	X	B	A	B	A	0	1
0	0	1	X	A	A	B	A	1	0
1	1	1	Y	B	B	A	B	0	1
1	0	0	1	1	0	1	1	0	1
1	1	0	0	1	1	0	1	1	1
0	1	0	0	0	1	1	1	1	0

Right array:

1	0	1	0	1	0	0	1	0	1
1	0	0	1	0	1	0	1	0	0
0	0	1	1	0	1	1	0	1	0
0	1	0	A	A	B	B	X	0	1
1	1	1	B	A	B	A	Y	0	1
0	0	1	A	A	B	A	Y	1	0
1	1	1	B	B	A	B	X	0	1
1	0	0	1	1	0	1	1	0	1
1	1	0	0	1	1	0	1	1	1
0	1	0	0	0	1	1	1	1	0

zontal shift is kept always an integral multiple of the cell size, then no cell of the background will ever be partly covered by the shifted areas, and thus no monocular cue will be present. Indeed, when figure 2.4-1* is viewed monocularly, it gives a homogeneously random impression without any global shape or contour. However, when binocularly fused (either by crossing eyes or by using a prism for the stereogram; or by viewing its anaglyph) a center square-shaped area jumps out in vivid depth above the background. Figure 2.4-4*, *A* and *B,* shows another random-dot stereo-

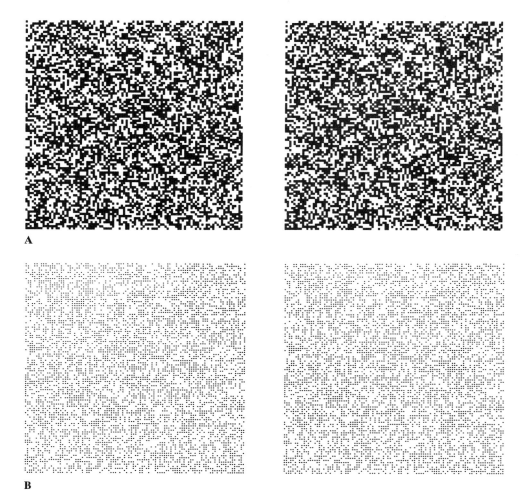

A

B

2.4-4* Random-dot stereogram
portraying a T-shaped area in depth:
(*A*) portrayed by 100 × 100 black
and white square-shaped cells;
(*B*) portrayed by 100 × 100 dots.

gram in which the cyclopean image portrays a T-shaped area (in *B* the random-dot texture is less dense).

This book contains a great number of examples of this stimulus class. In most cases they are random-dot stereograms in the strictest sense—that is, the monocular images are random-dot textures. However, sometimes the monocular images are not composed of dots but of line segments, and in other instances the monocular images are not random but possess some global structure. Nevertheless, in each case the binocularly portrayed information is independent of the monocular information. Perhaps "cyclopean stereograms" might be a better name, but for historic reasons the name "random-dot stereograms" is preferred.

It must be emphasized that cyclopean stimulation by random-dot stereograms is neither monocular nor binocular stimulation. Since real-life objects appear to be the same whether viewed monocularly or binocularly, cyclopean stimulation is a new condition in which the monocular and binocular information is separated. The only condition under which cyclopean stimulation is approximated by nature is in animal camouflage. Here the animal develops a pigmentation that blends with the texture of the background. Very often monocularly camouflaged objects are seen as separated in depth when viewed stereoscopically. However ideally camouflaged an object might be, it will cover some parts of the background that might have some periodic or familiar texture; and, because of this, some monocular cues might be produced. Furthermore, if we were to wrap objects having identically patterned textured covers with the background, the camouflage would still not be perfect. Owing to the "retinal gradient of texture" (Gibson 1950), surfaces closer to the viewer would have textures with larger granularity than the ones farther behind. It is only with the aid of computers that an ideally camouflaged image can be produced for the monocular views.

Random-dot stereograms need not portray the binocular information by depth changes. For example, if figure 2.3-5 is viewed stereoscopically, the identical areas (T-shaped texture) will give rise to binocular fusion, while the uncorrelated areas will yield binocular rivalry. In such cases the fused areas often will appear in front while areas in rivalry will be seen floating behind, but this is a more central depth effect than stereopsis (this will be discussed in § 7.8). Nevertheless, the cyclopean information is conveyed by the static appearance of certain areas versus the dynamic appearance of less stable areas for which dominance of the eyes alternates.

At this point let us briefly review some facts about binocular vision. If the left and right retinal images are identical or similar in the sense that they are the projections of a three-dimensional object from slightly different views, fusion of the two images into a single binocular view will inevitably take place. If the disparity between the retinal projections increases, as is the case for very near objects, the images are seen as double images, but still in depth. This is the range of patent stereopsis. For random-dot stereograms patent stereopsis probably does not exist. Certainly prior to fusion no double image can occur, since before fusion there is no global image. After fusion one can try to increase disparity by pulling retinally stabilized images apart, but the random-dot stereograms suddenly break and lose their depth (Fender and Julesz 1967a). That no double image can exist prior to fusion is important to the theory of stereopsis, since Hering and his followers assumed that the crossed or uncrossed nature of double images would serve for the stereopsis mechanism as a cue to increase or decrease disparity. If the images on the retinae are not similar or if the disparity increases above the limit of patent stereopsis, binocular rivalry is experienced. Under this condition parts of one image dominate parts of the other image, and they alternate dominance in time.

With random-dot stereograms it is possible to create images that give rise to fusion but, however, where the monocular global information is suppressed (Julesz 1966a, 1967a). While figure 2.4-1* has demonstrated that it is possible to portray binocular shapes with monocularly shapeless images, it is possible to reverse this paradigm. In figure 2.4-5* the monocularly visible global organization of a text (the word YES) is scrambled by the fusion of its constituent random-dot elements. In the stereoscopic

2.4-5*   Random-dot stereogram in which a monocularly visible global organization of a text (the word YES) is scrambled when stereoscopically viewed. If the reader has a strongly dominant left eye, he should reverse the viewer, that is, view the anaglyphs with the green filter over the right eye. (From Julesz 1967a.)

view a transparent lacelike plane is seen above a background. These planes in depth appear quite random, and the text is unreadable. The right image of figure 2.4-5* is identical to the left image, except that every even row (of the 100×100 cell array) is shifted to the left by four cells. This nontopological transformation scrambles the monocularly apparent text in the binocular view and gives rise to two depth planes, as shown by Julesz (1967a). One might argue that it is no wonder that an already poorly readable text becomes unreadable when some additional "noise" is added in the other channel. In fact, just the opposite is true. This random pattern is not uncorrelated noise which masks the text in the other display, but a totally correlated pattern which gives rise to stereopsis. It is an interesting paradox that when the two images are uncorrelated, binocular rivalry occurs. One might expect in this case the largest masking of the text by the competing uncorrelated noise; the opposite is in fact the case, since during binocular rivalry the text is quite often visible as dominance alternates. For the type of stimuli demonstrated here, the local fusion is perfect and stable; therefore, if the text is masked in the binocular view, it remains so indefinitely.

This technique can be used to suppress other monocularly perceivable global organizations, particularly symmetries of various types (Julesz 1966a, 1967a). The results obtained suggest that whenever binocular fusion occurs, this process precedes or dominates the recognition of monocularly perceivable global organizations. This technique therefore augments the previous ones. From now on I will call this technique "inverse cyclopean stimulation."

The reader might come to the conclusion that the only known cyclopean stimuli in the strong sense are random-dot stereograms and cinematograms. This is not the case. There are other cyclopean stimuli in the strong sense. For instance, Stromeyer's eidetiker has a cyclopean retina somewhere above the retinal site but before Area 18. It is possible to portray a figure as the monocular combination of an eidetic image and a physical image such that neither of them alone contains the figure. Thus this eidetic-cyclopean stimulus is cyclopean in the strong sense. Unfortunately, this cyclopean retina may only exist for a few select individuals. Furthermore, if our criterion is generalized to include stimulation with high spatial resolution and at a site that is *after* the cyclopean retina of movement perception or stereopsis, then our possibilities become very limited.

One might assume that the Zöllner-anorthoscopic-illusion (Zöllner 1862), resurrected as the Parks-effect (Parks 1965), qualifies for being regarded as a strong cyclopean stimulus. This effect is obtained under appropriate conditions if an object is hidden behind an opaque screen and then moved past a stationary narrow slit in the screen. The object is seen in its totality by an observer even when the object is much wider than the slit, so that only a small portion of the object is presented at any one time. The same effect occurs regardless of whether it is the slit or figure that moves. The Zöllner-anorthoscopic-illusion is really an "illusion" since the perceived figure is foreshortened in the direction of movement. The amount of foreshortening depends on the speed of the moving figure behind the screen. Wenzel (1926) studied the effect of speed, slit-width, and slit-shape on the illusion and found that it was possible to reduce the foreshortening but never to eliminate it completely.

Obviously, by predistorting the stimulus (to compensate for the foreshortening), one could easily portray any desired figure on a cyclopean retina.

But is the Zöllner-anorthoscopic-illusion a cyclopean stimulus? Helmholtz (1909) was convinced that this illusion was the result of eye-movements. Indeed, in the case of tracking eye-movements the entire stimulus could be "smeared" across neighboring areas of the retina. Thus, the stimulus would exist on the retina and then would be an ordinary stimulus (not even weakly cyclopean). Parks (1965) believed that no such smearing on the retina takes place since the same effect occurs if two figures are moved simultaneously and in opposite directions past a single slit. Obviously, the eye cannot be tracking both figures at once. However, Anstis and Atkinson (1967) pointed out that one of the two figures when oppositely moving becomes its own mirror image as expected from the geometry that results when eye-movements are considered. When eye-movement reverses, the other figure becomes the mirror image. This finding and some direct measurements of eye-movement during the experiencing of the anorthoscopic illusion by Anstis and Atkinson suggest that Helmholtz was right and that this illusion is not even cyclopean in the weak sense.

Hochberg (1968) tried the Zöllner-anorthoscopic-illusion with random-dot stereograms. He obtained a similar effect, however the exposure duration is of a different order of magnitude from the classical version. Furthermore, Hochberg explicitly notes the absence of sensory quality to either the form or the background that certainly characterizes the Zöllner (Parks) effect. Therefore, the cyclopean Zöllner (Parks) effect is clearly different from its classical counterpart and is only cyclopean because of the random-dot stereogram techniques that were used.

Even if the Zöllner-anorthoscopic-illusion may not be cyclopean, it raises an important point: the visual system may assemble a set of partial views that fall on the same retinal area over a period of time into a single percept. Such an experiment was reported by Rock and Halper (1969). They asked the subjects to track a moving luminous spot. The subjects perceived the path accurately despite the absence of an extended retinal image and presumably did so on the basis of information via sensing the eye position. If the tracking of the luminous spot could be accomplished without errors and delay, this stimulation would be strongly cyclopean (in spite of the fact that subjects perceived only the path of the moving spot, but could not perceive a line figure). However, the slightest error in tracking would paint a retinal image, and in this very probable case the effect would not yield a strong cyclopean stimulus.

I included these examples to show that in order to extend the class of strongly cyclopean stimuli beyond random-dot correlograms one must be very careful. The investigator must be absolutely sure that no hidden factor may portray the stimulus on the anatomical retina.

2.5 Classical versus Cyclopean Perception: Three Outcomes

Classical studies of visual perception used outline drawings or patterns in which the information was portrayed by brightness gradients as shown in figure 2.3-1. However, every experiment in the textbooks on visual perception can be repeated under cyclopean stimulation (using random-dot correlograms), except for those which study brightness and color perception for its own sake.[3] As a result of these stimuli, a new, cyclopean psychology begins to emerge. Since the information is portrayed by depth or velocity gradients of monocularly identical textures, there are many cyclopean phenomena that differ from the classical ones. The relation between classical and cyclopean perception depends on the similarity or difference between classical and cyclopean phenomena as given in table 1.

TABLE 1. *Relations between Classical and Cyclopean Perception*

Outcome 1. Classical phenomenon = cyclopean phenomenon (implies central processing)

Outcome 1*. Classical phenomenon ≠ cyclopean phenomenon (implies both central and peripheral processing)

Outcome 2. Only classical phenomenon exists (implies peripheral processing or central processes not adequately stimulated by cyclopean stimuli)

Outcome 3. Only cyclopean phenomenon exists (implies central processing usually overshadowed by peripheral processes or central processes not adequately stimulated by classical stimuli)

Before these three outcomes (if we regard Outcome 1* a special case of Outcome 1) are illustrated by examples let me discuss the argument on which these implications

---

3. If we use the "inverse cyclopean techniques" discussed with fig. 2.4-5*, random-dot stereograms can also be used for certain brightness and color studies. Furthermore, if we go beyond random-dot correlograms and regard cyclopean stimulation in the general sense as stimuli that affect a cyclopean retina beyond the anatomical retina, then techniques exist for studying color and brightness phenomena. We will see how eidetic imagery (§ 7.4) or the McCollough effect (§ 7.9) can be exploited in the study of central color and brightness processes. However, in this chapter I restrict the discussion to random-dot stereograms and cinematograms.

are based. Figure 2.5-1 is the "flow chart" of the visual system. The classical input stimulus portrays the information on the retina, and the first black box in the pathway represents all the peripheral processings that are performed prior to a certain stage where the cyclopean stimulation takes place. The location of this cyclopean retina depends on the particular method to be used and the organism in question. For random-dot stereograms and human subjects, for instance, one can safely assume that this cyclopean stimulation occurs after the LGN, somewhere in the visual cortex. The second box in figure 2.5-1 represents all the remaining processes—the central processes. Although the cyclopean stimuli are also presented on the retinae and some local constituents are processed by the peripheral mechanisms, this processing has no effect on the cyclopean image since the desired information is portrayed by the counterpoint, the beating between two (or more) peripherally processed patterns. Thus the peripheral pathways are (operationally) skipped under cyclopean stimulation.

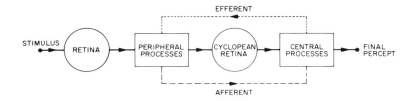

The dotted information flow indicates a possible feedback loop from the central processors to the peripheral processors. Whether neurophysiological evidence for such an efferent (or centrifugal) information flow exists is an interesting question, but it is of no concern to us at the moment. The dashed information flow indicates a possible direct feed-forward loop from the peripheral processors to some central processing stage (thereby skipping the central input stage).

All the black boxes can contain passive or active, linear or nonlinear, and parallel or serial processes. If we exclude a most unlikely case that for certain stimuli the peripheral processes are identical to the central ones, we can make the following assumption: If a classical image produces a certain perceptual phenomenon and the corresponding cyclopean image produces the same perceptual phenomenon both qualitatively and quantitatively, we can conclude that the phenomenon was the result of central processings. It is important that the classical and cyclopean phenomenon agree exactly. Were the cyclopean phenomenon only qualitatively similar to the classical one but exhibited a reduced amount of the classical finding (say, a certain classical figural aftereffect would be reduced to 60% under cyclopean stimulation), one would conclude that both peripheral and central processes were at play. However, because of the feedback channel,[4] one cannot conclude that 40% of the phe-

4. There is no neurophysiological evidence of any efferent pathway from the cerebral sites back to the retina. However, there are many cortical pathways from higher centers back to the LGN and to Area 17. Since the cyclopean stimulation is most likely to affect sites above Area 17, the feedback to the LGN or Area 17 is a physiological possibility.

nomenon was due to peripheral processes and 60% to central processes. Without cutting the feedback loop (or establishing by neurophysiological and neuroanatomical methods the absence of such efferent pathways) in this case, we can only conclude that the processing underlying the phenomenon started at peripheral levels. In this book our primary interest will be either those phenomena that are exactly identical under classical and cyclopean stimulation or phenomena that have either no cyclopean or no classical counterparts.

Researchers, who work on peripheral processes, sometimes encounter cascaded stages that act on the same stimulus parameter (e.g., various stages that enhance contrast). They may argue that the quantitative identity of classical and cyclopean phenomena is not conclusive in telling central processes apart from peripheral ones. After all, the peripheral and central processes might act in cascade such that if the periphery had an opportunity to enhance a stimulus parameter, the center would do little, and vice versa. Thus, the quantitative identity of a classical and cyclopean phenomenon amounts to nothing more than a situation in which the peripheral and central processes together (under classical stimulation) have the same effect as the central process alone has (under cyclopean stimulation). However, it is a very rare case when hierarchical processing stages are similar and work on the same stimulus parameters (e.g., the concentric Kuffler-units in the retina and LGN that are similarly organized, except that the LGN units have smaller receptive field diameters than the retinal units). Usually, at each successive processing stage a stimulus transformation occurs that creates a *new* entity that cannot be compared to earlier stimulus parameters. For instance, the edge detectors in the cortex get their inputs from many concentric Kuffler- units in the LGN, and a new stimulus feature, orientation, is extracted, that does not emerge at any earlier processing stage. This hierarchy of increasingly complex processing levels is not restricted to the well-known simple, complex, hypercomplex, and so on, units in areas 17, 18, and 19 of the visual cortex, but occurs at the earliest levels. For instance, the receptors in the retina extract the illumination information from the stimulus, whereas the bipolars at the next level (according to Werblin and Dowling 1969) are contrast extractors. I cannot see how a brightness or contrast detector stage could compensate for, say, the angular resolution of the orientation detectors. Therefore, it is safe to assume that Outcome 1 usually implies a central process that acts after the cyclopean retina.

If only the classical phenomenon exists and it is impossible to evoke the same phenomenon under corresponding cyclopean stimulation, we can assume peripheral processings alone. At this point one could argue that here is a logical jump. After all, classical stimulation uses brightness gradients, while cyclopean stimulation uses depth, velocity, texture, and other changes. Could it be that certain classical phenomena cannot be obtained under cyclopean conditions simply because central processes are evoked only by brightness changes? Perhaps the dashed flow of figure 2.5-1 exists. Indeed, if we were unable to evoke classically known perceptual phenomena under cyclopean stimulation, this objection would be seriously considered. However, as we will soon see, there are many instances in which cyclopean stimulation will evoke the classical phenomena, thus indicating that in most instances the central processes can operate on more general classes of stimulus parameters than brightness or color. Yet theoretically, if we are purists, the second class of table 1 should imply either peripheral processes at work or the possibility of central processes also operating but processing only global brightness (or color) changes.

At this point I must draw the reader's attention to the inverse cyclopean stimulation discussed in § 2.4. This method will be elaborated in detail, yet it should suffice now to say that it can be used for investigating Outcome 2 of table 1. The way to proceed is as follows: If we have established that a certain classical phenomenon cannot be evoked by cyclopean stimulation, we try the inverse cyclopean stimulation. We present the classical stimulus using brightness gradients (composed of random-dot textures of various densities) for, say, one eye. We present to the other eye a corresponding random-dot stereo image such that we scramble certain global parameters in the binocular percept rendering them unperceivable. If the perceptual phenomenon still holds, the underlying processes must be the result of peripheral processings.

In the next few paragraphs I will illustrate these techniques by concrete examples. Before this illustration, let me briefly discuss Outcome 3 of table 1. The fact that only the cyclopean phenomenon exists for a certain stimulus certainly implies its central origin, yet there are several possible explanations why under classical conditions it cannot be evoked. One plausible reason might be that the powerful peripheral processes interfere with the central processes. Another explanation might be that the classical stimulation is not effective enough to stimulate adequately the central processors. This is the converse of the previous case where we assumed that for certain stimuli only brightness or color gradients might adequately stimulate the central processors.

With these provisos in mind the assumptions of table 1 are good working hypotheses. Even if we cannot be absolutely sure, these techniques will enable us to separate peripheral and central phenomena with high probability.[5] After all, one must bear in mind that since we do this without opening the black box (except for the a priori knowledge of where the cyclopean retina is approximately located), it is quite impressive that purely psychological methods enable us to perform such a process localization, even to this extent. The final verification must come from the neurophysiologists. But such neurophysiological evidence accounting for complex perceptual phenomena is far away; until then we will have to be satisfied with available techniques.

In §§ 2.6, 2.7, and 2.8 I will illustrate the three outcomes of table 1 by a few characteristic examples. Furthermore, the majority of the experiments and demonstrations in this book belong to the three outcomes of table 1. In a way, many experiments of textbook psychology will be retested under cyclopean stimulation in order to localize the stage at which they might be processed. Throughout the book under "localization" of a phenomenon we will simply mean the separation of central processes from the peripheral ones.

5. For Outcomes 1 and 3 in table 1 the implications could be regarded as more than assumptions; they might be promoted to the status of "proofs."

The reader should note that "peripheral" or "central" does not necessarily mean topographic localization. Rather, they refer to the directionality of the information flow. If peripheral denotes process A and central denotes process B, I simply mean that process B utilizes the output of process A. To say that process A occurs before process B might be a better way to indicate different levels of hierarchical order than to call them peripheral or central. Thus, if we find that a figural aftereffect belongs to Outcome 3 when produced by random-dot stereograms, we might say that the aftereffect in question is processed after stereopsis. Whenever we call a phenomenon under study peripheral or central, we always mean with respect to that local phenomenon that portrays the phenomenon under study.

This chapter familiarizes the reader with the two procedures of comparing classical and cyclopean phenomena and the inverse cyclopean stimulation. Under favorable conditions there is a third, even more powerful, procedure that enables us to trace the information flow by purely psychological means. This technique will be introduced in §§ 3.6 and 3.7, where I will return to the problem of process localization.

Before I go into the subtleties of these psychoanatomical investigations let us again review these ideas in a historical framework. The use of binocular stimulation as a tool for telling retinal processes apart from cortical ones is as old as experimental psychology itself. Helmholtz in his *Physiological Optics* (1909) reviewed a large literature on binocular color mixture. Although Helmholtz could not experience binocular colors himself, he accepted the reports that affirmed them and was well aware of their importance to the localization of color processing. Sherrington (1906) studied flicker fusion with binocular stimuli and found slight elevation (decrease) in critical flicker frequency for in-phase (out of phase) stimulation. The first figural aftereffect reported by Gibson (1933) was tested by him for binocular transfer. After presenting a subject with a slightly curved line for a few minutes, a straight line of the same orientation was shown which then appeared to be curved in the opposite way. Though the effect was limited to the particular retinal region of the adapting stimulus, it was evidently a cerebral rather than a purely retinal process. For when one eye was closed during adaptation and then tested alone, the aftereffect was present, in half strength at least. Many other aftereffects from prism adaptation to movement aftereffects show similar transfer to the other eye. All these binocular control experiments established some cerebral (cortical) component for color mixing, for flicker, and for a gamut of various aftereffects.

The limitations of these binocular techniques severely curtail their applicability. First of all, since both retinal and cerebral processes are simultaneously stimulated, one can establish only whether some cerebral interaction takes place, but not its extent. But of even more importance is the use of dichoptic stimulation in binocular control experiments.[6] For the interocular-transfer experiments a positive result with

6. Dichoptic conditions refer to a special class of binocular presentations that do not yield binocular fusion. Namely, for some rates of intermittent stimulation it is possible to present to the left and right eyes different images (which cannot be fused) and still avoid binocular rivalry. However, most frequently, binocular rivalry is unavoidable; thus it is impossible to separate the stimulus figure into parts and present certain parts to different eyes.

measurable transfer to the unexposed eye is conclusive. However, in case of a negative result one cannot be sure whether a weak transfer has been disguised by binocular rivalry (since during testing the unexposed eye, the other eye is shut off). Some
other limitations of binocular transfer techniques will be discussed in § 7.9 on aftereffects and in § 10.1 on other psychoanatomical techniques.

Thus we have several techniques at our disposal both old and new for separating
retinal phenomena from cerebral ones. Our criterion for using random-dot stereograms, or cinematograms, or dichoptic stimulation, or interocular transfer techniques,
depends solely on one consideration: which of these methods gives a positive result.
The majority of perceptual phenomena do not lend themselves to dichoptic stimulation and do not show transfer to the other eye. Many of these phenomena, however,
can be studied under cyclopean stimulation. There are phenomena which in turn have
no cyclopean counterparts, but yield positive results under classical binocular stimulation. Of course, we should be content if this is the case. If a phenomenon can be
studied by both classical methods and cyclopean techniques, then the latter will often
yield additional insight. For instance, as will be shown in § 7.2 on optical illusions,
dichoptic techniques of presenting the test figure to one eye and the inducing figure to
the other established central interaction for some illusory figures. But because of
retinal rivalry, the illusory effect was greatly reduced in some cases. On the other
hand, illusory figures when portrayed by random-dot stereograms or cinematograms
often give rise to the same illusory effect as classical line drawings. In such cases we
can conclude that the process underlying the illusion is entirely central and not partly
central. Furthermore, cyclopean techniques are not restricted to binocular presentations but can stimulate cyclopean retinae at more central levels than the level where
binocular signals interact.

2.6 Outcome 1: Cyclopean Phenomenon Identical to Classical Phenomenon

The first outcome of table 1 is separated into two subclasses. Outcome 1 includes the
results of experiments in which the observed perceptual phenomenon is identical
under both classical and corresponding cyclopean stimulation. Here "identical" means
not only a qualitatively but a quantitatively exact agreement. This first case is illustrated by figures 2.6-1 and 2.6-2*, which show the well-known Müller-Lyer illusion
by the usual brightness gradients and by depth gradients. This cyclopean portrayal of
optical illusions using random-dot stereograms was first applied by Papert (1961) and
extended by Hochberg (1963), who concluded that the ones he tried—including the
Müller-Lyer illusion—are the same under classical and cyclopean stimulation.[7]

---

7. About the same time Ohwaki (1960) and Springbett (1961), elaborating on the early studies
by Witasek (1899) and Lau (1922), showed that the illusory effect of geometric illusions is
greatly reduced when stereoscopically presented (with the "test" figures of the illusion shown to
one eye and the "inducing" figures shown to the other eye). They concluded that the illusory
effects are mainly the result of peripheral processes. The correct interpretation of these results
is that the reason for the diminished illusions is binocular rivalry. Since all the conventional
binocular techniques produce unwanted rivalry, the advantage of random-dot stereograms can
be appreciated.

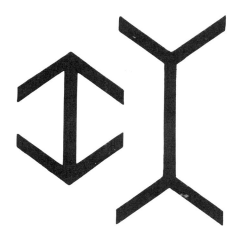

2.6-1 The classical Müller-Lyer
illusion.

2.6-2* The cyclopean Müller-Lyer
illusion.

Were the agreement only qualitative—that is, the left arrow would seem shorter
than the right one in figure 2.6-2*, but the illusion would be quantitatively different
from its classical counterpart—we might assume that the illusion is the result of both
central and peripheral processes. However, it can be shown by careful experimentation that the agreement between figures 2.6-1 and 2.6-2* is quantitatively the same;

therefore, we can conclude that the Müller-Lyer illusion is the result of central processings. Although many researchers suspected that optical illusions are related to higher mental functions, in the case of the Müller-Lyer illusion there is a hypothesis that the changes in brightness gradients at the arrowheads might cause the illusion. This, of course, was disproved (as discussed in § 7.2). Later in the book there will be many other optical illusions which appear identical to the classical ones under cyclopean stimulation; the reader will be able to compare the two stimulations in each case for himself.

Besides optical illusions, there are a large number of classical phenomena that are qualitatively identical under classical and cyclopean conditions. For example, the "waterfall illusion" under cyclopean stimulation appears to be qualitatively identical to the corresponding classical results. When a three-dimensional random-dot stereogram movie is produced such that the frames contain correlated left and right images but the random-dot textures in successive frames are statistically independent, monocularly dynamic noise is experienced. When stereoscopically viewed, however, the dynamic noise is segregated into separate depth planes. For instance, one can portray a moving (or static) bar in front of a surround such that both the bar and the surround are composed of dynamic noise (whirling dots). Since 1964 I have produced several such random-dot movies; and a summary of these efforts is given in Julesz (1966b) and Julesz and Payne (1968). Similar random-dot stereogram movies were produced by Papert (1964) at M.I.T. He observed that a moving cyclopean bar or stripe after prolonged viewing would cause an aftereffect of opposite movement. This cyclopean waterfall illusion is qualitatively identical to its classical counterpart. This preliminary report by Papert has never been followed up by careful quantification. However, if the cyclopean waterfall illusion is qualitatively identical to the classical, one can conjecture that it is partly central, occurring after stereopsis. Since we know from neurophysiological findings that for the cat and the monkey only the cortex contains movement-sensitive detectors, it is not surprising that the "adaptation" of motion extractors of a given orientation occurs centrally. The only nontrivial fact is that we now have techniques to show this by pure psychological means.

Thus, if careful experimentation were to disclose some quantitative difference between the classical and cyclopean perceptual effect, one would have to assume that in addition to central processes some peripheral processes are also at work. In most cases where the classical and cyclopean phenomena were qualitatively identical they were also quantitatively in agreement. Nevertheless, this does not have to be the case; and, therefore, we include Outcome 1* as a separate category in table 1.

2.7 Outcome 2: Only Classical Phenomenon Exists

The second outcome of table 1 is the most controversial case. While Outcomes 1 and 3 refer to positive results, Outcome 2 refers to a negative response. That we are *not* able to obtain a certain percept of a cyclopean stimulus can have countless reasons, out of which the conclusion that the process occurs peripherally is only one conjecture. I illustrate Outcome 2 only for didactic reasons and note that for this case the technique of inverse cyclopean stimulation, when applicable, should be used.

Obvious examples for Outcome 2 are afterimages. Afterimages are assumed to be caused by chemical changes in retinal receptors. The brighter the stimuli the longer afterimages last. For sharp afterimages the retina must not move during stimulation; this is usually achieved by presenting the stimulus for brief durations. After a bright and brief flash followed by a secondary stimulus (e.g., a dark field), a sequence of afterimage phases can be observed with various latency times in which the brightness and color distribution of the perceived image appears to change. Craik (1940) demonstrated that light which was not perceived during the pressure blinding of the retina gave rise to an afterimage when pressure was released. This result was interpreted to indicate that the site of the afterimage is retinal and not central. Kohlraush (1925) regarded the persisting electroretinogram potentials following intense brief flashes as a likely indication of the retinal origin of afterimages, an opinion shared by Brindley (1962).[8] In the light of these findings the retinal origin of afterimages is well-established, and it is not surprising that no cyclopean afterimages have been produced yet, and it seems doubtful that they ever will be.

Another example for Outcome 2 of table 1 is illustrated by the famous Hermann-Hering grid as shown in figures 2.7-1 and 2.7-2*. The cyclopean Hermann-Hering grid has been studied by Julesz (1965b), and it appears that the classical inhibitory effects at the intersections in figure 2.7-1 are completely absent in figure 2.7-2*. In the latter presentation there is no change of depth or average texture brightness at the intersections.

One possible conjecture is that the processors at the cyclopean retina have no inhibitory-type receptive fields. One might argue that the cyclopean Hermann-Hering grid is only superficially similar to the classical one since there is no reason to believe

2.7-1   The classical Hermann-
Hering grid.

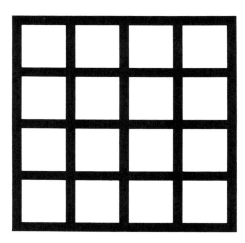

8. There are some claims by the Popovs (1954), however, who reported to have obtained afterimages as conditioned responses to auditory stimuli. This result, if verified, might suggest also a central component of the afterimages although in this case it might be more proper to regard them as aftereffects.

 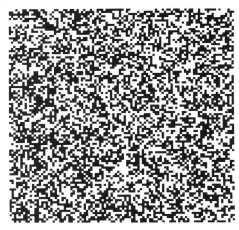

2.7-2* The cyclopean Hermann-Hering grid. Unlike the classical case, no inhibitory effects are experienced at the intersections.

that brightness gradients can be mimicked by depth gradients. Nevertheless, one could argue as follows: When the black grid over the white surround in figure 2.7-1 is presented on the retina, "off" units stimulated by the black grid will increase their firing rates and "on" units on which the white surround casts its image will do likewise. Somewhere in the cortex at a stage where the left and right retinal projections of the same binocular disparity are combined, a similar neural firing rate change takes place. Binocular disparity units tuned to the disparity value of the cyclopean grid in figure 2.7-2* will change their firing rate; similarly, neural units tuned to the disparity value of the background area will do likewise. From a neural point of view the difference between classical and cyclopean stimulation is minimal. Classical stimulation generates two sets of firing patterns in the peripheral pathways conveyed on "on" and "off" units. Cyclopean stimulation generates many sets of firing patterns in the cortex conveyed by a multitude of neural sets that are extracting various binocular disparity values. For the presentation in figure 2.7-2* we have only two disparity values; therefore, only two sets of neurons (corresponding to the two depth levels used) will encounter larger firing rate changes. In the peripheral system at ganglion and LGN levels the sets of neural firing patterns are further processed by lateral inhibition accentuating large spatial gradients. This lateral inhibition by the next hierarchical level of neurons that process the firing patterns of retinal receptors can cause the illusion of dimming at the intersections. One might expect that neural processing above the cyclopean retina would also emphasize rapid spatial changes in neural firing patterns. Therefore, one could argue that it is not a naive idea to look for some perceptual changes at the intersections of the cyclopean grid.

There are, however, severe inadequacies to this argument. From the identity of neural firing patterns one could argue that the corresponding sensations ought to be identical, an assumption that is contrary to the doctrine of the specificity of nerve energies! Even if this doctrine were defunct, there is no reason to assume that depth or movement changes ought to produce similar percepts to brightness changes. However, the converse outcome (Outcome 1) is different. If a depth or movement gradient produces the same phenomenon as a brightness gradient it is very unlikely that this is a sheer coincidence and not the result of a common central process. It might be the case that lateral inhibition may still exist at cyclopean levels, yet this inhibition does not cause depth or texture changes, but some other effects not yet noticed.

The problem of whether the Hermann-Hering grid phenomenon is peripheral or central cannot be decided from the experiments of figures 2.7-1 and 2.7-2*. Baumgartner (1960) assumed that this phenomenon is the result of concentric receptive fields with excitatory centers and inhibitory surrounds as discovered in the retina of the cat by Kuffler (1953) and in the monkey by Hubel and Wiesel (1960). Ronchi and Bottai (1964) reported (and Ronchi and Salvi, 1965, further elaborated) that the largest inhibition is perceived when the Hermann-Hering grid intersects at $32°$ of angle. This result can be explained both by peripheral symmetric Kuffler units or by assuming some cortical complex corner detectors at work. Some recent experiments on an eidetiker (to be reported in § 7.4) even suggest that (at least for this subject) the inhibitory effects of the Hermann-Hering grid occur more centrally than the retina.

Many of the phenomena of sensory psychology that are evoked by brightness and color gradients belong to Outcome 2. Indeed, it would be very difficult even to imagine a random-dot correlogram counterpart of a Mach-band or some other simultaneous contrast phenomena. Yet, there are other cyclopean techniques that can be used in order to study these phenomena.

The problem of determining whether certain phenomena that are attributed to peripheral processes alone might be also centrally processed is of great interest. After all, one might question whether simultaneous contrast phenomena are the result of lateral inhibition in the retina and other subcortical stages or whether lateral inhibition further shapes the information in the cortex. The negative outcome of the experiment shown in figure 2.7-2* does not prove that brightness enhancement could not take place cortically.

However, under favorable conditions (i.e., being able to produce a positive outcome) there is a powerful dichoptic method that can also be used for localizing the site of simple brightness and color interactions. Such a dichoptic study was carried out by Diamond (1958), who was able to evoke the Pulfrich phenomenon by simultaneous contrast effects. While in the classical Pulfrich phenomenon an attenuator over one eye produces a depth change in the appearance of a moving object, attenuation of brightness in one eye induced by a neighboring monocular bright field yields an inverse depth change. Yet we know from experiments (reviewed in § 7.6) that attenuation in the Pulfrich phenomenon causes a time delay that occurs before stereopsis, therefore the site of the simultaneous contrast process must occur (partly or totally) before stereopsis. Unfortunately, binocular rivalry limits this method to the localization of very simple spatial interactions.

Another possibility of localizing peripheral processes by regular psychological techniques is provided by inverse cyclopean stimulation, as described earlier. The idea is as follows: We generate a stereogram in which the left field portrays the Hermann-

Hering grid as shown in figure 2.7-3*. In order to provide stereoscopic cues, the horizontal and vertical black bars are randomly speckled with a few white dots. The right image of figure 2.7-3* is generated such that when stereoscopically fused with the left image the horizontal bars appear in front of the vertical bars. At the intersection of the grid there are no changes in luminance; only cyclopean edges produced by binocular disparity changes exist. Stereoscopic inspection of figure 2.7-3*, however, yields a strong perceived brightness change at the intersections, similar to the classical effect. Since the intersecting bars appear at a distance from each other in the three-dimensional perceptual space, one can assume that lateral inhibition should be negligible between the corresponding binocular disparity detectors in the neural-space. The fact that the perceived brightness changes at the intersections are not affected by the depth changes, however, strongly suggests that this lateral inhibition is a peripheral process, occurring before stereopsis.

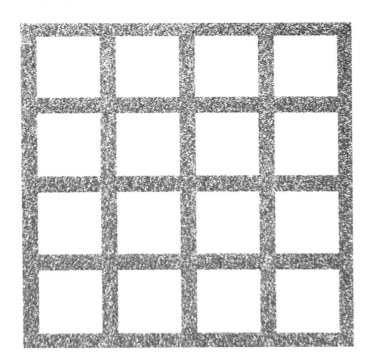 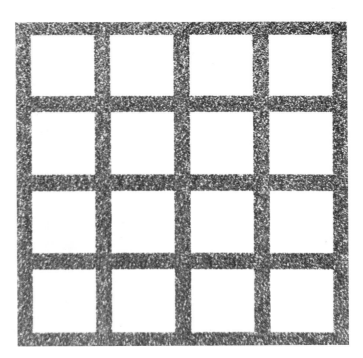

Another example that is based on inverse cyclopean stimulation is as follows: We generate a random-dot stereogram in which the left field portrays some well-known simultaneous contrast effect. For instance, in the left field of figure 2.7-4* two medium-gray squares of the same size and brightness are presented, one of them surrounded by a dark-gray area, the other surrounded by a bright-gray area. These three brightness values are portrayed by different texture densities. The right field of figure 2.7-4* is generated such that the picture elements of the two gray squares will

2.7-3* The inverse cyclopean Hermann-Hering grid.

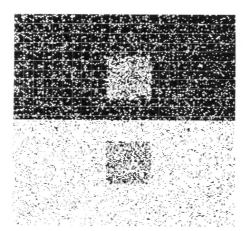 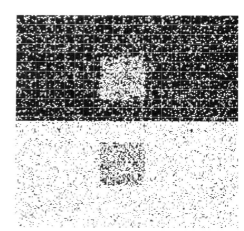

2.7-4*   Simultaneous contrast
phenomena under cyclopean
stimulation.

stand out in depth above the dark and bright surrounds when stereoscopically viewed. When the monocularly viewed gray squares in the left (or right) field are compared with each other and psychophysically scaled, the square in the bright surround will appear considerably darker than the square in the dark surround. Will the ratio of this darkening effect due to simultaneous contrast change under stereoscopic fusion? If it will, one can conclude that some lateral inhibition operates after stereopsis, which is interfered with when neighboring areas are "broken" in depth.

On the other hand, if the simultaneous contrast effect remains the same when the gray squares are perceived in front of the dark and bright surrounds, we can conjecture that lateral inhibition operates before stereopsis. In viewing figure 2.7-4* first monocularly, then stereoscopically, it is very hard to notice any change in the ratio of the darkening effect. If any change appears at all it is a slight increase in the ratio under stereoscopic viewing. Since the lifting of squares above the inhibitory surrounds does not reduce simultaneous contrast, we can assume that the bulk of lateral inhibition occurs before stereopsis.

2.8 Outcome 3: Only Cyclopean Phenomenon Exists

The third outcome, probably the most interesting one, consists of some new, hitherto unobserved phenomena that are evoked under cyclopean stimulation. Since no peripheral information exists, the phenomena have to be central. On the other hand, the reason why they have never been observed previously is probably the overshadowing influence of monocular cues. We have to produce an "eclipse" of these powerful peripherally processed cues in order to reveal a new psychology that is qualitatively different from the classical discipline.

In a sense, the basic random-dot stereoscopic phenomenon demonstrated in figure 2.4-1* belongs to Outcome 3. The fact that, contrary to common belief (e.g., see Ogle 1962), no monocular contours are necessary to obtain stereopsis is qualitatively different from the findings on stereopsis under classical conditions. I will devote a chapter to some new, hitherto unobserved phenomena that are evoked by random-dot stereograms.

Before I review some of these entirely new phenomena let me give an example of some simpler kinds of new phenomena that resemble classical phenomena in general, but differ from them in some important way.

Such is the case of figure-ground reversal. Figure 2.8-1 shows the classical "vase–two faces," ambiguous figure. In 2.8-1 one can see the vase as figure, and the two faces become the ground and hence are unnoticed; or the two faces become attended to and the vase merges into a meaningless background. It has been noted by Rubin (1915) that the figure becomes a "thing" and seems to be localized in front of the ground, which becomes "nothing." The weakly cyclopean version of figure 2.8-1 is shown in figure 2.8-2. Here the information is portrayed by small line segments in various orientations such that one class of line segments has orientations close to horizontal while the other class has orientations that are close to vertical. Perceptual reversal is easy to obtain.

2.8-1 Classical figure and ground reversal. The vase, or two faces. (From Wyburn, Pickford, and Hirst 1964.)

2.8-1

2.8-2 Weakly cyclopean version of figure 2.8-1. Figure-ground reversal is similar to the classical case.

Figure 2.8-3* is a random-dot stereogram version of figure 2.8-1. When 2.8-3* is viewed stereoscopically so that the cyclopean vase is in front, it is impossible to shift attention to the two faces in the background and to perceive them as a figure. Similarly, when the left and right images are reversed and the two faces jump in front, it is again impossible to perceive the vase as a figure. The two faces will always be seen as figure (Julesz 1969a). This is a phenomenon which illustrates that stereopsis is incorporated in figure-ground perception and that the importance of an object close to the observer cannot be overcome by voluntary shifts in attention.

There are phenomena belonging to Outcome 3, however, that do not exist at all under classical conditions. One has been discovered by Julesz and Payne (1968). Figures 2.8-4 and 2.8-5* show the classical and cyclopean grid used as stimuli. Figure 2.8-6 illustrates schematically how such a pair of grids is shown in rapid temporal alternation. For classical stimuli this stroboscopic presentation may result in succession, movement, and simultaneity depending on the luminance, duration, and temporal pause between stimuli according to Korte's laws (Korte 1915). However, for cyclopean stimulation there is a fourth response between the classical optimal movement and simultaneity region. In this new region a single grid is seen at a standstill. This cyclopean standstill should be contrasted with simultaneity in which the two alternate stimuli are seen as superimposed. This new region can be experienced only when the two alternate random-dot stereograms have uncorrelated textures. (Of course the corresponding left and right images are correlated.) If the random-dot stereograms that are alternately shown are composed of the same textures, then in addition to the cyclopean bars (global information) being seen as swinging back and forth, the texture elements (local information) will also follow the same movement. Under such conditions the presentation will not be cyclopean since, when monocularly viewed, the correlated textures in temporal succession will portray the bars in the same way as had been illustrated by figure 2.3-5. Indeed, in this case, the

2.8-3* Strongly cyclopean version of figure 2.8-1 in the form of a random-dot stereogram. Only the area closest to the viewer can be perceived as the figure. For figure-ground reversal one has to reverse the red-green viewer as well.

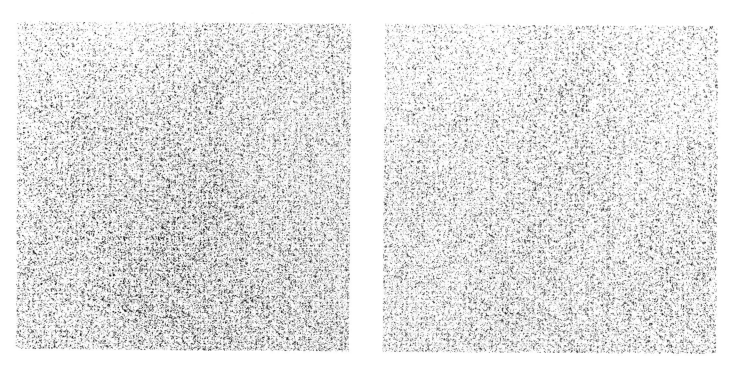

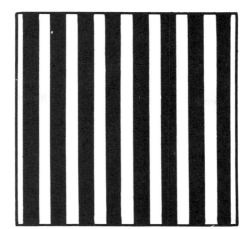

2.8-4   Classical stimulus portraying vertical stripes.

2.8-5*   Random-dot stereogram portraying figure 2.8-4 cyclopeanly in depth. (From Julesz and Payne 1968.)

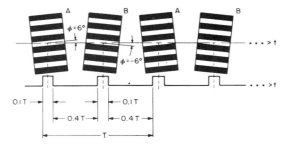

2.8-6   Schematic illustration of how a pair of cyclopean grids (such as fig. 2.8-5*) is presented in rapid succession in order to yield stroboscopic movement. (From Julesz and Payne 1968.)

cyclopean standstill cannot be perceived. On the other hand, when the global information in the form of the bars is correlated in the alternate stereograms, although the textures are uncorrelated, one can separate contours of moving objects from the movement of their covering textures. While the bars in depth are seen as oscillating back and forth (in the optimal movement region), the texture of the bars and their surroundings appear as dynamic noise. In more poetic terms it seems as if a snowstorm segregates into two different patterns in depth that perform some simple movement in a specified trajectory. As the rate of alternations is increased, the back-and-forth oscillation of the bars in depth appears reduced and finally comes to a standstill, while the covering texture continues to appear as dynamic noise. This experience is even more surprising since the frames of the stereograms (which, of course, are monocularly perceivable) appear to be oscillating at the same time. This phenomenon has been described in this introduction in such detail only because it shows how the slightest monocular cues can conceal a cyclopean phenomenon. It is interesting from another point of view as well. In the monkey, and probably in man also, monocular movement detectors reside in the visual cortex. Our findings suggest that these monocular movement detectors precede, or at least dominate the binocular movement detectors. In everyday life both the contours and the surface texture of objects move together, and the monocular and binocular movement detectors are presented with compatible information. Only in this special cyclopean condition, when the two movements become independent, is the central nervous system confronted with contradictory information, at which time the existence of monocular and binocular movement detectors is revealed.

To this third case of new cyclopean phenomena belongs a class of phenomena that are not eclipsed by the peripheral process but are not adequately stimulated under classical conditions. Indeed, the simple two dots and lines of classical stimuli will excite only a few dot or edge detectors of the CNS. On the other hand, the random-dot stereograms contain ten thousand dots (or line segments) with controlled features that stimulate almost all the available feature extractors of the CNS. Since many of the perceptual phenomena are phenomena that the physicist would call "cooperative" (in which many local processes participate simultaneously), it is not surprising that only effective stimulation can produce them. For instance, figures 2.8-7* and 2.8-8*

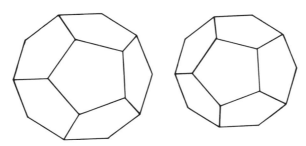

2.8-7* Classical stereogram in which one of the images is 15% expanded. Stereopsis is difficult to obtain.

2.8-8* Random-dot stereogram in which one of the images is 15% expanded. Stereopsis can be easily obtained.

show the same 15% size difference (in the linear dimension) between the left and right images of a stereogram. The classical stereogram of figure 2.8-7* composed of a simple line drawing will yield weak fusion, while the same amount of aniseikonia still gives rise to strong stereopsis and fusion for figure 2.8-8*.

Another even more striking example of a cooperative phenomenon has been discovered by Fender and Julesz (1967a). They showed that under binocular retinal stabilization, after a random-dot stereogram is fused (i.e., the left and right images are brought within Panum's fusional area of 6 min of arc disparity), the percept of the two images resists the effect of a horizontal shift by as much as 120 min of arc. The image stays fused. After the shift exceeds this amount the fused images suddenly break apart and have to be brought within 6 min arc alignment for fusion. This hysteresis, a lag between cause and effect, is practically nonexistent for a simple vertical line target. This finding implies that stereoscopic fusion is a cooperative phenomenon requiring the simultaneous interaction of thousands of local matches of corresponding points. As in physics, no single atom of a ferromagnetic material would exhibit a Curie point or hysteresis, but only after the simultaneous coupling among large numbers of atoms will such phenomena occur.

The foregoing was intended to convey to the reader the essence of cyclopean perception, the separation of peripheral processes from central ones. In the following chapters we will discuss in great detail many examples of cyclopean phenomena belonging to the three cases.

2.9 Some Limitations of Cyclopean Techniques

The basic limitations of cyclopean reasoning are identical to the limitations of experimental findings in general. It is impossible to prove the null hypothesis; it can be only disproved. Thus our inability to evoke a certain classical or cyclopean phenomenon is much less convincing than the opposite—that is, the successful creation of an existing or a new phenomenon. It is particularly difficult to localize the origin of perceptual phenomena that are based on brightness or color gradients. In § 2.7 I discussed how the inverse cyclopean technique can be used to study simultaneous contrast phenomena. In the example of figure 2.7-4* (that portrays the simultaneous contrast phenomenon by inverse cyclopean technique) I assume that no lateral inhibition exists among areas that lie in distant depth planes. This was the assumption on which I based the conjecture that simultaneous contrast is a peripheral process. Were there lateral inhibition between areas at different depth levels, lifting the gray squares above the dark and bright surrounds would accomplish nothing.

From these limitations the reader might get the impression that only Outcome 1 and Outcome 3 of table 1 reveal something definite, besides a few positive outcomes obtained under inverse cyclopean stimulation. For Outcomes 1 and 3 we can surely assume central processes at work; and for Outcome 3 we have an additional bonus, the creation of a new phenomenon. Thus, if a cyclopean contour in depth acts as a classical contour, we can localize the phenomenon in question. Everything else gives only hints. For complex perceptual phenomena this appraisal of the procedure introduced in this chapter is correct. However, another procedure lends itself for certain simple perceptual phenomena, such as flicker and color, superresolution (vernier acuity), stroboscopic movement perception, and so on, for which there is a possibility that they might occur prior to stereopsis. Instead of portraying the monocular patterns of a stereogram or cinematogram by randomly selected brightness values, one can use for picture elements various colors, edges of various orientations or movements, tiny breaks in lines, and so on, which are binocularly (or successively in time) correlated. If these correlograms can product a cyclopean image (e.g., a center area in depth), we have proven that the phenomenon in question occurs prior to stereopsis (or movement perception). I will discuss this technique in §§ 3.5, 3.6, and 3.7 after I have reviewed some neurophysiological background material.

We are now in a position to face the basic opposition to cyclopean methodology. Is it really meaningful to compare percepts of similar forms that are portrayed by different perceptual entities? Why should the perception of an illusory figure portrayed by brightness gradients have anything in common with the perception of a similar illusory figure portrayed by movement or depth gradients? Is not such a comparison as arbitrary as the comparison of perceptual processes in two different modalities? Let us try to answer the objection by citing an example.

Hirsh and Fraisse (1964) compared the temporal thresholds for simultaneity and succession between two auditory, visual, and tactile impulses, respectively. They obtained different critical durations for the simultaneity threshold for all three modalities. However, the critical duration for determining which of the two stimuli came first (a more difficult task than reporting simultaneity) was identical for all three modalities. These results suggested to them that simultaneity detection is a relatively early process, dependent on the modality in question, whereas successivity detection is a highly central process that evaluates the output of all three modalities. The purist might object to this implication. He could argue that successivity detection occurs independently for all three modalities, and it is merely by chance that their critical durations coincide. Although this view is theoretically possible, it is very unlikely. Similarly, the finding that the classical Müller-Lyer illusion holds for its cyclopean counterpart could be regarded as a mere coincidence. However, the fact that all the dozen different illusions tried hold for both classical and cyclopean conditions and, what is more, can be evoked by either depth gradients or movement gradients, is an outcome that is most unlikely to be a chance coincidence.

In answering the objection we can go a step further. Cyclopean stimuli are not stimuli of a different modality from the classical counterparts. We know from neurophysiology the hierarchical chain of processors in the visual system. We will see how monocular edges, binocular edges, and moving edges are extracted in the visual cortex. Usually, an object in real-life simultaneously stimulates all these extractors. With cyclopean stimuli it now appears possible to stimulate selectively only one of these extractors. If only one of these stimulated extractors will not evoke a certain perceptual phenomenon, we cannot infer very much. However, if the global percept is similar to that which can be obtained when all the extractors are simultaneously affected, we must conclude that the global process may utilize the outputs of these extractors. This, according to our definition, is equivalent of saying that the global process in question occurs after the very extractor we were able to stimulate by cyclopean methods.

2.10 An Example of Studying Brightness-Contrast with Cyclopean Techniques

From the foregoing the reader might think that the localization of all brightness (color) phenomena requires hypothetical assumptions. Are there phenomena using brightness or color gradients that can be localized without any additional assumptions? Such a case is illustrated by a recent experiment to study the well-known Benussi (Koffka)-ring phenomenon under classical and cyclopean conditions as shown in figures 2.10-1 and 2.10-2* (Julesz 1969a). In figure 2.10-1 the gray ring is uniformly (or almost uniformly) of the same shade, appearing to resist the enhancements of contrast due to the surrounding dark and gray areas. However, if a thin line is placed at the vertical boundaries between the black and white half-fields so that the ring is also cut in half, the brightness of the ring changes dramatically. That half of the ring which lies on the white background appears dark-gray while the other half of the ring seems light-gray. The Gestaltists assigned particular significance to this phenomenon and regarded it as an illustration of how the global principles of "good-gestalt" and "uniqueness" overcome the local processes of contour enhancement.

In figure 2.10-2* Julesz (1969a) tried to halve the Koffka-ring *in depth*. The gray ring is composed of random-dot textures with an average brightness between the black and white half-fields. The black half-field contains randomly speckled white dots of 10% density (in order to help the depth localization) and the white half-field has a random-dot density of 90%, the gray ring has a 50% random-dot density. The left image of figure 2.10-2* is identical in geometry with the classical display of

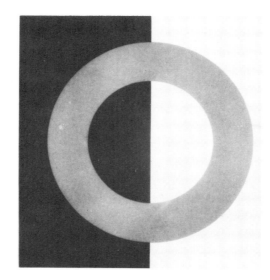

2.10-1 The classical Benussi
(Koffka)-ring phenomenon.

figure 2.10-1. The left and right images differ in the shifting of the left half of the
random-dot ring in the temporal direction, while the right half of the ring has a
disparity in the nasal direction. When figure 2.10-2* is stereoscopically viewed, the
right half of the ring appears in depth and closer to the reader than the left half of
the ring that lies behind the depth plane of the printed page containing the black
and white half-fields.

The cyclopean demonstration of figure 2.10-2* gives only a slight contrast enhancement when the ring is "cut in depth." The reason why the binocular effect is
small can be explained by observing that figure 2.10-2* when monocularly viewed
does not seem to contain a uniform gray ring. Obviously, it is more difficult to notice

2.10-2* The cyclopean Benussi
(Koffka)-ring phenomenon.

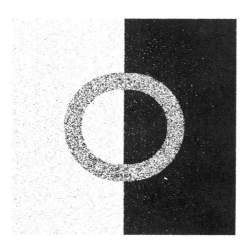 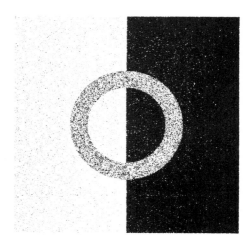

an additional increase of contrast than to observe a departure from uniformity. The main factor of not being able to portray a ring that would appear uniformly gray in a black and white half-field is that the gray was achieved by speckled black and white dots. It seems that random textures are less effective in counteracting contrast enhancement than uniformly shaded areas. Furthermore, a careful inspection of the classical case shown in figure 2.10-1 reveals that even in this case the gray ring is not entirely uniform in appearance. Cohen, Bill, and Gilinsky (1968) studied Benussi (Koffka)-rings of various shapes (e.g., square and diamond) and observed that the uniformity of the grayness of the rings did not vary with their shapes. However, they manipulated the amount of contact between the two halves of the ring (for instance, by increasing the width of the ring) and concluded that this is the primary cue. They rank-ordered the uniformity of the ring from zero (such a wide ring that it becomes a disk) to four (a ring cut in half) and obtained for the ring shape of figure 2.10-1 a scale of 1.4. This departure from uniformity during monocular viewing conditions makes the increased contrast under stereoscopic conditions less dramatic.

It would be much more revealing to cyclopeanly cut a gray shape that is monocularly the most uniform. In the study of Cohen, Bill, and Gilinsky (1968) one of the most uniformly gray shapes was obtained by a filled-in gray disk over a black and white half-field. Figure 2.10-3* shows such a configuration portrayed by random-dot textures; and, when the disks are monocularly viewed, they seem rather uniformly gray. However, the randomly dotted disk reveals an important fact that is less ap-

2.10-3* Similar to figure 2.10-4* except that a cyclopean edge in depth cuts the disk into two halves.

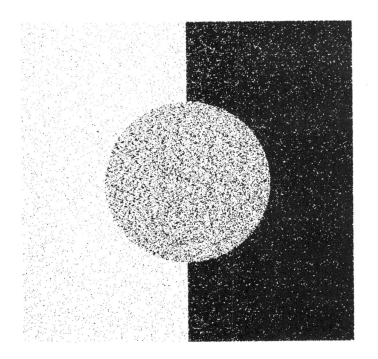
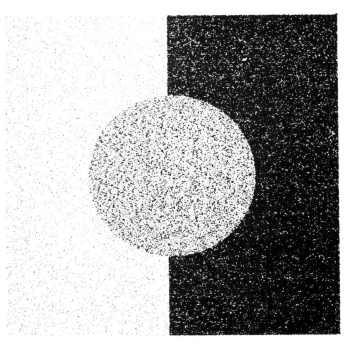

parent for a homogeneous gray disk. As long as we attend to the center of the disk, it appears very uniformly gray. However, as our attention shifts to the left or the right border of the disk there appears a crescent-moon-shaped brighter or darker area, respectively. Thus, lateral inhibition of the surround is acting on the disk; however, the contact at the center counteracts simultaneous contrast for a large area. When figure 2.10-3* is stereoscopically viewed, the disk appears to be split in half by a cyclopean edge (composed of a randomly textured vertical bar above the surround). The crescent-moon-shaped brightness discontinuities spread to the cyclopean border, and two half-disks of different brightness are perceived. The cutting of the disk by a monocularly perceivable edge (brightness gradient) results in an even stronger contrast enhancement. The cyclopean cut of the ring is thus almost as effective as the classical edge. From this we can conclude that the contrast reducing mechanism is partly central.

Two variants of this cyclopean experiment are of interest. In the first one, shown in figure 2.10-4*, the two halves of the randomly dotted disk are portrayed in two different depth planes, one half-disk above and the other half-disk behind the surround. In the second variant, shown in figure 2.10-5*, the left half of the dotted disk is identical in the two fields, while the right half is uncorrelated. This pair is not intended as a stereogram, but as a cinematogram portraying monocularly stroboscopic movement as the two frames alternate in temporal succession. The left half of the disk will appear as static noise, while the right half gives rise to dynamic noise.

There are two methods to demonstrate a random-dot cinematogram. The simpler method is as follows: Figure 2.10-5* is viewed through the red and green viewer, while the reader tries to cover each of his eyes alternately by rapidly moving his palm back and forth in front of his eyes. A better but more elaborate method is as follows: If the two images of figure 2.10-5* are Xeroxed, folded at the dark vertical midline,

2.10-4*   The cyclopean modified version of figure 2.10-1 using a gray disk and a cyclopean cut of the two halves of the disk in depth.

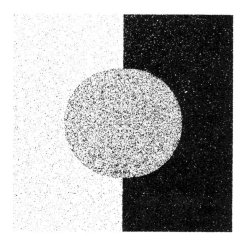
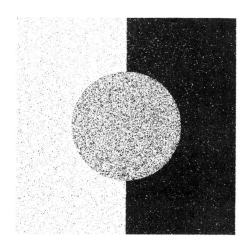

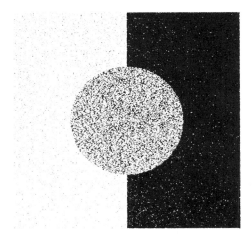

2.10-5* Similar to figure 2.10-4*
except the cyclopean cut is obtained by
dynamic and static stroboscopic
movement. This is not a stereogram.
For its viewing see text.

and glued together with the center of a rubber-band between them, the two images
are then superimposed in rapid succession as the rubber-band is twisted.

Neither figure 2.10-4* nor figure 2.10-5* are as effective stimuli as figure 2.10-3*,
though both yield a contrast enhancement. I have shown these variants in order to
demonstrate that cyclopean methods permit the cutting of an area in more than one
way. Why a thin cyclopean bar in depth is more effective in cutting the disk than
entire half-fields at different depth or motion is unknown. However, the important
fact is that one of the cyclopean cuts is quite effective.

These findings indicate that Cohen et al. (1968) were partly correct when they
assumed that a cross-contact effect reduces simultaneous contrast, and the contrast
reduction increases with increased cross-contact. However, they assumed that this
cross-contact effect acts at the same level as the simultaneous contrast mechanism.
Since the latter is a peripheral process, their conclusion was opposed to Gestaltist
views. Furthermore, their findings that squares, diamonds, and other shapes having
the same cross-contact resulted in identical uniformity of grayness was regarded as
another evidence against Gestaltist claims. Indeed, since the Gestaltists regarded a
circular ring as a better gestalt than a square- or diamond-shaped ring, it seems clear
from these results that the uniformity of the ring is not caused by gestalt organization. Yet, our experiments prove conclusively that the cross-contact effect is partly
central (after stereopsis), while the simultaneous-contrast effect is probably peripheral (prior to stereopsis). The fact that a brightness contour disrupts uniformity more
effectively than a cyclopean contour does not mean that the cross-contact effect is
partly peripheral. It could be entirely central since a cyclopean contour is not as
effective in influencing this mechanism as a brightness gradient is.

We have elaborated the foregoing in order to show how certain postulated processes
responsible for the cross-contact phenomenon can be localized, while other simultaneous contrast processes are not as conclusively localizable.

2.11 Auditory versus Visual Cyclopean Stimulation

Although this book deals with visual perception, a short review of auditory analogies is revealing. These analogies are not very deep, as pointed out by Julesz and Hirsh (1968). If we define the study of perception as the science of those patterns of energy distributions in space and time that are perceivable and the relations between these percepts, this definition then is broad enough to hold for all modalities. However, visual perception is primarily concerned with spatial patterns that characterize *visible objects,* while auditory perception is restricted to temporal patterns that describe *auditory events.* It is this preoccupation of the visual processes with the spatial dimension and of the auditory processes with the temporal dimension which makes the comparison between the two modalities quite arbitrary. Of course, there are *visual events* as well as *auditory objects;* however, these constructs are rather artificial and of secondary importance. The few examples of pure visual events such as flicker phenomena or Morse-code transmission by light signals (tasks for which the spatial extent of the object is unimportant) occur under very unnatural conditions and human performance is notoriously poor when compared with similar auditory tasks. (For instance, flicker fusion threshold is at most 50 Hz, while a 20 kHz sound wave still gives rise to pitch.) One might argue that the velocity of objects is important in visual perception; however, the vectorial quantity of velocity is a mere change in spatial coordinates and, as such, is a spatial derivative and not a temporal event.

It is even more difficult to find an example of auditory objects. Except for bats and some other echolocating animals, auditory processes extract temporal changes in the stimuli. Even for binaural hearing the azimuthal information is conveyed mainly by temporal delays between the two ears. Though no real auditory object exists, some of the binaural phenomena give rise to spatial sensations. For instance, two identical auditory patterns presented in phase to the two ears give rise to the sensation that the event is emanating from a point in the center of the head. By changing the ratio of amplitudes, or the time delay between the left and right channels, one can shift this perceived point to any place along a line between the two ears. Licklider (1948) presented correlated white noise to the two ears and produced various localizations by temporal delay. Cramer and Huggins (1958) noticed that a delay T between the identical noise sequences presented to the two ears would give rise to a faint pitch of $1/T$ Hz. Both of these binaural experiments, which use Gaussian noise for inputs and produce a new pattern of sensation, might be mistaken for cyclopean stimulation. In the visual case the retinal images were two-dimensional spatial patterns, and a different cyclopean image was portrayed along the same spatial dimensions. However, Licklider's input was a temporal pattern, and the fused pattern was the same temporal pattern, only perceived at a different locale. "Huggins' pitch" is still not a truly cyclopean phenomenon since the input variable is a single time delay, while the perceived variable is a single pitch. If we want to produce changes in perceptual dimensions, then practically all perceptual phenomena might fall into this category. For instance, ordinary stereopsis using classical stimulations might be the analogy of the auditory examples given above. The perceived binocular pattern is identical to the monocular ones, except that it acquires an additional attribute, such as three-dimensionality. Even in the tactile domain, where spatial stimulation evokes spatial percepts, the many ingenious bitactile stimuli devised by Békésy (1960) are of the classical kind, although it is the tactile domain in which deep analogies to visual cyclopean stimulation could be found.

In this section I would like to clarify another point. The development of a scientific discipline often follows a helical path. The historically inclined mind is usually impressed by the cyclical nature of this progress as expressed by the maxim "there is nothing new under the sun." The researcher, however, concentrates his attention on the climbing pitch of the helix, to the few radically new ideas that are based on the slowly accumulating new findings. He hopes that some of the new paradigms will result in more substance than the mere sum of previous knowledge. This difference between the historical and research-motivated approaches can be illustrated by the technique of cyclopean stimulation. When I first published my results with random-dot stereograms, the findings were well-received, but several scholars quickly pointed out that the basic idea was a mere extension to vision of the technique of binaural noise stimulation. As was discussed, this is only very superficially so. Although the invention of counterpoint took place centuries ago in the auditory domain, strangely enough no auditory investigations used cyclopean counterpoints as stimuli prior to 1960. This is the more unexplainable since auditory perception is free from the aesthetic constraints of musical harmony. The essence of cyclopean stimulation is not in the complexity of the stimulation but in the production for the perceiving mind of a pattern that is independent in shape of the patterns at the transducer levels. If one were to use the Cramer and Huggins technique and, instead of a single pitch, produced a series of pitches, one would obtain a "cyclopean melody." This would be cyclopean since a pattern of temporal delays in time (that exist only as a relation between the left and right noise inputs) produces a pattern of perceived pitches in time. Similarly if one were to vary the temporal delay as a pattern in time in Licklider's experiment and produce a changing spatial pattern of perceived localization in time, this latter pattern would be strongly cyclopean.

Let me conclude this fundamental chapter with a final auditory example that will further clarify the notion of cyclopean stimulation. This example shows that in audition even a full analogy to random-dot stereograms does not yield a genuine cyclopean stimulus. Guttman and Julesz (1963) and Julesz and Guttman (1963) played repetitive segments of white noise, prepared by a computer so that the end of the segment blended smoothly into the beginning of its next occurrence. The results showed that with very short segments, a fraction of a second long, the listener experiences "motorboating," and with segments that are short enough for pitch sensation the listener hears auditory pitch. When the cycle is many seconds in length, the repetition is not noticed at all, except by deliberate search for particularly codable fragments of sound. The longest segments heard as repetitive without such an effort were about one second long—what Guttman and Julesz estimated as the limit of "auditory memory." Neisser (1967) called this "echoic memory." Obviously, in the range of mechanical Fourier analysis the heard pitch (or rather a melody produced in this way) is not cyclopean since it already exists on cochlear levels. However, in the range of 1–16 Hz (1 sec to 60 msec segment duration) the spontaneous perception of repetition is probably the result of a central process; therefore, stimulation in this range is quasi-cyclopean. It is only quasi-cyclopean, since the percept of a constant repetition—the monaural counterpart of Huggins-pitch—is too simple to be regarded as a cyclopean phenomenon. However, a slow variation of the repetition rate could portray a cyclopean rhythmic pattern (similar to a cyclopean melody produced by a sequence of Huggins-pitches). Since at repetition rates longer than 1 sec the perception of reoccurrence requires deliberation and scrutiny (although possible up to about 4 sec), according to our definition,[9] we will regard this stimulus condition as beyond the realm of our concern. This technique, however, can be used for studying auditory memory. Since the perception of repetition in the motorboating range requires neural mechanisms whose anatomical locations are not fully known, we must await further elucidation of the localization of the "cyclopean ear." Of course the same dependence on neurophysiology occurs in vision, but in the last decade cortical neurophysiology of vision has made much greater progress than neurophysiology of the auditory cortex.

These great advances in cortical visual neurophysiology, particularly the epoch-making discoveries of Hubel and Wiesel, make the study of a physiological psychology of visual perception possible. The same sophistication has not yet been reached in audition, and to write a similar book on a cyclopean auditory perception has to be postponed for the present.

It might well be that the deeper neurophysiological insight into the visual system as compared to that of the auditory system reflects a fundamental difference in the anatomy of the two sensory modalities. In mammalian vision a new geniculate-cortical system developed and became dominant over a parallel archaic visual system. This neo-system has a precise structure with well-specified connections, whereas the archaic visual system is diffuse and less specific. Almost all phenomena of visual perception take place in the geniculate-cortical system. There are very few visual phenomena that are attributed to the archaic system such as vergence,[10] accommodation, and pupillary control by the superior colliculus. However, in audition this neo-pathway has not developed; consequently, the auditory system is basically an archaic, diffuse structure that is hard to probe.

The use of cyclopean techniques in audition is further handicapped by the great gap between pitch perception and linguistic-cognitive processes. For vision, the cyclopean retina at the level of stereopsis or movement perception is conveniently located in the middle of a chain of perceptual processors. It is not at all trivial that, say, super-resolution processing of vernier acuity occurs before stereopsis or that apparent movement perception may occur after this level. However, in auditory perception it is very difficult to invent cyclopean techniques that stimulate a cyclopean cochlea which is conveniently located in the processing chain and is not too close to the input or too high.

9. See § 3.2.
10. Vergence will denote convergence-divergence motions of the eyes.

Let me illustrate this difficulty in audition by a visual example. In figures 2.3-3, 2.3-4, 2.3-5 and 2.4-4* the global image of the letters "T" or "V" has been portrayed by various cyclopean techniques. This merely served as a demonstration to show that perceptual phenomena based on orientation, vernier acuity, movement, and depth gradients can be used as local elements for portraying a global image. However, to conclude from these experiments that letter recognition occurs after orientation, vernier acuity, movement, or depth perception is so obviously true that it borders on the comical. Similarly, speech recognition or enjoyment of music are such highly central processes that most auditory processes which could be used as local elements to portray them occur at much earlier levels. For instance, we can transmit linguistic information by phonemic utterances or by dots and dashes in the Morse code, which in turn can be changed to two pitch or timbre values. The cognitive processes are so much more elevated in the hierarchy than the perceptual processes of pitch, timbre, or pulse duration that for the study of the cognitive processes these cyclopean stimuli are too peripheral.

There are two ways to avoid trivial localizations. One is to look for perceptual phenomena that are simpler than the cognitive ones and thus might reside in the vicinity of a given cyclopean cochlea. The other, more difficult way is to find cyclopean techniques that stimulate a cyclopean cochlea at some highly central level. Such a visual example is given by figure 3.1-1 wherein two textures are composed of English and nonsense words of equal length. Since texture discrimination cannot be experienced, one can conclude that after word recognition no spatial cluster finding operation exists that would organize the half-fields into a higher entity. A similar cyclopean experiment was conducted at cognitive levels by Treisman (1964) and will be further discussed in § 4.9. She showed that bilingual subjects noticed that two messages (one of which was delayed with respect to the other) were identical when the attended and unattended ears received messages in different languages. In this example the global information is the semantics of the text while the actual languages, say, English or Hungarian are "only" the local elements for conveying the global information.

These few cognitive examples show the possibility that cyclopean reasoning can be applied on this level as well. Much of art, especially music and poetry, is basically cyclopean since the basic elements are themselves complex cognitive processes that portray an even more global phenomenon. The success of an artistic style can be measured by its ability to evoke in the reader, listener, or viewer the same global message. If the global process cannot utilize the perceptual or cognitive phenomena that were used to portray the global message, then the artistic attempt is barren; it reduces to an empty numerology—playing with symbols and relations.

I return now to visual perception; this brief departure into auditory perception and cognition was intended only to deepen the reader's familiarity with cyclopean reasoning.

3

Cyclopean Stimulation and Physiological Psychology

3.1 Perception as a Hierarchy of Pattern Matchings

Neurophysiology cannot restrict its explanations to strictly physical, chemical, and anatomical terms. A large volume of findings exists, obtained by sensory experiments, in which an essential element is the subject's report of his own sensations. No physiochemical theory would be able to predict the sensations that a certain radiant energy might evoke. These subjective experiences are of immense value to the neurophysiologist; and, except for a few purists, the majority pays attention to psychological findings. However, there are several neurophysiologists working in sensory physiology who would restrict the use of psychological findings to a subclass of phenomena, called "Class A" by Brindley (1960). Class A experiments are based solely on observations that can be experienced as the identity or nonidentity of two sensations (e.g., the case in sensory matches and threshold determinations). The rest he called "Class B" experiments. Only if some of the Class B experiments can be converted into Class A experiments would they be accepted as having the same impact as physiochemical and anatomical evidence has. This restriction seems to be too pedantic. There are countless perceptual findings of Class B that yield great insight into the visual mechanisms. Most experiments reported in this book, however, are Class A. Of course the sensations which have to yield the same or different experience are here of a much more complex kind than brightness or color. Typically, in the experiments of interest to us the subject must decide whether two complex spatial patterns are identical in their textures or whether they appear at two different depth planes.

There is one difficulty with these more complex pattern matchings which is less cumbersome for simple brightness and color matchings. For the patterned stimuli an operation is necessary that selectively attends to certain aspects of the stimuli and normalizes them prior to or during the pattern-matching operation. When, for instance, two triangles having different sizes and positions are drawn over a complex background and are compared for their similarities, they first have to be attended to while the rest of the stimulus material is ignored. Second, the figures effectively must be zoomed and superimposed by some normalizing operation. The selective attention process will be discussed in chapter 4, while the normalization process is dealt with in chapter 9. These processes are so basic that they will be discussed in great detail even if many relevant findings are of the Class B variety. However, for texture discrimination, random-dot stereograms and cinematograms, the most explored subjects in this book, selective attention and normalization problems are of a much simpler kind.

Next I shall discuss visual perception as a hierarchy of pattern-matching processes. When discussing pattern matchings I will restrict myself to cases where at least one (if not both) of the patterns to be matched are visually presented. Thus I will never compare mental images with each other as they occur during dreams, hallucinations, or pictorial recall. On the other hand, one of the images can be a mental image stored in memory, but it must always be compared with a physically presented stimulus. Most of the shape recognition problems belong to this category. An input image, say of a familiar object, is compared with some stored representation of it. The stored image undergoes some drastic information reduction and the input image can be greatly distorted without affecting the recall. The ease with which we recognize cartoons demonstrates to what extent a few stimulus features can be used to retrieve the memorized representation. The perceptual weakness of such processes has been demonstrated by Julesz (1962a), as shown in figure 3.1-1. Here the left half-field of the image contains seven-letter English words, while the right half contains nonsense words of the same length. We can see that this is so by scrutinizing the field word by word, but it is impossible for us to perceive the two halves as separate entities. The discrimination in this example is based to some extent on cognitive processes, not on perceptual ones; and, therefore, we will ignore them in this book.

A simpler level of pattern matching occurs when two patterns are physically presented side-by-side to subjects who are asked to compare them. Such a process operates in visual texture discrimination, where the basis of comparison is not point-by-point but rather, some extracted features are matched. Figure 3.1-2 shows two random-dot

```
METHODS SCIENCE COLUMNS NIATREC YLKCIUQ DEHCNUP
SUBJECT MERCURY GOVERNS ECNEICS YFICEPS ESICERP
COLUMNS CERTAIN QUICKLY SDOHTEM SDROCER EZIDIXO
RECORDS EXAMPLE SCIENCE STCIPED HSILGNE NIATREC
MERCURY SPECIFY PRECISE TCEJBUS DEHCNUP SNREVOG
METHODS COLUMNS MERCURY ELPMAXE YLKCIUQ YFICEPS
CERTAIN DEPICTS ENGLISH ECNEICS ESICERP ELPMAXE
QUICKLY METHODS EXAMPLE YFICEPS YRUCREM DEHCNUP
ENGLISH SUBJECT RECORDS ELPMAXE SNREVOG EZIDIXO
MERCURY PUNCHED CERTAIN SNMULOC TCEJBUS ESICERP
SUBJECT OXIDIZE GOVERNS HSILGNE SDROCER ELPMAXE
METHODS ENGLISH COLUMNS NIATREC ESICERP DEHCNUP
DEPICTS SPECIFY PRECISE EZIDIXO YLKCIUQ ECNEICS
CERTAIN RECORDS SCIENCE STCIPED ELPMAXE HSILGNE
PRECISE QUICKLY METHODS YFICEPS YRUCREM SNREVOG
```

3.1-1

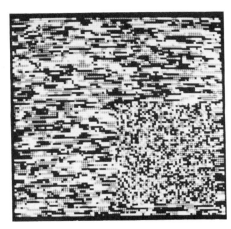

3.1-2

3.1-1 Nonspontaneous discrimination of two half-fields containing seven-letter English and nonsense words, respectively. (From Julesz 1962a.)

3.1-2 Spontaneous discrimination between two areas having the same first-order but different second-order statistics. (From Julesz 1962a.)

textures with the same first-order statistics but different second-order statistics (Julesz 1962a). Inspection immediately reveals a difference in the granularity of the two patterns, and the textures are spontaneously perceived as two distinct entities. In figure 3.1-2 the separation between the two textures is visually quite apparent. In contrast to shape recognition, no memory is required. Furthermore, since these textures are stochastically generated, the only processes that can operate here are some relatively simple feature extractors that are inherently built into the visual system. Such discriminable textures based on changes in some stochastic or topological parameters may or may not be used as cyclopean stimuli, as shown in figures 2.3-2 and 2.3-3. The only criterion is whether the peripheral processes are equipped to extract the very features that are the cues for discrimination or whether some central processes are evoked.

A third, even more primitive level of pattern matching occurs when two stimuli, presented simultaneously or in rapid temporal succession, are compared with each other point-by-point. This process is still much more complex than those sensory processes by means of which brightness or color matchings are performed. Stereoscopic depth perception and stroboscopic (apparent) movement perception are examples of such point-by-point matchings for which anatomically and neurophysiologically each corresponding point is functionally connected in the two eyes' views or in two successive frames. Such pattern matchings have been achieved by the random-dot cinematograms and stereograms as already shown in figure 2.3-5 and 2.4-1*. These strongly cyclopean stimuli give rise to a vivid sensation of movement or depth, respectively; and the experienced separation between textures is much more compelling than in texture discrimination.

These examples suggest that as we go from shape recognition to point-by-point matchings, the less memory and preprocessing is required. At the same time, slight changes in the patterns become more apparent. While we can recognize a face in a blurred snapshot, the slightest scratch in a stereoscopic image pair or motion picture frame is immediately detected. Thus stereopsis and stroboscopic movement perception of textures are the most visually vivid and concrete experiences, while shape recognition is a much more abstract process with less visual imagery. But even shape recognition is perhaps not as abstract as the verbal cognitive processes.

Since visual texture discrimination and point-by-point matchings can be studied without touching upon the many enigmatic problems of memory, most of our efforts will be devoted to these, while problems of shape recognition, which are closely linked to memory, will be pursued in lesser detail. However, some of the principles that have been clarified for stereopsis and texture discrimination can be applied to shape recognition after some generalization. I will discuss such generalizations under § 9.5.

3.2 Perception versus Scrutiny

In the previous sections we have already assumed that perception is a rapid, effortless, and spontaneous process in contrast to slow deliberation and scrutiny of the stimulus. Indeed after the expenditure of sufficient effort and time, any difference between two patterns of space, time, and energy distributions can be detected. The cognitive processes of language, logic, and mathematical manipulations, in particular, permit such a slow, thorough evaluation. However, our interest is in the effortless sensing of certain patterns, which will be regarded by us as perception. Any internal structure in the stimulus that is not spontaneously perceivable becomes empty numerology. For instance we know that in music no temporal symmetry in the melodical sequence is perceivable by the ear. Yet many baroque composers liked to use symmetrical melodies because they looked pretty in the score. Thus, in audition, symmetry is nonperceivable, requiring scrutiny by another modality. On the other hand, the periodicity so readily perceivable by the auditory system is less perceivable by the visual system for very complex patterns. This is illustrated in figure 3.2-1. Here the quadrants of a 100×100 random-dot array are identically repeated. This can be verified by looking consciously for certain micropatterns that will be found four times; yet during a brief presentation this pattern appears to be like the random-dot array in figure 3.2-2. However, if we take figure 3.2-1 with ¾ redundancy but "mirror" the same four quadrants across a horizontal and a vertical axis of symmetry as shown in figure

3.2-1

3.2-2

3.2-1 Nonspontaneous perception of repetition. The four quadrants are identical yet can be perceived only with scrutiny.

3.2-2 Random-dot texture.

3.2-3 Spontaneous perception of two-fold symmetry. The four identical quadrants are "mirrored" and yield a strong percept of symmetry.

3.2-3, the two-fold symmetry of the pattern is quickly and effortlessly perceivable (Julesz 1966a, 1967a). A two-or-four-fold symmetry can be perceived within 40 msec frame rates in random-dot movies portraying dynamic noise, as demonstrated by Julesz and Bosche (Bosche [1967] describes one of our movies). Here each frame consists of uncorrelated random-dot textures having an octal symmetry. When presented in temporal succession, in addition to seeing dynamic noise, the global symmetries are well perceived in each frame. That perception is not related to mathematical constraints in the stimulus is well conveyed in figures 3.2-4 and 3.2-5 where the random-dot patterns with one-fold symmetry are identical in the two displays, except that in 3.2-4 the axis of symmetry is horizontal while in 3.2-5 it is vertical. Inspection immediately reveals that the horizontal symmetry is much less perceivable than the vertical one, a result that was known to Mach (1886), who used simple

3.2-4   Random-dot pattern with
one-fold symmetry across the
horizontal axis. (From Julesz 1967a.)

3.2-5   Random-dot pattern with
one-fold symmetry across the
vertical axis. (From Julesz 1967a.)

3.2-4

3.2-5

3.2-6   Random-dot pattern with
centric symmetry. (From Julesz
1967a.)

amorphic shapes with various amounts of symmetry. While the horizontal one-fold symmetry is still perceivable, the centric symmetry (the reflection through a center point) of figure 3.2-6 requires scrutiny, as shown by Julesz (1967a,b). This result is different from Mach's findings with simple amorphic shapes. With the complex patterns shown here, only *n*-fold symmetries can be perceived, while centric symmetry cannot. A more detailed discussion on symmetry perception will be found in § 4.7. The reader might conclude that criteria for perceptibility may be somewhat arbitrary. This is true, as it is with most criteria, but the most important aspect is that they can be measured and rank-ordered.

3.3 Hierarchy of Neural Feature Extraction in Visual Perception

As the reader may have noted, frequent references have been made to some very recent neurophysiological findings in vision. These findings are so important that they will be referred to constantly, and it is therefore useful to summarize them now.

Since this book is written in the spirit of physiological psychology, let me first discuss the relation between psychology and neurophysiology. (Here I want to be brief since some of these problems involve the most sophisticated questions of epistemology and, as such, will be avoided.)

Perhaps the best analogy of the relation between neurophysiology and experimental psychology is that between quantum physics and chemistry. One might argue that, in principle, chemistry is no longer necessary since the solving of the Schrödinger equations for some complex atoms or molecules can determine all their chemical properties. There are two major difficulties with this approach—one practical the other theoretical. It would seem most impractical to solve the "many body" problems of differential equations after years of calculation by the fastest computers, when the same parameters could be determined in minutes by chemical methods. The theoretical difficulty goes much deeper. It is quite doubtful that a physicist without any chemical insight would ever be able to "dream up" such interesting macroscopic phenomena as, for instance, the solid or liquid states or the long-term order of ferromagnetic materials below the Curie point. It is most unlikely that any statistical average of microscopic events that is chosen ad hoc would be both interesting and revealing. Therefore, both the chemist and the quantum physicist follow each other's achievements with interest, often to the extent that there are specialists who work at the "interface." They try to close the gap between the local theory of quantum physics and the global discipline of chemistry; they regard themselves as chemical-physicists or physical-chemists, respectively.

A similar, heuristically fruitful interaction between neurophysiology and experimental psychology has been established for a century. The best understood problems belong to physiological psychology, and the recently discovered relations between neural cytoarchitecture and behavioral functions can be regarded as the beginning of a psychophysiology. Particularly, the new methodology of microelectrode probing of the nervous system in a local area and the use of computer-generated complex stimuli of global constraints make the previous analogy between a local discipline (neurophysiology) and a global discipline (experimental psychology) more fitting and make attempts to close the gap between the two more desirable.

A major difficulty we will encounter is that this book is devoted entirely to human visual perception, while most of the relevant neurophysiological evidence is collected on the horseshoe crab (*Limulus*), frog, rabbit, cat, and monkey. Despite Hartline's epoch-making discoveries of lateral inhibition in the *Limulus* (1949) and the discovery of feature extractors early in the frog's visual system by Lettvin, Maturana, McCulloch, and Pitts (1959), I will allude to these findings only because of their great heuristic value. Even the movement detectors in the rabbit's retina as revealed by Barlow and Hill (1963) will be of secondary interest to us, since the visual system of the rabbit is quite different from ours. The most important findings, on which we will base certain arguments, were obtained in the visual system of the cat and especially of the monkey. Hubel and Wiesel (1968) have shown that their earlier findings in the visual cortex, lateral geniculate nucleus, and optic nerve of the cat are similar in some important organizational principles to the findings obtained in corresponding organs of the monkey; thus, we can be confident that a very general structure has been revealed. However, the cat has no fovea and minimal color vision; furthermore, its visual resolution is poor compared to that of the primates. Therefore primary emphasis will be given to findings obtained in the monkey, regardless of their paucity and recency.

One of the most important notions of early feature extraction is the concept of a receptive field. Kuffler (1953), placing his microelectrodes near retinal ganglion cells of the cat, found that certain areas of the retina when stimulated by spots of light caused the ganglion cells to fire, while other areas inhibited firing. The shape of the excitatory areas for retinal ganglion cells was shown to be a small disk surrounded by an inhibitory annulus or vice versa. Similar concentric receptive fields have been found in the optic nerve and the LGN of the cat and monkey by Hubel (1960) and Wiesel and Hubel (1966). The only difference for LGN cells is that the receptive fields are smaller, and the receptive fields in the monkey are proportionately smaller than those in the cat. Furthermore, in the monkey LGN units having concentric receptive fields (called Kuffler-units) are also segregated into color-sensitive areas where the excitatory (or inhibitory) center is responsive to one color, while its surrounding inhibitory (or excitatory) annulus is responsive to another color as seen in both the monkey and in the goldfish (MacNichol and Svaetichin 1958). These concentric receptive fields also have a characteristic temporal behavior: If the center fires for "on" responses, then the annulus fires for "off" responses or vice versa.[1]

It is obvious that these units perform the earliest pattern analysis or feature extraction in the visual system. Only spatial changes evoke responses, while homogeneous illumination, however strong, influences very little the firing of these units. In functional terms these are "dot detectors" or more specifically "bright dot in a dark surround detectors" or "dark dot in a bright surround detectors." Of course there are "Ganzfeld detectors" in the retina, responsive to average brightness of a large region, which regulate the pupillary mechanism through the superior colliculus, but we are only interested in neural units that participate in processing patterned stimuli.

The really revolutionary discovery was the relation between receptive field geometry and the cytoarchitecture of the cortex. Mountcastle (1957) discovered the columnar organization of the cat's somatosensory cortex. This vertical modular arrangement in the somesthetic cortex means that units along a track perpendicular to the cortical surface all give rise to the same sensory discharge—for example, the successive cells along one track respond to skin touch-pressure, and the cells along another track to joint rotation in the monkey (Powell and Mountcastle 1959). The beauty of this correlation between cortical organization and functional organization became fully apparent in the monumental findings by Hubel and Wiesel (1960, 1962) in the visual cortex of the cat. They found feature extractors of hierarchically increasing complexity. However, as one goes from the so-called simple units having elongated

---

1. There are other types of chromatic units, for instance, one which may respond at "on" over its whole receptive field to red light; and "off" over its whole receptive field to green light (De Valois 1960). A detailed account of all these neurophysiological findings goes beyond the scope of this section.

receptive fields with antagonistic surroundings—also called slit or edge detectors (though Hubel and Wiesel have deliberately avoided such terms)—to complex and hypercomplex units that respond to highly special features (like movement in a certain direction or the end of a line), one notices that despite their diversity all of these feature extractors have a common characteristic: they respond optimally to a certain orientation. In a vertical module (column) perpendicular to the cortical surface of the cat and the monkey (Hubel and Wiesel 1960, 1962, 1968) there are several types of units, from the simple and complex kind in Area 17 (and even hypercomplex units in the monkey) to the complex and hypercomplex kind in Areas 18 and 19. But in a given column, all detectors have the same preferred orientation. In addition to this meticulously precise mapping of the orientational information, the retinal position is also maintained, and units with receptive fields in neighboring retinal positions tend to lie in close proximity.

Another exciting finding by Hubel and Wiesel is the hierarchy of feature extraction. It seems as if each unit in the hierarchy results from combining the outputs of units of lower complexity using both excitatory and inhibitory connections. Thus the simple units of the slit or edge detector type are found in the input layer (fourth layer) of Area 17 and may be built from Kuffler-units in the LGN by summing several adjacent Kuffler-units that fall on a line of a given orientation. This summation results in a narrow elongated receptive field having an elongated elliptical excitatory (inhibitory) area surrounded by an antagonistic neighborhood. Such cortical units fire optimally for line segments (slits or edges) that fall on the proper location on the retina and have the preferred orientation. These simple units are the only ones (in addition to Kuffler-units) whose receptive fields can be plotted by luminous dots and segregated into inhibitory and excitatory areas. The complex and hypercomplex units, on the other hand, respond to such complex features as movement of an edge in a certain orientation and direction or the perpendicularity of two intersecting line segments.

It would take too long to review the great variety of these complex and hypercomplex feature extractors, and the reader is referred to the original publications by Hubel and Wiesel and others. Here it should suffice to note that the notions of straightness, orientation, velocity, position, parallelism, perpendicularity, abrupt ending of a line, corners, and so on, appear to be innate; it is most likely that they are genetically programmed into our CNS (central nervous system).

In the light of these discoveries the old controversy between the structuralists and adherents of the Gestalt school has been resolved in a higher synthesis. There are elementary building blocks of perception—contrary to the Gestaltists; however, these are not the simple "atoms" of sensations of the structuralists, but complex "perceptual molecules."

It is also interesting that these perceptual molecules are obtained by parallel processes (i.e., many units are simultaneously evaluated). This parallel, built-in feature extraction probably does not extend much beyond these relatively simple features. Some of the more complex percepts are probably obtained by serial (sequential) processes. Even such relatively simple percepts as a pentagon or a hexagon need not be extracted by a parallel process. When the sides (and corners) of a pentagon are presented in sequential or nonsequential order, the sequential presentation requires slower repetition rates to produce unitary percept of the pentagon (Julesz 1964a). If the detection of the pentagon were a parallel process, then within the temporal inertia of the system the order of the presentation should be immaterial. One might argue that this conclusion is valid only if the essential features are presented, and it is not certain that edges and corners are the essential features for the "pentagon detector." Although I do not see what other more appropriate features a pentagon may possess besides edges and corners, this criticism is philosophically correct. After all, one can never prove the null hypothesis. However, I think believers in parallel processing would have rejoiced if sequence had made no difference. Probably the truth is that the visual processes are spatiotemporal: under conditions where the spatial aspect dominates, we think of the process as parallel; when the temporal aspect is more important, the process appears to us as serial.

Later, I will review many experiments that show serial processes at work for complex patterns, a sort of pictorial syntax. Yet it is important and revealing that a considerable amount of the input information undergoes a parallel preprocessing.

Another important discovery by Hubel and Wiesel (1962, 1968) is the preponderance of binocular units in the cortex. The majority of the units can be stimulated by the left or right eye to various extents. In the cat this binocularity occurs in Area 17, while in Area 17 in the monkey the monocular units are in the majority; but in areas higher up the majority of units are binocular. Hubel and Wiesel (1965b) and Wiesel and Hubel (1965a,b) showed that these genetically determined binocularly driven units require patterned stimuli to keep them functional. They showed that kittens with one eye sutured from birth, or with artificially introduced strabismus, grow up with only monocularly driven cortical units, and that this deterioration of cortical binocular function is irreversible. However, recent study by Chow (1969) indicates that the sutured eye that becomes nonfunctional can be made functional again by prolonged forced training. As a result of this forced training deteriorated layers of the LGN that received their projections from the sutured eye become undistinguishable from the other layers. There is no evidence yet that cortical damage could be reversed by forced stimulation. The essence of these findings is that the influence of stimulation in an early period of life is necessary to preserve innate abilities, and the old controversy between nativist and empiricist views is significantly modified. These neurophysiological discoveries by microelectrode recordings thus have settled both the structuralist-Gestaltist and the nativist-empiricist debates in unexpected ways.

An even finer distinction among binocular cortical units in the cat has been made by Barlow, Blakemore, and Pettigrew (1967) and by Pettigrew, Nikara, and Bishop (1968). They found units in Area 17 that responded optimally to certain binocular disparities and as such, can be regarded as representing the various depth planes of the model of stereopsis that has been proposed by Julesz (1960a, b). The units they discovered fire optimally when the orientations of the two line segments are identical in both retinae and the binocular disparity (horizontal displacement) between the corresponding retinal coordinates is of a certain value.

As we will see later this local fusion of units having the same orientation is prone to ambiguities, and one must look for the next higher neural level that represents a global solution. This psychologically postulated global mechanism has not been found yet by the neurophysiologists. Furthermore, the cat is not a visually sophisticated animal, and perhaps its relatively coarse retinal patterns have few ambiguities and do not require a global mechanism in order for the stereopsis mechanism to function satisfactorily. Until behavioral experiments with random-dot stereograms reveal that the cat can perceive the global binocular patterns, there is no need to assume that stereopsis in the cat resembles that of the monkey, ape, or man. At this time several neurophysiologists are trying to train cats and monkeys using random-dot stereograms. Bough (1970) has successfully established stereopsis in the macaque monkey in a training experiment with random-dot stereograms. The stage is thus set for exploring the neurophysiological bases of stereopsis in the monkey, with all of their implications for human stereopsis.

Let us mention another fact about binocularly driven neural units. In the cat, Hubel and Wiesel (1961, 1962) found no binocularly activated units in the lateral geniculate body (in contrast to the finding that 84% of the single units in the striate cortex were fired by light stimulation to either eye). However, Bishop et al. (1959) found that 8.5% of the LGN units were binocularly driven. Similar results were obtained in the cat's LGN by Erulkar and Fillenz (1960). In the LGN of the monkey DeValois (1960) found no binocular units. Recent findings by Chow (1969) in cats whose visual cortex had been largely removed show that after degeneration the LGN becomes greatly enriched with binocular units. Thus, it seems that there may be binocular units in the LGN, but their role is clearly secondary compared to the binocular units in the visual cortex. This does not mean that we regard the LGN as a passive "switchboard terminal" as was commonly assumed. Bizzi (1968) showed that the convergence angle of the eyes (monitored through the superior colliculus) changes the firing rate of LGN units. Richards (1968) regards this as the mechanism for size constancy—the site for a "perceptual zooming" process, a hypothesis to be discussed later.

I highly recommend careful reading of a recent paper by Hubel and Wiesel (1968) entitled "Receptive fields and functional architecture of monkey striate cortex." The following is a condensed version of their summary: After studying the striate cortex in lightly anaesthetized macaque and spider monkeys by recording extracellularly from single neurons while stimulating the retinae with spots or patterns of light, they found that most cells can be categorized into simple, complex, or hypercomplex classes with feature extraction capabilities very similar to those previously described by them in the cat. However, receptive fields usually are smaller than in the cat and are more sensitive to changes in stimulus orientation. A small proportion of the cells is color-coded. There are at least two independent systems of columns extending vertically from surface to white matter. The first type of columnar module contains cells with common receptive-field orientation (similar to those in the cat, but narrower). In the second modular organization cells are aggregated into columns according to eye dominance. The ocular dominance columns are larger than the orientation columns and their boundaries seem to be independent. There is also a tendency for cells to be grouped together according to anisotropy of responses to movement; in some regions the cells respond equally to the two opposite directions of movement of a line, but other regions contain a mixture of cells preferring one direction and cells preferring the other.

In addition to these vertical organizations a horizontal organization corresponding to cortical striation was also found by Hubel and Wiesel. The upper layers (II and the upper two-thirds of III) contain complex and hypercomplex cells, while simple cells are virtually absent.[2] These cells are primarily binocularly driven. Simple cells were found deep in Layer III and in IV-A and particularly in IV-B. In layers IV-A and IV-B they found units lacking orientation specificity and conjectured that they might have their origin in the LGN. In Layer IV most cells are monocular and are segregated into mosaics with cells of some regions responding to one eye only, those of other regions responding to the other eye. Layers V and VI contain mostly complex and hypercomplex cells, binocularly driven. Thus the cortex is seen as a system organized vertically and horizontally in entirely different ways. In the vertical columnar organization stimulus features such as retinal position, line orientation, ocular dominance, color, and perhaps directionality of movement, are mapped in sets of superimposed but independent mosaics. The horizontal system segregates cells into layers by hierarchical orders of feature complexity, the lowest orders (monocularly driven simple cells) are located in and near Layer IV, the higher orders are found in the upper and lower layers.

Let us quote the concluding sentences of this article by Hubel and Wiesel (1968):

> Specialized as the cells of 17 are, compared with rods and cones, they must, nevertheless, still represent a very elementary stage in handling complex forms, occupied as they are with a relatively simple region-by-region analysis of retinal contours. How this information is used at later stages in the visual path is far from clear, and represents one of the most tantalizing problems for the future.

Of course, the striate cortex is not the only site of complex visual processes. Recently Gross, Bender, and Rocha-Miranda (1969) have found very large receptive fields (greater than 10×10 degrees) in the inferotemporal cortex. They found several units that responded most strongly to very complex figures. For example, to quote the authors, "One unit that responded to dark rectangles responded much more strongly to a cutout of a monkey hand, and the more the stimulus looked like a hand, the more strongly the unit responded to it." Furthermore, the authors believe that the activity of units in the inferotemporal cortex might depend on other variables besides the features of the stimulus. This conjecture was based on the fact that in many units a fast EEG was necessary to demonstrate receptive fields and also on the fact that removal of the inferotemporal cortex results in visual-learning deficits. Thus inferotemporal units may require that the stimulus characteristic should be proper and of significance to the animal.

---

2. The definition of the various layers in the visual cortex is supplied in Hubel and Wiesel's papers. Here it should suffice to know that the input to the cortex is Layer IV.

Although I have summarized a good deal of physiological data above, let me note again that this book focuses on findings obtained solely by psychological means. That is, instead of opening the black box, we try, rather, to determine its properties by studying its output as a function of the input stimuli. Nevertheless, I will offer "peeks" into the black box whenever opportunity permits. More important, I hope that the psychological findings reviewed in this book will guide the neurophysiologist in his research, particularly in his search for global structure. Despite the great heuristic value of these neurophysiological findings discussed above we know at present only a very limited number of perceptual phenomena that might be direct consequences of them. For instance, the Mach-bands as conjectured by Ernst Mach (1886) in the last century, might be the result of lateral inhibition. The neurophysiological evidence was only found half a century later in the *Limulus* eye by Hartline (1949). However, the extensive literature on Mach-bands—reviewed in a comprehensive book by Ratliff (1965)—only indirectly suggests the role of lateral inhibition. Pertaining to concrete details, direct evidence is less convincing. We must conclude that much work has to be done by both experimental psychologists and neurophysiologists before the physiological findings of early feature-extraction will have indisputable psychological significance.

In spite of the epoch-making discoveries in recent years by neurophysiologists, there are sceptics who contend that the brain is such a complex organ that according to the stimulus class presented, its observable architecture will follow suit. They argue that if in place of line segments one were to use, for instance, curved arcs, a completely different cytoarchitecture might be revealed, provided that the researchers were fortunate enough to place their probes into the proper cells. It is as if the shape of the holes in a fisherman's net determined the types of creatures the sea contains.[3] However, until someone is able to show a different cortical organization from Hubel and Wiesel's by finding and using other "canonical" sets of stimuli, such arguments have only epistemological value. Furthermore, lines can be regarded as tangential segments of curves and, as such, are the first-order approximation of functions (in a Taylor series expansion). Thus the orientation-sensitive feature extractors of Hubel and Wiesel may correspond to the most significant term in the approximation; more

---

3. Recent neurophysiological studies reported by Hirsch and Spinelli (1970) and Blakemore and Cooper (1970) of kittens raised in specially patterned environments raise further questions. In a normal cat, the orientation sensitive cortical edge detectors are about equally represented with respect to orientation. However, early visual experience can change this organization. Hirsch and Spinelli reared kittens with one eye viewing vertical stripes, the other horizontal, and found only monocular cortical units with elongated receptive fields whose orientations matched the pattern experienced by that eye. Blakemore and Cooper allowed their kittens normal binocular vision, but from birth on raised them in a vertically (or horizontally) striped environment. After five months, these kittens were both behaviorally and neurophysiologically tested. They had "normal" vision for vertical (horizontal) contours, but were blind for horizontal (vertical) edges that were absent in their early experience. If this deficit were merely a passive degeneration of certain cortical neurones because of under-activity, the findings would be only a generalization of Hubel and Wiesel's squint experiments. However, Blakemore and Cooper note that they did not find large regions of "silent" cortex and assume that the under-activated cortical units reorganized and changed their preferred orientation. If this assumption is true, is this reorganization restricted to orientation changes or can an edge detector change into a curvature detector?

complex units such as corner and hypothetical curvature detectors extract second and higher spatial derivatives, and thus they belong to higher-order terms. If this is true, then Hubel and Wiesel's findings reveal the basic, most elementary structures, and the rest can be regarded as additional refinements.

3.4 Psychological Studies of Feature Extraction

In reviewing the neurophysiological findings obtained by single microelectrode recording, we might have concluded that the perceptual molecules are the province of neurophysiology, while the resulting complex shapes belong to psychology. This view is only partly justified. After all, the ultimate goal of neurophysiology should be the explanation of all mental events, however distant such a goal may be. Similarly, experimental psychologists cannot expect the workings of early feature extraction to be relevant to perception until these extractors are also demonstrated by psychological means.

Such a psychological demonstration of the existence of local feature extraction is by no means simple. If one satisfies oneself with some general idea, say neural crispening by lateral inhibition (without specifying a detailed mechanism) then such nonspecific neural operators have been proposed by psychologists as early as the last century. However, to demonstrate the existence of units responsible for Mach-bands or to show the existence of detectors of chromatic, moving edges, the psychological demonstration requires great sophistication. There are many perceptual phenomena that one might think are explained by the latest neurophysiological evidence, but it is difficult to find a close neurophysiological parallel. Take, for instance, the *orthogonal afterimages* of MacKay (1957, 1961) that were first reported by Hunter (1915). After adapting the viewer's visual system with a pattern consisting of concentric circles or radial lines, a field of white noise presented afterwards looks like a family of lines orthogonal to the adapting pattern. Thus radial lines will evoke concentric circles and vice versa. This phenomenon is obviously related to the extraction of orientation information. But it is not at all clear why the adaptation (perhaps satiation is a better word) to a certain set of curves will produce the perpendicular set of curves. It would have been a much clearer finding if the noise were perceived as the sum of all other possible patterns *but* the adapting one. In audition this is indeed the case. When Zwicker (1964) presented subjects with auditory white noise in which a very narrow frequency band had been filtered out, the subjects heard a faint pitch (corresponding to the center of the narrow filter) for a brief period after the termination of the noise. Such a paradigm is a much more direct illustration of the intactness of a certain critical band (an auditory perceptual molecule) and the satiation of all the rest than is the dual experimental paradigm of MacKay.

Unfortunately, visual patterns are two-dimensional, and the possible visual feature extractors are of a much richer variety than the auditory pitch detectors (and some of their combinations). Therefore a visual counterpart of the Zwicker phenomenon is not easy to produce. One possible analog would involve a random-dot movie, each frame of which consists of uncorrelated random-dot textures. Each frame should be generated by a Gaussian process such that first its two-dimensional Fourier transform is taken, then a two-dimensional spatial frequency component is filtered out, and finally the inverse Fourier transform is generated. The backtransform will look practically indistinguishable from the original visual noise. Since the same spatial frequency component is filtered out of each random frame, one might expect that after some time the visual dynamic noise has displayed all spatial frequencies except one and will have occurred often enough to satiate the visual system for all frequencies except that one. This experiment assumes that, first, the visual system has built-in spatial Fourier analyzers, and second, that all the spatial frequencies occur often enough by chance to satiate these postulated analyzers. (Both of these assumptions are questionable, although Enroth-Cugel and Robson [1966] and Campbell and Robson [1968] claim to have found neurophysiological evidence for spatial frequency analyzers.) Special care must be exerted to keep both the amplitude and phase of the removed component constant. One then might expect to see a $f(x,y) = \cos(2\pi \, mx)\cos(2\pi \, ny)$–shaped function after the removal of dynamic noise, where m,n are the spatial frequency components of the filtered out frequency. One could try to filter out some more interesting shapes from the dynamic noise (for example the noise could contain all possible patterns except the letter "A"). But then the probability of occurrence in each random-dot image of the complex combination of all those spatial frequencies that constitute the letter "A" is very low. Therefore, the unfiltered and filtered version of the visual noise would not differ adequately to produce satiation.

The Zwicker effect is quite weak in the auditory domain; and, therefore, it is not likely that its visual analog will be discovered. However, there is a strong effect in audition, which demonstrates the existence of critical bands (broadly tuned frequency analyzers) and which can be demonstrated in the visual domain as well. If auditory white noise passes through a band-rejection filter, then the masking of a single pitch by the noise depends upon whether its frequency falls inside or outside the rejection band. Masking experiments in the visual domain were recently carried out by Pollehn and Roehrig (1970), who used filtered two-dimensional visual noise and spatial sinusoidal gratings; but they were not studying the question of whether critical bands exist in vision. Also, two-dimensional visual noise is not the proper way of masking a vertical grating; therefore, Julesz and Stromeyer (1970) used one-dimensional filtered noise for masking. The noise consisted of vertical stripes whose amplitude along the horizontal axis of a visual monitor was determined by a Gaussian process. The visibility of a sinusoidal grating strongly depended on the frequency of the grating. If this frequency fell in the rejection band, the grating was visible, otherwise it was masked. However, the rejection band had to be on either side at least an octave wide, and two octaves wide for best results. Thus, the frequency analyzers in vision are very poor filters.

The finding that the critical band in vision is more than an octave wide has been shown by Blakemore and Campbell (1969a) using an adaptation technique (to be discussed in the next paragraph). Our finding confirms their results. It seems that frequency analysis occurs in the visual domain as well. However, the frequency analyzers have such a shallow characteristic that the analogy to Fourier analysis is rather remote.

Our findings with masking one-dimensional noise are merely a verification (although based on strong effects 30 dB above threshold or even higher) of the original findings of Pantle and Sekuler (1968c) and Campbell and Robson (1968) who first demonstrated the existence of frequency analyzing channels in vision using gratings at threshold. Historically, the first hint of spatial frequency analysis was made by Campbell and Kulikowski (1966) who investigated the visibility threshold of a test grating as a function of the orientation of a masking grating. They briefly mentioned that largest threshold increase occurs when the masking and test gratings have similar geometry. That spectrum analysis actually occurs in the visual system was demonstrated by Pantle and Sekuler (1968) using adaptation and test gratings of different frequencies and by Campbell and Robson (1968) who noted that a square grating appears as a sinusoidal grating until the higher harmonics reach their visibility thresholds. Blakemore and Campbell (1969b) demonstrated that prolonged inspection of a grating with specified grid constant (spatial frequency) will elevate the detection threshold for grids which have the same periodicity but not for grids which have different spatial frequencies. In an amusing demonstration, they pictured a human figure holding an umbrella on a rainy day. The rain is portrayed by sporadic tiny line segments that make up a vertical grating. After adapting to another grating, with high contrast but the same grid constant as the rain droplets, the rain in the test picture disappears while the figure with the umbrella appears unchanged.

A recent study by Nachmias, et al. (1969) showed that at the threshold of visibility the various spatial frequency analyzers are statistically independent of each other (as long as the various spectral components have frequency ratios in excess of 5:4). This experiment finally established—after a hundred years delay—that the visual system is similar in its workings to the auditory system as proposed by Helmholtz. The finding by Nachmias, et al. (1969) indicates that at threshold the phase of the signals is ignored. This is also very similar to the auditory case. However, for perception above threshold phase information is used by the higher processors. This is also true in audition. After all, both an impulse function (Dirac delta function) and white noise have the same flat amplitude spectrum, but very different phase spectra. The fact that they are seen and heard as being very different shows that ultimately the phase information is utilized. However, the similarities between visual and auditory frequency analysis are not deeply rooted. First, the auditory frequency analysis is the result of the mechanical properties of the basilar membrane and cochlear geometry, while the visual frequency analysis is the result of neural computation.[4] Second, auditory signals are one-dimensional, and the one-dimensional frequency analysis (approximated by one-dimensional Fourier analysis) is a general operation; the visual signals are two-dimensional in general, and only for one-dimensionally

4. Maybe the proper auditory analogy to visual frequency analysis is the residue pitch. Schouten (1940) showed that the finding of the missing fundamental in a complex sound (residue) is not due to nonlinear distortions of the ear, but rather to some neural frequency analysis. Perhaps residue pitch analysis and visual frequency analysis would show more similarity to each other than ordinary pitch and visual grating perception do. However, residue pitch can be perceived only below 1.5 kHz, while visual frequency analysis appears to operate as high as the visual acuity limit.

varying spatial gratings does the spectral analysis model hold. It must be a very complex mechanism that determines the two-dimensional Fourier spectrum from one-dimensional Fourier spectra that are taken from all possible orientations. It can be shown that the sum of a rotating one-dimensional Fourier spectrum is equivalent to $F(m,n) / \sqrt{(m^2 + n^2)}$, where $F(m,n)$ is the two-dimensional Fourier transform of an $f(x,y)$ spatial distribution and m,n are the spatial frequencies in the horizontal and vertical directions (Julesz, Slepian, and Sondhi 1969). Thus, if we assume that edge detectors are responsible for the one-dimensional spectrum analysis and assume that the outputs of various such edge detectors in all possible orientations are summed, we get back the original stimulus except that the low frequencies are preemphasized. This corresponds to a spatial integration. In order to obtain the exact input stimulus the summed output has to be differentiated [or the spectrum has to be multiplied by $\sqrt{(m^2 + n^2)}$ before the inverse Fourier transform]. Whether such a low frequency preemphasis and deemphasis actually occurs is an interesting question and remains to be seen.

This seeming paradox that the outputs of threshold analyzers are not combined, but at higher levels the percept is the combination of the output of the constituent analyzers is probably a general organizational principle in the CNS. For instance, in color perception we perceive only mixed colors and are not conscious of whether a perceived yellow was a spectral yellow or some combination of red and green components. However, at threshold the component color analyzers can be demonstrated as achieved by Krauskopf and Srebro (1965). They presented, tachistoscopically, dim, narrow spots of monochromatic light to human observers. There was a high probability that *only* one or two color receptors would be stimulated. Indeed, subjects often report red or green sensations (in spite of the fact that the spot is of a different wavelength). Krauskopf also collected evidence that at threshold the output of the individual color analyzers is not combined by the detecting mechanism but is independently evaluated.

These examples might give the false impression that in every instance only the complex combined percept can be consciously perceived and the workings of the early analyzers always take place on a subconscious level. However, in a musical chord we are able to perceive the chord as an entity in itself, and yet we are also able to perceive each individual pitch separately.

All these speculations assume that the visual system is linear at least at threshold. However, this assumption might be questionable. In an experiment with random-dot patterns of various contrast and density (area fraction of black and white) Fender and Julesz (1967b) found a nonlinearity at the threshold of seeing. The contrast as a function of area fraction of black and white for detection of the texture, was *not a symmetrical* function at low and high area fractions. Thus the visibility of, say, 20% black dots in a white surround required a different contrast from 20% white dots in a black surround. The entire shape of the curve was different for area fractions below 20% and above 80%, although it appeared quite symmetrical within this range. This departure from symmetry and thus from linearity *at threshold* depends on the stimulus shape.

Finally, the linearity of the auditory system holds in a wide range above threshold, while the Fourier techniques (also called modulation transfer function technique) in vision was restricted to visibility threshold. I bring up these differences to point out that the analogy between audition and vision might be quite superficial. The fact that both systems specialize in extracting temporal or spatial changes is rather trivial.

In addition to the demonstration that adaptation to a certain grating can raise the threshold of a test grating (having the same spatial frequency), there are some new aftereffects that bear on spatial frequency analysis. Blakemore and Sutton (1969) showed that after prolonged adaptation to a grating, if one views a grating of the same orientation with somewhat narrower bars (higher frequency), then the bars seem even thinner than in fact they are. (The perceived spatial frequency appears elevated.) Broader bars seem broader still. This perceptual shift of perceived spatial frequency after prolonged adaptation is similar to the pitch shifts in the auditory domain. Yet one has to be extremely cautious with such analogies.

Most of this section was written before our experiments with masking filtered noise. In the light of these latest findings it appears that spectrum analysis occurs orders of magnitudes above threshold without affecting the critical bandwidth. It is possible even to take a complex signal (e.g., a squarewave grating) and selectively mask its first harmonic so that only the third harmonic of the signal becomes visible. Thus the analogies to auditory spectrum analysis appear deeper than had been suspected. Perhaps this transformation of spatial (temporal) patterns into the frequency domain has the same purpose both for visual and auditory processings: it renders the patterns invariant under any translation. Indeed, both a visual pattern and an auditory melody appear constant under a uniform positional or pitch shift, respectively.

The most interesting demonstration to date of the role of feature extraction in perception was given by Celeste McCollough (1965), who demonstrated a color aftereffect that depended on the orientation of stationary edges. The aftereffect was built up by alternately presenting for several minutes a pattern of black and colored (e.g., red) vertical stripes, and a pattern of horizontal black and green (e.g.) stripes. After adaptation a vertical black and white grid appeared pale green; an adjacent horizontal grid, pink. The aftereffect depended on the retinal orientation of the edges, for the colors switched grids when the head was tilted 90° to the side. McCollough linked the aftereffect to orientation-specific visual mechanisms.

In a refined study Stromeyer (1969) showed that only one colored grating is needed to produce the aftereffect; for example, alternating a red (green) vertical adaptation grating and a white horizontal grating gives a saturated green (red) aftereffect on a vertical (horizontal) test grating.

Harris and Gibson (1968) asked whether this aftereffect is actually the result of color adaptation of edge detectors in the cortex or could be simply due to ordinary negative afterimages on the retina. They said that the facts that the McCollough effect is linked to test pattern orientation, whereas afterimages are best seen on homogeneous surfaces, and that afterimages require fixation of the adapting stimulus (while the McCollough effects do not) can be explained in terms of afterimages if we make one assumption. The assumption is that observers tend to look predominantly at some part of the colored adapting grid and thereby build up a "plaid" colored afterimage. However, Harris and Gibson found that even when they used tachistoscopic illumination too short to permit systematic eye movements and random placements of the adapting grids, they still obtained the aftereffect. In addition, the long persistence of the aftereffect (days) disallows some simple chemical bleaching or retinal adaptation. They concluded that the aftereffect cannot be the result of afterimages and proposed a model postulating hypothetical neural "dipoles"—essentially contrast-detectors with two, nonconcentric opponent regions. These experimental results, however, are also compatible with the findings by Hubel and Wiesel (1968) of color and orientation sensitive edge detectors in the monkey's cortex. Yet the fact that even a retinal afterimage could explain most characteristics of the orientation-specific aftereffect shows that very careful psychological studies are necessary to establish that certain perceptual phenomena are the result of the feature extractors described by neurophysiologists.

I quote this work to make the reader aware that the McCollough phenomenon can produce strong and lasting aftereffects. Although it has not been conclusively proven yet, the site of this phenomenon is probably cortical, but certainly it is not retinal. Because of this most likely conjecture, the McCollough effect can be regarded as a cyclopean stimulus. Since the color aftereffect can be perceived without presenting a color to the retina and the color originates more centrally than the anatomical retina, the aftereffect is cyclopean. (However, because the color of the aftereffect is determined by the orientation of the grating presented on the retina, therefore, the phenomenon is only cyclopean in the weak sense.) It is then possible to exploit this aftereffect and try to produce other phenomena by utilizing it. Such an experiment will be discussed in § 7.9: Land colors can be produced by McCollough color aftereffects, showing that Land color phenomena occur after the McCollough effect.

An even more novel colored aftereffect has been produced with moving edges by Hepler (1968) and by Stromeyer and Mansfield (1970). Knowing about the McCollough aftereffect and the classical waterfall illusion (to be discussed later), it is perhaps not so surprising to find that after adaptation to oppositely moving bars of red and green color one might get a colored aftereffect in which a white-black grid will be seen as oppositely colored, depending on the direction of movement. The really startling discovery is that these aftereffects lasted as long as 6 weeks (after only a short adaptation period)! A variation of the experiment by Stromeyer and Mansfield was even more interesting. Subjects fixated at the center of a rotating spiral. The spiral rotated alternately in opposite directions, paired with a red square in a green surround for one rotation and a green square in a red surround for the opposite rotation. After 50 minutes of adaptation, colored images of the squares were seen when the center of the spiral was fixated and the direction of rotation sharply reversed. Further results are even more revealing. Stromeyer and Mansfield found that color aftereffects do not occur unless during adaptation the edges move in both directions, each direction paired with a different color. The colored square aftereffects with the spiral showed that the aftereffect did not change retinal locus. The movement-color aftereffects did not transfer to the other eye. This monocularity is also true for the

McCollough phenomenon (McCollough 1965). This fits well the monkey cortex findings of Hubel and Wiesel (1968) who claimed that the simple units were mainly monocular and some of them were color specific, whereas the complex and hypercomplex units were binocular and the majority of cells were not sensitive to color. If these preliminary neurophysiological findings remain true, these color aftereffects should be basically monocular.

I have devoted so much space to these color aftereffects because several complex features such as orientation, velocity in a given direction, and color have to be simultaneously presented in order to evoke them. As we mentioned, these aftereffects are sensitive to edge orientation. Some aftereffects less sensitive to orientation were studied by Sekuler and Pantle (1967) and by Pantle and Sekuler (1968a) using black and white moving stripes for adaptation and measuring visibility thresholds of a test bar of various orientations and directions of motion. The threshold was the highest when both the adapting and test stripes had the same orientation and direction of motion. The threshold decreased as orientation differences increased, and even at 60° difference the threshold was considerably elevated. However, Pantle and Sekuler (1968b) found that the movement aftereffect does not affect orthogonal edges. They found that adaptation to horizontal stripes moving vertically raised the visibility threshold of stationary horizontal stripes, but did not affect vertical stripes. While orthogonal directions of edge movement appear independent, opposite directions affect the produced aftereffect of each other. Wohlgemuth (1911) showed that an alternating waterfall illusion had considerably less duration than the original one with one -directional movement.

This independence between orthogonal edges in spite of strong interaction at even 60° angles might explain why, in the MacKay phenomenon, a set of edges causes adaptation to all orientations but the orthogonal set. But this explanation is psychological and not neurophysiological. Except for a few "corner," "parallelness," and "end-of-line" detectors in which two edges of perpendicular or identical orientations, or at some other angle, excite or inhibit each other, we have no evidence of how edge detectors are connected into higher organizations.

There are many indications that the visual system is sensitive to curvatures as well. That means that two adjacent edge detectors at various angles must be able to report this angle information to a next level (if it is true that the earliest cortical sites contain only edge detectors). The frog's retina has such curvature detectors as shown by Lettvin, et al. (1959), and it is likely that in higher forms the cytoarchitecture of the cortex will also reflect curvature extraction. The importance of curvatures is simply demonstrated by figure 3.4-1 in which only points of local maxima of curvature are drawn and the rest is filled in by straight-line segments (Attneave 1954). Surprisingly, these few points portray an intricate drawing exceedingly well. However, as Green and Courtiss (1966) pointed out, it is not the points of maximum curvature alone that convey information, since figure 3.4-2 containing only these points is unrecognizable. The direction, the order of connections must also be known. I bring this up to indicate that many local features can be predicted by the psychologists which neurophysiologists could search for.

Some of the simplest neural units do not seem to explain adequately certain perceptual findings which were attributed to them. We have already noted that according to many researchers, the Hermann-Hering grid illusion is peripheral or occurs prior to stereopsis and is believed to be the result of Kuffler-units. However, some corner, edge, or curvature detectors in the cortex might be other candidates for explaining the illusion (although these units must occur before the level of stereopsis).

We will return once more to the problem of the relation between neurophysiological evidence and perceptual findings in the chapter on models. Here let us note, that as long as the system is linear, any orthonormal decomposition (e.g., sinusoids, Bessel functions, etc.) of the input stimulus is compatible with the facts of perception, and only the opening of the black-box or the selective adaptation (masking) of a stimulus feature can decide which analysis is actually used.

The possibility that each feature extractor reports its findings both to the next hierarchical level and directly to some higher center is very attractive. After all, at the output of the edge detectors the shape of a dot is lost; at the output of a complex unit the exact location of an edge is lost, and so on. However, when we view an edge we know not only that it is parallel to another edge of the same orientation, but we perceive its exact location and are able to perceive individual dots. The structure is similar to transformational grammars for which it is inadequate to have access to a certain syntactic level, but all lower and higher levels in the hierarchy have to be reachable. This simultaneous perception of all hierarchical levels and our ability to shift attention to any particular level is characteristic of our visual system. We will discuss some of these problems in § 4.9 on selective attention. Here we note only that this enormous complexity of all features acting individually or in unison makes psychological verification of feature extractors so difficult. What does it mean that from a long list of words one can detect one's own name almost immediately? Does this mean that a neural feature extractor has been built in for certain very familiar patterns? Does one look for a certain face or name detector by simple neural probing? As one goes to increasing complexity, it becomes more unlikely that one could stimulate a given cortical unit with the specific stimulus feature. This is the reason why the probing of complex and hypercomplex units gets more and more difficult.

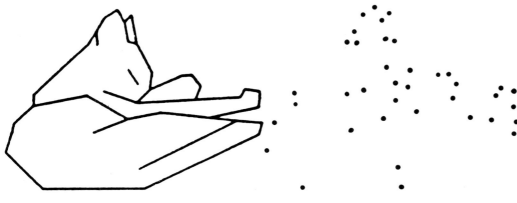

3.4-1

3.4-2

3.4-1 Figure portrayed by connecting points of maximum curvature by Attneave. (From Attneave 1954.)

3.4-2 Only points of maximum curvature of figure 3.4-1 are shown. This illustrates that, besides points of maximum curvature, direction has to be given as well. (From Green and Courtis 1966.)

This hierarchical structure can be interpreted in another way. The Kuffler-units detect dots, while the edge detectors merely "name" the event in which several adjacent Kuffler-units falling on a line are simultaneously stimulated by calling this occurrence an "edge." Similarly if several edge detectors of the same orientation are simultaneously presented, a complex unit will fire, indicating that a "parallel" event occurred. I assume that the direct (artificial) stimulation of a cortical edge detector might not evoke a vivid image of an edge, but just a faint notion of an edge. This faint notion of an edge (or parallelism, or perpendicularity) might be no more visual than tactile or auditory cognitive information to this effect. However, I was greatly surprised by the response to cortical stimulation of a blind woman by Brindley and Lewin (1968). The reported cortically induced phosphenes had the shape of small bright spots, but a few elongated edge-shaped phosphenes were also observed. This finding might still be reconciled with my original assumption that the Kuffler-units not only communicate with their next "superior" but can synapse directly to some "executive" levels. The macroelectrode stimulation of the cortical surface might have fired some central representations of the Kuffler-units.

Instead of assuming a direct access to each extracting level, there is another possibility of neural organization. After all, one can reconstruct a dot as the intersection of several edge detectors at a higher stage. It is only strange that when a dot is extracted by a Kuffler-unit, this information is thrown away to be reconstructed anew by a complex process. Of course, theoretically this is possible since a very large set of edges (or slits) in all positions, orientations, and width-to-length ratios can contain the original input information. However cumbersome it may be to reconstruct a complex texture of dots from edges, present neurophysiological evidence suggests such a solution, since almost every cortical unit has elongated receptive fields.

There is also some anecdotal psychological evidence for the preponderance of edge extractions in the visual cortex. Some vivid parallel edges and corners can be experienced at the boundaries of a scotoma prior to the onset of a migraine attack. These phenomena were known for a long time, yet had to wait for present neurophysiological sophistication to be interpreted. It is interesting to note that frequently only edges, scintillating parallel edges, and two edges forming a corner are reported.

How closely related the neurophysiological findings will be to the psychological observations as we move to increasingly complex phenomena remains to be seen. In spite of the great gap which still exists between the two disciplines, we can say with some satisfaction that in the last decade some foundation has been laid for a psychological physiology.

3.5 Further Classes of Cyclopean Stimulation
Based on Emmert's Law and Color Phenomena

In the previous chapter we discussed several classes of cyclopean stimuli both in the weak and strong sense. Let us add a few more classes of cyclopean stimulation in the strong sense. First of all we prefer for vision cyclopean stimuli varying in the spatial dimensions with high spatial resolution (e.g., random-dot stereograms and cinematograms) to cyclopean stimuli varying in time. An example of a cyclopean temporal stimulus can be obtained by exploiting Emmert's law. According to this law, the perceived sizes of objects (or afterimages) cast on the retinae increase with decreasing convergence angle (increasing perceived distance). (This law is the result of a binocular size constancy mechanism.) We could present an object binocularly and change its size according to some temporal sequence and simultaneously change its distance according to some temporal sequence. The two independent temporal events can produce a desired third event: the apparent size variations in time of the perceived object. But we would rather generate spatial patterns on some cyclopean level.

This size-zooming with convergence was studied by Richards (1968). He used random-dot stereograms which contained a stripe of uncorrelated textures while the remainder of the images was identical in the two views, similar to figure 2.3-5. He varied the area of the stripe (its width) until the size was sufficient to cause binocular rivalry. He reported, that for near targets (increased convergence) the stripe width had to be increased to produce the same amount of rivalry that is experienced when viewing the stereogram from a distance. This stripe width for near and far vision was inversely proportional to retinal size change. He concluded that there is a perceptual zooming mechanism which operates prior to binocular combination. Whether this zooming occurs in the LGN as Richards suggested, or in the cortex, cannot be decided from this experiment. However, it is interesting that the zooming according to Emmert's law might operate before binocular combination. Richards also believes that the retinal boundary size of scotomas in patients with cortical lesions increases with increased convergence. This is in agreement with Emmert's law. Usually, according to Emmert's law the retinal image of a stimulus appears diminished to the subject when seen from a near distance. If this size scaling occurs between the retina and the site of the scotoma, the probing dots on the retina corresponding to the boundary of the scotoma cover an increased area with increased convergence. The first results reported by Richards (1969) on one patient showed such a zooming scotoma but only for the contralateral field, while the ipsilateral part of the scotoma remained constant during vergence movements.

I have given examples of two cyclopean stimulus classes in the strong sense: the random-dot stereograms and cinematograms. The random-dot stereograms were further subdivided into stimuli for which the information has been portrayed by changes in velocity, orientation, or position of the picture elements. A class of cyclopean stimulation in the strong sense can be produced by binocular color mixing. Under favorable conditions two different colors presented to the left and right eyes can produce a third mixed color. People's abilities differ very much in this respect, as was noted by Helmholtz (1909), who himself was unable to obtain color mixing but quoted a list of illustrious researchers who were able to obtain it. For complex stereograms, complementary colors, and low luminance levels it is possible to obtain color mixtures and avoid binocular rivalry, a fact exploited in the anaglyph technique of separating the left and right eye's view by red and green presentation. For most readers the anaglyph of figure 2.4-1* will appear achromatic or yellowish-white after stereoscopic fusion. Even if the only binocular color mixture someone can experience is of this limited kind, that is of experiencing an achromatic hue from red and green dots presented to the two eyes, it is adequate for obtaining a new cyclopean stimulation class.

Imagine that we want to portray an achromatic T-shaped area amidst red, green, and black dots. If we produce a random-dot stereogram of red, green, and black randomly chosen dots such that the red and green dots of the T-shaped area are -100% correlated while the surround is $+100\%$ correlated and the black dots in both the T-shaped area and in the surround are also $+100\%$ correlated, the desired result is obtained. Monocularly the two images are composed of red and green randomly chosen dots; however, the correlated dots after fusion will retain their color, while the complemented areas will produce achromatic colors. This method is really an extension of a type of random-dot stereogram in which the information is portrayed not by depth changes but by fusion and rivalry. The difference there is that the positive and negative areas (complemented brightness dots) in the two images cannot be fused, while in the present case they give rise to mixed colors. We have devised here a cyclopean Ishihara color test. The classical Ishihara color test chart is a cyclopean stimulus in the weak sense, since the global information is physically contained in the stimulus; however, the retinal receptors of the color deficient person cannot extract this information, or conversely, only color-blind subjects can see certain global organizations that are hidden for the color normals. The cyclopean color test chart in the strong sense makes possible an objective test for evaluating binocular color mixing capabilities.

This cyclopean stimulation based on color mixing should not be mistaken for random color-dot stereograms based on ordinary fusion of similar hues. In 1961 I produced random-dot stereograms in two colors such that the hues in the left and right images were correlated but their brightness values were negatively correlated. Since positive and negative random-dot stereoscopic half-pairs cannot be fused (Julesz 1960a), the fusion of these color stereograms is the result of the correlated hues. The main finding was that the more different the two constituent hue values were, the more opposite brightness contrast could be tolerated before rivalry ensued. Thus for random-dot stereograms of, say, red and orange dots, a few percent of opposite brightness modulation in the two fields destroyed fusion, while for red and blue dots the two fields could be modulated with large contrast difference such that a bright gray dot in one field would correspond to a dark gray in the other field without losing fusion. I have never quantified the distance between hue values as a function of opposite brightness contrast at the threshold of fusion except for a pilot study. But the main result is that the hue information contributes to stereopsis in the same way that brightness does. Thus stereopsis by randomly colored dot stereograms is just a special case of random-dot stereograms, while the method based on binocular color mixing exploits an entirely different psychological phenomenon.

The recent discovery of some long lasting color aftereffects induced by edges discovered by McCollough (1965) and an even longer persisting color aftereffect of movement by Stromeyer and Mansfield (1970) that lasted as long as 6 weeks poses the problem of whether cyclopean stimuli could be created with the help of these phenomena. While eidetic children are rare and their images last usually only for a few minutes, these color aftereffects consist of relatively detailed patterns that persist for hours, days, and even weeks for the majority of people. Furthermore, these aftereffects are not afterimages and their possible site may be cortical. Therefore, it would be most revealing to try producing cyclopean aftereffects. For instance, the left and right fields of a random-line stereogram could be presented (perhaps with lower resolution than in figure 7.9-1) alternately through colored filters, say, at 10-min intervals. Since the two images are never simultaneously presented, no stereopsis is obtained. But perhaps after a long period of such adaptation, monocular aftereffects could be built up that might binocularly interact, and a cyclopean image (e.g., a center square in depth or in different color from its surround) might be experienced. Such a stereoscopic aftereffect would be most informative and would help in locating the site of the aftereffect, that is whether it occurs prior to stereopsis or not. An unsuccessful experiment along these lines will be discussed in § 7.9.

Another kind of cyclopean stimulation could exploit Benham's top (Benham 1894) originally reported by Fechner (1838). Benham and also Bidwell (1897) investigated this phenomenon very thoroughly. As areas of the retina are stimulated by alternate black and white contoured fields sweeping over them and as the time interval between black and white increases to 200 msec, 400 msec, and 600 msec, respectively, red, green, and blue color sensations can be experienced. Thus it is possible to create color sensations by proper spatiotemporal manipulations of black and white patterns. Would it be possible to generate cyclopean stimuli by correlating such induced colors in a random-dot stereogram or cinematogram (and avoid correlated flicker frequencies)? If the experiment succeeded, the location of Benham's color mechanism could be placed before stereopsis. It is relatively easy to generate a computer-movie in which certain areas appear colored by using Benham's top. One must only generate two such film sequences in which the corresponding areas are of the same Benham's color and present them stereoscopically.

Some authors (Hess 1952, Campenhausen 1968) used dichoptic stimulation by presenting sections of a Benham's disk to the two eyes, respectively, such that their binocular superposition would become a Benham's disk. They failed to produce color sensations so they concluded that Benham's colors must be of retinal (peripheral) origin.

However, as I discussed in § 2.5, negative results with dichoptic stimulation are not conclusive. Because of binocular rivalry, it is impossible to superimpose a left and a right part-image into a binocular whole-image. Only random-dot stereograms (in which every element is binocularly fused) portray a binocular whole-image; and, therefore, the controversial conjectures obtained by dichoptic techniques must be rechecked by cyclopean methods.[5]

5. Benham's colors are regarded by some as being produced by a central process that encodes a temporal code of color information. However, there might be simpler explanations than this —even if Benham's colors could be evoked under scotopic illumination, or in color-blind subjects. For example, rods might be connected to different cone-mechanisms at some retinal stage constituting color detectors with different time constants that could yield color sensations.

This is one of the few sections that proposed some experiments to be tried in addition to reporting the results of experiments already performed. My reason for speculating on a few hypothetical experiments in such an early chapter is merely didactic. The purpose of this book is to teach the reader a certain virtuosity in using "counterpoint" in his psychological studies. Familiarity with already known cyclopean stimuli and the results obtained by their application is crucial. However, for the active researcher it is at least as important to master the spirit of cyclopean techniques to the extent that he could invent new methods for a particular research problem.

3.6 Localization of Vernier Acuity Perception

In §§ 2.5, 2.6, 2.7, and 2.8 we discussed a procedure that compared classical and corresponding cyclopean phenomena and from the outcome inferred their central or peripheral origin. For certain phenomena that are complex enough to be regarded as perceptual yet simple enough to be expected to occur prior to stereopsis, one can devise another powerful procedure. Such early phenomena are some of the color effects discussed in the previous section. Now we turn to some other simple perceptual processes.

Take for example superresolution, our ability to distinguish minor breaks (dislocations) in a line that are fractions of the visual resolution of the retinal receptor mosaic. This is called "vernier acuity," whereas the resolution of the retinal mosaic is called "visual acuity" and for optimal conditions (using line gratings) is reported to be 20 sec of arc (Shlaer, et al. 1942). This ordinary visual acuity is actually given as the logarithm of the reciprocal spacing between the lines of the grid at the limit of visibility. It is well-correlated with anatomical evidence based on the geometry of the retinal mosaic. The limit of superresolution is called "vernier acuity" when the lines break in a plane and "stereoscopic acuity" when the break occurs in depth. Optimal vernier and stereoscopic acuity thresholds are about 2 sec of arc, depending on the type of test targets used (Berry 1948, ten Doesschate 1955). The longer the line, the smaller is the break that can be detected (Andersen and Weymouth 1923). That vernier and stereoscopic acuity are much better than the inter-cone separation and that they improve with increased line length indicate central nervous system processings. Julesz and Spivack (1967) demonstrated that this superresolution process occurs prior to stereopsis. They generated random-line stereograms by a computer in which the only monocular cues were minute breaks occurring at random in thin vertical (or horizontal) line grids. The breaks were binocularly correlated. The picture elements of figure 3.6-1* are composed of vertical line segments 1 dot wide and 10 dots long, placed within an area 10×10 dots in size. Two types of such picture elements were selected with equal probability. The picture elements of type A have the vertical line segment in the 5-dot position, while those of type B have it displaced horizontally in the 7-dot position. Figure 3.6-2 illustrates by an array of 5×4 picture elements the way figure 3.6-1* was generated. A center square of 2×2 picture elements (shown by the dotted lines) is horizontally shifted by one picture element to the left in the right array, and the uncovered 1×2 area of the surround is randomly covered by the

3.6-1* Random-line stereogram. The minute breaks in the vertical lines are correlated in the two images, and a center square is seen in depth. (From Julesz and Spivack 1967.)

two types of picture elements. Otherwise the left and right images are identical.

Actually, the left and right images of figure 3.6-1* are similar to figure 3.6-2 except that they are composed of a 100×100 array of picture elements ($1,000 \times 1,000$ dots). The center square is 40×40 picture elements. In one image, the center square is displaced by one picture element to the left of its location in the other image (10 dots). When the images (as printed here) are viewed from a comfortable distance of 30 cm, they subtend about $10°$; thus the one-picture-element disparity of the center square is 6 min of arc, and the 2 dot breaks in the line segments span 72 sec of arc. When figure 3.6-1* is stereoscopically viewed, stereopsis can be obtained, and the center square is seen in front of the surround. This stereopsis is still maintained when viewing distance is increased to 140 cm so that the fused image extends $2.2°$, and the breaks in the line amount to only 16 sec of arc. This value is in the range of vernier acuity. It is somewhat larger than the optimal vernier acuity threshold because half of the breaks have a line length between two breaks only five times the break size, which would be too short to obtain maximum vernier acuity. Stereopsis can still be obtained as we further decrease the break size, but the square loses its dense texture and becomes transparent (lacelike) although seen in depth.

Since the only cue in figure 3.6-1* is the correlation of the breaks and since these breaks are too small for the retinal mosaic to be resolved, we have a double cyclopean

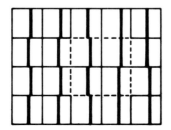 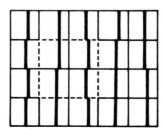

3.6-2 Illustration of how figure 3.6-1* has been generated. (From Julesz and Spivack 1967.)

stimulus. The monocular information is already cyclopean in the weak sense, while the binocular information is strongly cyclopean. That stereopsis can be experienced indicates that superresolution occurs prior to the binocular matching of the monocular images.

A similar display to figure 3.6-1* is shown in figure 3.6-3* except that the line segments are horizontal and the breaks are vertical. Stereopsis can be easily obtained

3.6-3* Random-line stereogram. Similar to figure 3.6-1* except that lines are horizontal. (From Julesz and Spivack 1967.)

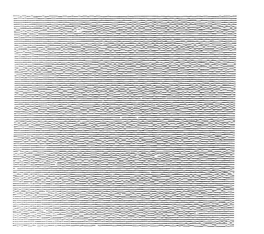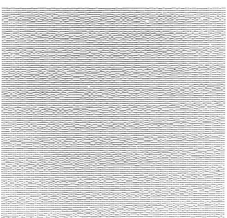

as long as the breaks are monocularly resolved. This demonstration is particularly interesting. It shows that horizontal line segments can give rise to stereopsis; but, more importantly, it separates vernier acuity from stereoscopic acuity. In the case of figure 3.6-1* one could have argued that it was only shown that local stereoscopic acuity perception is utilized by the global stereopsis mechanism, a trivial finding. However, horizontal line segments do not yield local stereopsis and, therefore, the superresolution of the vertical breaks is strictly due to vernier acuity perception. Thus, the result of this demonstration shows that vernier acuity perception occurs before global stereopsis.

It is particularly interesting that figure 3.6-1* yields global stereopsis (i.e., the square and its surround). After all, 50% of the picture elements are identical in the left and right images owing to chance alone (A-A and B-B)[6]; these elements could have been perceived in a depth plane with zero disparity. Moreover, 25% of the picture elements (A-B) could easily have been fused with 16 sec of arc disparity as a lacelike transparent plane slightly in front of the zero-disparity plane, while 25% (B-A) could have been fused with a —16 sec of arc disparity behind the transparent zero-disparity plane. This possible organization would have been an easy solution to the problem of fusing the two images. That one perceives another global organization (even when the center square has 10 times more disparity than the local breaks) indicates that the CNS searches for a particular solution, even if it is more unlikely.

6. A and B refer to the two types of picture elements that are used in these stereograms.

Thus the fusion process apparently favors a solid (densely packed) organization with large disparities in preference to a transparent (lacelike) organization, even if the latter has much smaller disparities. It seems likely that this weighting process among various organizations occurs higher than the local binocular fusion of similar picture elements. I will discuss this remarkable ability of the fusional process to find the "true" solution from a gamut of possible "false" solutions in § 4.5.

Before we proceed with our main task of process localization let us note an interesting property of figures 3.6-1* and 3.6-3*. The center square (or any other cyclopean figure) can be resolved only in the center of the fovea. Since vernier acuity resolution drops off rapidly towards the periphery, these stimuli can be resolved only by foveal vision. Hochberg and Strickland (reported by Hochberg 1970) made use of this property of random-dot stereograms in their research. They used, however, a simpler method. They portrayed the cyclopean figure by an ordinary random-dot stereogram, except that they used only 9 sec of arc disparity. Since stereoscopic acuity diminishes rapidly toward the periphery as well, these stimuli can be resolved only by the fovea, while the periphery "sees" only a flat random plane. Thus, for figural studies in which the stimulus has to be restricted to the fovea, this technique can be used to avoid tachistoscopic flashes or stabilization techniques with contact lenses.

We now need to clarify the meaning of such statements as, for instance, "vernier acuity is processed at a stage *prior* to stereopsis." First of all, in this technique we use the phenomenon under study in such a way as to produce the local picture elements for the global pattern matching process by which correspondence is established between areas in the two retinal projections. By "stereopsis," I always refer to this global matching process. Even if vernier acuity were a special case of stereoscopic acuity—and I regard the latter a local matching process—it would occur before global stereopsis.

Let us clarify another notion. Words like "before" or "after" can have a different connotation for the psychologist than for the neurophysiologist. The psychologist is interested in the information flow. If the output of process A is the input of another process, B, or in other words, process B utilizes the outcome of process A, we regard process A as occurring before process B. The antenna input of a TV set certainly occurs prior to the output of the video amplifier, which feeds in signals to the picture tube. This direction of information flow will not be influenced by the chance of the antenna input and video output being adjacently placed in the actual layout. Neurophysiologists using single microelectrode recordings often find a variety of feature extractors in close proximity to each other and may incorrectly conclude that these features are processed at the same hierarchical stage.

Before using multiple simultaneous recordings that may reveal interdependencies, the neurophysiologist has to rely on neuroanatomical tracings of connectives (histology of degenerative neurons or electron-micrographs of serial sections). In some cases, such as Hubel and Wiesel's complex and simple units, the relation between the extracted features leads to the conjecture that the complex units are constructed from adjacent simple units by excitatory and inhibitory synapses. I cite these examples to show that modern neurophysiological thinking is also flow chart oriented. When Hubel and Wiesel found a few units with concentric receptive fields in the fourth layer of the visual cortex (Area 17), they conjectured that these units were of LGN origin prior to any histological evidence.

In summary, a process (or neural net) A is before another process B if B utilizes A. This directionality in the information flow is not affected if a process C is after process B and the output of C is utilized by A (feedback). Even if A and B are connected in a closed loop, B thus utilizing A and vice versa, we will regard A as being before B if A (and not B) utilizes a process I that utilizes neither A nor B. It is obvious at this point that this directionality of information flow is a specific phenomenon. For instance, a vestibular process H which determines head tilt may occur at a highly central level. However, this information may be utilized by a visual process V at some level. For this visual process the head position extractor H is a "peripheral" process since it occurs before the visual process. On the other hand, when process H is studied we might find that the visual process V is utilized by process H, and in this context V occurs before H.

3.7 Localization of Stroboscopic Movement Perception

We know from neurophysiological evidence that the visual cortex contains binocular units tuned to various velocities, orientations, and directions. However, we still have no idea, particularly in humans, whether stereopsis or movement perception comes first.

Would it be possible to apply the previous idea and generate random-element stereograms in which the picture elements are line segments of, say, two apparent (stroboscopic) movements in opposite directions? Figure 3.7-1 gives a schematic

3.7-1 Schematic illustration of a stereogram in which the left-to-right, right-to-left motion of vertical line segments is correlated, respectively.

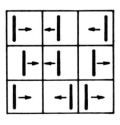
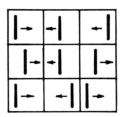
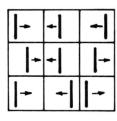
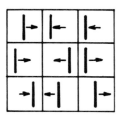

3.7-2 Illustration of how the local movement of figure 3.7-1 is out of phase for corresponding line segments in the two views.

illustration of this idea, using two types of picture elements, in this case vertical line segments that move left to right or right to left, respectively. The corresponding directions of movement are binocularly correlated. Similarly to the previous experiments, a center square is shifted horizontally in order to produce a depth effect after binocular fusion. When Spivack and I made a computer-movie, we produced these moving picture elements by apparent movement perception in four successive frames. Figure 3.7-2 shows the movements and positions of vertical line segments at corresponding locations in the left and right views. Movement was produced in four successive frames. After four frames the same sequence was identically repeated several hundred times. The most important characteristic of this experiment is also shown in figure 3.7-2. Whenever line segment A in the left images moves into positions 1, 2, 3, 4, 1, 2, 3, 4, . . . the corresponding line segment A in the right image takes the positions 3, 4, 1, 2, 3, 4, 1, 2, ... Similarly, an oppositely moving line segment B in the left image cycles as 4, 3, 2, 1, 4, 3, 2, 1, . . . while the corresponding line segment B in the right image is shifted in phase by two positions and cycles as 2, 1, 4, 3, 2, 1, 4, 3, ... The important property is this phase shift between corresponding moving line segments. Were these line segments in phase, stereopsis would result not because correlated movement detectors extract the same velocity information but because in each frame line segments have the same orientation and position in the two views and can be fused as static stereograms. However, if the corresponding line segments are out of phase, the individual stereograms of any frame in the movie cannot be fused. Would the dynamic presentation of these movies produce local movements whose binocular correlation would yield stereopsis? The results were negative! Although we were able to produce excellent perceived movement and the local line segments moved from left to right or vice versa in the left and right images under stereoscopic viewing, no stereopsis could be experienced. At the speeds we used, vivid stereopsis could be experienced when the left and right images contained moving line segments in phase. This control experiment indicated that the selected speeds were well chosen, and the reason we were not able to evoke stereopsis was something else.

We also noticed that adjacent line segments moving in the same orientation grouped together, forming monocular clusters. These movement clusters were not quite correlated in the left and right views because of the two-dot phase delay. In order to eliminate this cluster formation, we selected instead of line segments, dots that were perceived to move clockwise or counterclockwise. These whirling dots did not interfere with their neighbors, and no monocularly apparent groupings were noticed. Yet stereopsis could not be experienced. Again the control (in which the corresponding rotating dots in the left and right images were in phase) resulted in strong stereopsis; stereopsis ceased only when the frame rate reached the simultaneity region and both the clockwise and counterclockwise rotating dots appeared as static rings.

The failure of these experiments, of course, is not as convincing as a successful attempt might have been. However, it suggests that stereopsis might occur prior to movement perception or that movement perception may precede stereopsis but cannot be used as a cue by the stereopsis mechanism.

There is, however, more convincing evidence that stereopsis occurs before movement perception. In a series of random-dot stereogram movies (produced in 1966 by Julesz and Bosche) one of the scenes consisted of alternate random-dot stereograms in which every odd frame portrayed a center square above the surround while each even frame portrayed the same square behind the surround. The random dots in each frame were uncorrelated from those in the successive frames. The binocular percept was a dynamic noise segregated in three depth planes: the surround, the center square above it, and another square behind the surround. Although in each frame 100% of the dots were correlated in the left and right images, the obtained percept was that of a transparent square in the foreground through which the square behind the surround could be seen. (We will discuss this finding later in another context.)

This result suggests that movement perception must occur after stereopsis. Were movement processed prior to stereopsis and utilized by the global matching process, the local movements of the random dots that fell in the center areas would be uncorrelated in the left and right views. After all, the center of every left image in even frames is shifted in the temporal direction (while the right image is unchanged) and every left image in odd frames is shifted in the nasal direction (keeping the right image unchanged). Because of this, the monocularly apparent movement between successive random dots that are proximate in successive frames is different for the center area in each of the two eyes' views. Because of this uncorrelated motion stereopsis could not be obtained in the center if apparent movement perception were to occur prior to stereopsis. However stereopsis is experienced both for the center and for the surround. The center two squares are seen simultaneously in two depth planes which indicates that stereopsis is obtained by processing each frame individually. Only then after the dots are localized in depth are these dots processed by the apparent movement extractors. Of the two alternatives (namely, that stereopsis does not utilize movement perception or that stereopsis occurs before movement perception), the second is suggested.

Similar conclusions can be drawn from other stereoscopic movies in which the areas at various depth levels are portrayed by dynamic noise. When monocularly-viewed dots lie at the two sides of the boundaries (separating areas that have different binocular disparities) in successive frames, they will be perceived as moving across the boundaries. However, when stereoscopically viewed, no moving dot ever appears to cross the boundaries of two depth planes. Proximate dots in temporal succession yield apparent movement only if they are close to each other in three-dimensional space.

Another experiment is also relevant. Julesz and Hesse (1970) produced computer-movies in which tiny needles in all orientations rotated around their centers in a plane. In a rectangular area the needles rotated with twice the speed as those in the surround. This caused the global percept of the rectangle to differ from that of the surround; there appear to be sharp edges at the boundaries. John Merritt (a student from Harvard who attended my seminar) and I tried to use this film to produce stereopsis. One eye viewed the movie directly, while the other eye viewed it through a 90° rotating prism. Thus the corresponding needles in the two eyes were perpendicular to each other and the corresponding rectangles could be horizontally shifted by any amount. In spite of the fact that monocularly perceivable rectangles had a binocular disparity owing to binocular rivalry, no stereopsis could be experienced. It is true that under slow rotations (well below apparent movement threshold) only binocular rivalry is to be expected due to the orthogonality of corresponding needles in the two views. Yet, above apparent movement threshold one might have expected that the perceived rotation of needles having the same clockwise or counterclockwise direction would have compensated for the perpendicular orientation. The results, as we mentioned, were negative. Whether the result implied that stereopsis occurs before movement perception, or that binocular rivalry prevents stereopsis based on correlated movement extractors, seemed to be an open question.

However, a recent experiment by David N. Lee at Harvard in which I participated as a subject casts severe doubt on this conjecture. He performed the following: In static noise he slowly moved a vertical bar of static noise, using my usual techniques. However, he presented to the left and right eyes uncorrelated random dots; the static-noise-textured bar, when moving from left to right, had a nasal disparity, when moving from right to left it had a temporalward disparity. In spite of binocular rivalry, the movement contours of the cyclopean bars gave rise to vivid stereopsis. First the fused bar moving from left to right appeared in front of the surround, while the oppositely moving bar appeared behind the surround.

This experiment is quite similar to the previous one I conducted with Merritt, yet it gave a positive result. Probably, in our experiment the corresponding perpendicular line segments produced rivalry at each position (except for the coincident dots at the intersections), while in Lee's experiment only 50% of the dots were rivalrous and the cyclopean movement-contour for lateral movement yielded a stronger percept than for rotational movement of line segments.

Since then, this important study has been published (Lee 1971). E. Chiarucci and I verified this experiment in my laboratory, using random-dot textures with various densities. Since the left and right images are uncorrelated, it is crucial to provide a fusible reference frame that is identical in both images. The moving cyclopean edges (that have binocular disparity with respect to the reference frame) give rise to local stereopsis. Except for the fact that the edges are cyclopean (produced by monocular random-dot cinematograms), the stereogram, as such, is not cyclopean (since the reference frame and the moving bar can be seen by each eye separately). Therefore, this stereopsis is the ordinary kind, which we call "local stereopsis." Only if monocular cues are absent and the fusional process has to evaluate many local disparity values for a best fit, do we use the term "global stereopsis."

In Lee's experiment the local elements had identical movement gradient and thus gave rise to a single global motion. In the light of these experiments one could conjecture as follows: Local stereopsis can produce a local movement perception, while local motion cannot yield local stereopsis. However, global motion can give rise to local stereopsis, and of course, global stereopsis can yield local motion. While local stereopsis can occur before local movement perception, global movement perception may occur before local stereopsis. I will discuss some of these problems more thoroughly in § 7.5.

3.8 Perception of Local and Global Levels

As in all scientific disciplines there are at least two main approaches in our search for structure. One is the global approach which emphasizes that only in the large does a pattern make sense. The other is the local approach which searches for local constituents out of which the entire edifice is constructed. Until advances in technology permit meaningful experiments, these two viewpoints will be merely epistemological exercises. Premature local models were proposed by the Greek atomists and premature global models by the physicists who, a generation ago, unsuccessfully attempted to deduce all the laws of physics from an all-pervading ether. On the other hand, when technological advances open up new methods, some meaningful local or global models may be constructed. It is a further development in a field of science when relations between local and global features become apparent. A well-known example is the equivalence of energy and mass as predicted by the global theory of special relativity and so convincingly demonstrated by the local theory of nuclear physics a generation later.

As we have noted, psychology, particularly the study of perception, has followed a similar history. Structuralism versus the gestalt approach corresponds to the local and global viewpoints. This dualism already existed in early philosophies prior to becoming a banner for various schools of psychology. The huge amount of data collected by structuralists, and partly discredited by Gestaltists, was obtained by presenting simple line drawings. The stimuli by the former usually consisted of simple dots, lines, or large homogeneous areas having various brightness, color, or extent, while the Gestaltists used somewhat more complex outlined figures and showed that under their conditions the results differed from the ones used for simpler stimuli. Although adherents to the Gestalt school made their point that "the whole is more than its parts," they were unable to maintain the quantitative results of the structuralists.

I have already discussed how the recent findings of neurophysiology have resolved the structuralist-Gestaltist controversy in a higher synthesis. Perhaps it is closer to the truth that these results shifted the controversy to a higher level. Instead of the atomistic sensations of the structuralists, we now believe in perceptual molecules extracted from the stimulus by early feature analyzers in the CNS. However, the manner in which recognition of a complex pattern takes place, say of a face, from a gamut of such perceptual molecules is far from explained. One advantage of the technique of random-dot stereograms is the possibility of separating the local and global problems of organization. One can, for instance, stimulate the perceptual molecules by appropriate local picture elements in the monocular presentation, while the global organization is contained in the binocular view. These separated local and global cues can also be reversed. Sections 3.9 and 3.10 will be devoted to these methods. Of course, these sections will not shed light on local-global problems within the framework outlined in the first paragraph of this section. Yet, on a limited scale they might bridge the gap between two levels of complexity.

3.9 Monocular Local versus Binocular Global Information

The basic demonstration of figure 2.4-1* showed only that no monocular shape recognition is necessary in order to establish corresponding points and obtain stereopsis. Here shape recognition refers to some familiar macropatterns or Gestalten such as objects in our environment. The demonstration of figure 2.4-1* does not show that no monocular pattern recognition is necessary for stereopsis. One might argue that before stereopsis some micropatterns are found that resemble such simple signs as a T, +, O, and so on formed by chance and occurring in both the left and right fields. The finding of such similar micropatterns would solve the ambiguity problem and establish the correspondence between areas in the two fields.

This is such a basic point that in my first publication on random-dot stereograms (Julesz 1960b) I demonstrated an experiment shown in figure 3.9-1* in order to

3.9-1* Random-dot stereogram, derived from figure 2.4-1* except that in the left image the diagonal connectivity is broken. Eighty-four percent of the dots remain identical, and stereopsis is easy to obtain.

meet this criticism. Figure 3.9-1* is identical to figure 2.4-1* except that in the left field the diagonal connectivity is broken. Whenever 3 adjacent dots along the diagonal of +45° angle are all black or along the —45° diagonal are all white, the center dot is complemented to its opposite value (black becomes white and vice versa). This procedure changes only about 16% of the dots to their complements, thus 84% of the left and right fields of figure 3.9-1* are identical. Yet when monocularly viewed, the two fields look strikingly different. The left field of figure 3.9-1* seems to contain horizontal and vertical line segments of various lengths and about one dot width, while the right field looks speckled with random dots. In spite of this very great difference in monocular appearance, the stereoscopic image pair is easy to fuse since on a point-by-point basis the two images are binocu-

3.9-2* Random-dot stereogram in which, similar to figure 3.9-1*, 16% of the dots are complemented but at randomly chosen positions.

larly very similar. Although there is no similar micropattern in the two fields and on the average at least one point is different in each 2 × 3 array, the visual system easily takes advantage of the fact that 84% of the dots are identical. The monocularly apparent drastic difference between the monocular appearance of the textures is of no consequence, since in figure 3.9-2* the same 16% of the dots are complemented (and thus unavailable for fusion), but this time the complemented points are chosen at random (Julesz 1965b). As one can see, the textures of the two images appear monocularly very similar, but the goodness of fusion and stability of stereopsis (measured by the necessary perception time or sharpness of the percept) is identical to the previous case.

At this point one could still argue that only the highly detailed information (high

3.9-3 Various picture-element configurations of two kinds used in complex stereograms.

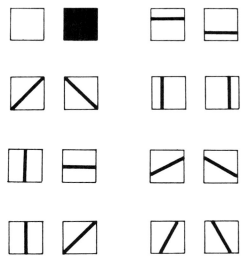

3.9-4* Random-line stereogram of correlated horizontal and diagonal line segments. When stereoscopically viewed, a hovering center square is seen.

spatial frequencies) were removed, but the patterns portrayed by the low spatial frequencies are identical in the two fields. In order to remove the low frequencies from the stereograms, random-dot stereograms were changed to random-line stereograms (in which clusters of many dots cannot form). In these displays instead of using two brightness values for picture elements, I have introduced line segments of various orientations to serve as the picture elements. Since these line segments are confined in square-shaped cells they cannot touch each other and no clusters (of low spatial frequencies) can form. Usually two kinds of picture elements were chosen differing in orientation or position as shown in figure 3.9-3. However, similar to random-dot stereograms that can differ in a large number of brightness values (in the limit about 100 levels to be noticed), the perceivable positional and orientational changes provide a large number of various random line segments to be used as picture elements. In figure 3.9-4* only two kinds of line segments are used (of 10 dot length and 1 dot width in an array of 100×100 picture elements) where each picture element is a small square array of 10×10 dots. The two kinds of picture elements contain line segments of horizontal or -45 degree orientations and were chosen randomly with equal probability. Of course, the left and right fields have been correlated, and in figure 3.9-4* a center square is seen in depth when stereoscopically fused (Julesz 1967a).

The local line segments do not have to be of exactly identical orientation in the two views as one of my students, the late Lloyd Marlowe, showed in his Ph.D. thesis (Marlowe 1968). He varied the orientation of the corresponding line segments by as much as $\pm 15°$ (i.e., a vertical line segment in the left field belonged to a $\pm 15°$ tilted segment in the right field) and could still get global fusion—that is, the center square could be perceived in depth. This is the more interesting since $15°$ rotation destroys the local fusion of the horizontal line segments (though they are within the limit of patent stereopsis). Marlowe's experiment will be reviewed again in § 5.3 and demonstrated by figures 5.3-1* and 5.3-2*.

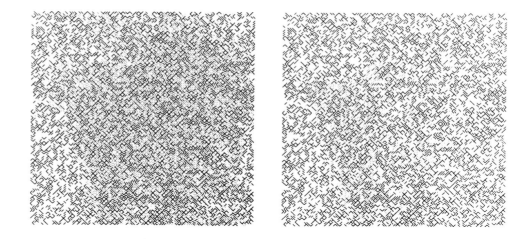

3.9-5* Random-line stereogram portraying a Kaufman effect. Orientation portrays a diamond in front, and brightness (line width) portrays a square behind the surround.

Somewhat prior to this, Julesz and Spivack (1967) used vernier acuity cues to portray correlated local information on the retinae. Figure 3.6-1* shows the display that was used. We discussed this finding in another context in § 3.6; here we note that the cyclopean square is portrayed by superresolution perception which in turn is carried by the high spatial frequencies, while the low frequency channel does not carry any global information.

It is also possible to juxtapose the low and high frequency information as shown by Kaufman (1964). He used a technique which Julesz and Miller (1962) introduced, where instead of local brightness changes, randomly chosen (binocularly correlated) alphanumeric characters were used. (However, Kaufman used these recognizable characters as markers in an original way.) He portrayed one center square of a given disparity by the correlation of the corresponding shapes of the characters, while he modulated the brightness of the characters and portrayed by corresponding brightness values another area of a different disparity. Instead of characters, it is easier to use line segments where their orientation is one parameter and their width another. Such a modified Kaufman pattern is shown in figure 3.9-5* generated by Marlowe and myself. The orientation information displays a center square in front, and the width information portrays a center square behind the surround. It is possible to obtain either one of the two organizations, though the percept is not of uniform fusion since half of the picture elements are inducing binocular rivalry due to the incompatible width or orientation information, respectively. However, if the images are blurred (by defocusing them, or by squinting) the high frequency information (carried by orientation) is filtered out, and the low frequency information (conveyed by the width modulation) dominates.

One can even go a step further. I made a display in 1967 in which the high frequency information is rivalrous. In figure 3.9-6*, ⅞ of the display is composed of diagonal line segments of ±45° orientations such that the corresponding line segments in the left and right fields are orthogonal to each other. The only correlated

3.9-6* Random-line stereogram
consisting of diagonal line segments
that are orthogonal to each other
in the two eyes' views except for
correlated black areas having ⅛
probability. In spite of binocular
rivalry, stereopsis is easily obtained.

information is conveyed by the blank areas of ⅛ probability that correspond to the
low frequency information. Yet fusion is easy in spite of the fact that 88% of the
perceived image is under binocular rivalry. It is not quite precise to regard the 12%
blank areas as corresponding to low frequencies. After all, the blank areas have intricate outlines. Yet they can differ only in one cell steps (which are 10×10 dot
size), while the line segments are portrayed by 10 times finer resolution.

These and similar demonstrations show that the pattern matching between the left
and right fields can occur separately using either the low, medium, or high frequency
information or any combination of these. Of course at least one of these frequency
channels must be correlated in the left and right fields, otherwise retinal rivalry would
ensue. What is interesting, however, is that certain frequency bands yield fusion while
the other frequency bands are in binocular rivalry. From the previous example we
have seen that a remarkably small amount of correlated frequency bands will yield
fusion. These few correlated spatial frequency components in the two monocular
images are not adequate to be individually detected and recognized.

In order to explain these findings one can think of a model in which, prior to
stereopsis, a frequency analysis is performed, say, by Kuffler-units of various diameters. Only the outputs of Kuffler-units in the left and right pathways that have the
same sizes are binocularly compared. Later, in § 4.5 and chapter 6 some detailed
models of stereopsis will be given. Here it should suffice to say that the global stereopsis process evaluates the outputs of many local stereopsis units that selectively fire
for a specific binocular disparity and spatial frequency. In the future when we will
discuss point-by-point comparison between the left and right fields, the reader should
bear in mind that the corresponding points to be compared not only have the same
coordinates but also receptive fields of the same dimensions.

The preprocessing in this model is purposely attributed to Kuffler-units and not
to edge or slit detectors. The vernier acuity stereograms of Julesz and Spivack (1967)
have breaks which are much less than the width of the elongated receptive fields.

Furthermore, a few seconds of arc tilt of a vertical line segment in one eye's view causes the stereoscopic percept to appear as a tilted line in depth. This slight tilt could not be resolved by any single binocular slit or edge detector and yet gives rise to perceivable depth sensation. Furthermore, Marlowe (1969) performed a most instructive experiment. He measured perception times for random-line stereograms where the corresponding line segments were randomly rotated by a few degrees in one view with respect to the other view. Then he replaced these random-line stereograms with the end points of the line segments alone. It turned out that these "dipoles" yielded the same perception time for stereopsis as was required for the entire edges. This finding is even more remarkable since the end points of a different adjacent line segment formed closer dipoles than the original ones and because the random tilts were often uncorrelated in the two views. Furthermore, the two points of any dipole were in slightly different depth planes because of the tilts. All this evidence strongly suggests that stereopsis is based neither on edge detectors nor even on Kuffler-units, but on some complex combination of all of these. Only such a complex stereo unit (not yet neurophysiologically discovered) can account for superresolution of vernier and stereo acuities or for the matching of different correlated frequency bands that are utilized by the global stereopsis process. In these stereo units several adjacent edge detectors of the same ad of different orientations must be combined in several ways in order to account for the phenomena discussed above. Because of this it is easier (less constrained) to describe the stereo units as aggregates of Kuffler-units connected in complex ways than to represent them as edge detectors that are themselves aggregates of Kuffler-units.

At present it is very difficult to verify the existence of orientation-independent global stereo units. The orientation and disparity sensitive local stereo units that were found in the cat and the monkey do not exclude the existence of global stereo units one hierarchical level above. Let me note that a very recent adaptation experiment by Julesz and Stromeyer, however, seems to indicate the existence of orientation independent disparity sensitive units in man. This experiment is a modification of the Blakemore and Julesz (1971) technique of evoking a cyclopean depth aftereffect (discussed in § 7.9). While Blakemore and Julesz used random-dot stereograms for both the adaptation and test stimuli, Julesz and Stromeyer used random-line stereograms, composed of tiny, vertical line segments or horizontal line segments, respectively. When the adaptation and test stereograms were portrayed by line segments of the same orientation, a sizable depth aftereffect was observed. When the line segments in the adaptation stereogram were perpendicular to the line segments of the test stereogram, the depth aftereffect was not noticeably reduced. This observation indicates that it is possible to adapt stereo units that are orientation insensitive but fire for a certain disparity.

The neurophysiological implications of these psychological experiments are manyfold. It is shown that the many local fusional possibilities are weighted by some global levels. The local stereopsis mechanism revealed by Barlow, et al. (1967) and Pettigrew, et al. (1968) is only a first physiological process that must be followed by global processes not yet revealed. The first neurophysiological evidence for the existence of a complex stereo unit in the cat cortex is reported by Blakemore (1968). He found, in addition to the binocular columns in Area 17 that contained edge detectors firing for the same binocular disparities, another columnar organization. Binocular units belonging to these columnar modules contained overlapping contralateral (left) receptive fields while the ipsilateral (right) receptive fields greatly varied in their locations. These columns thus contained units that "looked" at points which lay in the same direction (for the left eye) but at different depth. Blakemore suggests that depth localization is accomplished by the depth columns (disparity units) while positioning is performed by the direction columns. Thus there must be a higher process that combines both kinds of information which are necessary for global stereopsis. § 5.2 and figures 5.2-1 and 5.2-2 give more details.

Some of these psychological conjectures remain to be tested by neurophysiological evidence. However, several of the findings of random-dot stereograms had impact on neurophysiological research and were supported in the last few years. Before the evidence of random-dot stereograms it was widely believed that seeing with two eyes is more complex than seeing with one eye and the understanding of binocular vision must await the understanding of monocular vision. Sherrington (1906), for example, concluded "that during binocular regard . . . each uniocular mechanism develops independently a sensual image of considerable completeness. The singleness of binocular perception results from union of these elaborated uniocular sensations." Bishop (1969a,b) in two review articles on the neurophysiology of visual form and binocular depth perception starts with this Sherrington quotation and then emphasizes that paradoxically, the reverse is the case. Bishop, a neurophysiologist, then reviews the implications of random-dot stereograms as follows (1969b):

A number of important suggestions can be made as a result of Julesz's work, namely: 1) As stated above, the neural mechanisms for binocular depth discrimination are likely to be, in a sense, less complex than those required for form recognition and, in addition, antecedent to them in the brain. 2) The analyses by which depth discriminations are made have a mosaic point-by-point or feature-by-feature basis. Line contours of appreciable length are not required. 3) The fact that form perception can arise binocularly by viewing a pattern of dots which are merely random to monocular inspection suggests that the mechanisms concerned are an integral part of the chain of events leading up to pattern recognition.

I quote Bishop in order to show that the simplification that resulted from random-dot stereograms is well appreciated by some neurophysiologists. Much of the recent physiological evidence by Bishop and his students reveals many of the early processes of disparity extracting units. However, the global aspects of the paradigm of random-dot stereograms are less appreciated by the physiologists. How does the visual system resolve ambiguities inherent in random-dot stereograms? I will discuss this problem in great detail in chapters 5 and 6. The solution requires a next level of hierarchy, above the binocular cortical units of Barlow, et al. (1967) and Bishop (1969a,b).

I will review some further neurophysiological evidence of stereopsis in § 5.2. Let us now return to psychology and review the outcome of experiments where the global information is presented monocularly while the local information is given binocularly.

3.10-1* Inverse cyclopean stimulus. The monocularly apparent one-fold symmetry is suppressed in the binocular view. If the reader has a strongly dominant left eye, he should reverse the viewer, that is, view the anaglyphs with the green filter over the right eye. (From Julesz 1967a.)

3.10 Monocular Global versus Binocular Local Information

In §§ 2.4 and 2.5 I discussed briefly inverse cyclopean stimulation. This technique scrambles the monocular global organization in the binocular view. Figure 2.4-5* demonstrates how a monocularly readable text can be suppressed in the binocular view. I elaborated that this suppression is not due to binocular rivalry, but on the contrary, is the result of binocular fusion of each local element. Figure 3.10-1* illustrates this stimulus class by scrambling a monocularly apparent one-fold symmetry in the binocular percept (Julesz 1966b, 1967a). The fused image contains five picture-element wide, horizontal stripes at alternate depth levels, and the appearance of the covering texture is random. Whenever the i-th horizontal strip is shifted to the left in the right field the $(-i)$-th stripe is shifted in the opposite direction. This operation breaks the bilateral horizontal symmetry in the right field, and the cyclopean view is a percept between the left and right monocular percepts but still lacks symmetries.

After all, monocular vision is based on a topological space, while binocular vision is based on nontopological shearings and cuts. For monocular vision, except for some newly entering objects that might cover some old objects, most transformations are continuous motions. On the other hand, the monocular projections of stereoscopic views due to parallax shifts are discontinuous; there are always areas that project onto only one retina.

In one series of experiments, higher-order symmetries were scrambled binocularly as well (Julesz 1967a). The main finding was that the finer the resolution of the monocular global pattern (i.e., the more complex the global pattern is), the better it is binocularly suppressed. It appears that the coarser local patterns might stimulate monocular feature extractors which transmit the extracted information to higher centers. On the other hand, the finding that monocularly complex global patterns can be scrambled binocularly indicates that the high spatial-frequency units (with narrow receptive fields) are probably binocularly driven. It would be nice to confirm this conjecture by neurophysiological histograms. Perhaps the recent findings by Bishop (1969a,b) and his co-workers might be relevant. They found that many of the cortical units in Area 17 of the cat that were previously regarded as monocular are in fact binocularly driven. There is a possibility that the suppression of the monocular global information in figures 2.4-5* and 3.10-1* is simply due to the fact that the majority of the units are binocular, and only prolonged strabistic conditions render these binocular units monocular.

Although we will discuss problems of stereopsis in great detail in succeeding chapters let us summarize the essence of the findings obtained by the three experiments demonstrated by figures 2.4-1*, 3.9-1*, and 3.10-1*. Figure 2.4-1* demonstrated that stereopsis can occur in the absence of recognizable shapes. That this absence of shape recognition is not restricted to macropatterns of complex design alone but to simple micropatterns as well has been shown by figure 3.9-1*. Finally, figure 3.10-1* demonstrated that stereopsis not only *can* occur without monocularly recognizable shapes but actually *does* occur. That is, if monocular shapes exist, then stereopsis precedes their recognition and in the three-dimensional percept the monocular shape can be scrambled. These experiments lead to the flow diagram of figure 3.10-2. This schematic illustration indicates that instead of the thinly drawn monocular pathways, stereopsis is a simpler operation, as shown by the heavy lines. Form perception occurs after stereopsis and not before it as previously assumed. Some simple operations, such as low and high frequency filtering or edge detection do occur prior to binocular matching, and is assumed to be incorporated in the box called "binocular combination." This flow diagram was derived by psychological means and resembles the anatomically and neurophysiologically determined pathways of binocular vision.

Until now we discussed only one basic problem of stereopsis—that is, how two projections of an object give rise to a three-dimensional sensation. However, the second basic problem of how the combined binocular image is obtained has not been raised yet. If the two images are very similar, as is the case of fusible stereoscopic pairs, the derivation of the combined percept is rather simple (at least conceptually).

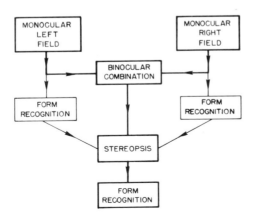

3.10-2   Flow diagram of stereopsis.

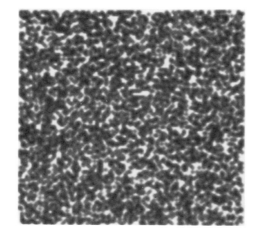 

3.10-3*   Random-dot stereogram, similar to figure 2.4-1* except that the left image is strongly blurred.

Except for some partly hidden areas seen by one eye alone and for brightness differences owing to highly reflective surfaces, it is immaterial which of the images is selected for the binocular percept. The only difference between the two projections are horizontal parallax shifts (binocular disparity) between corresponding areas which are shifted in alignment partly by convergence of the eyes and by a cortical shifting mechanism. This perceived image having the orientation between the corresponding left and right areas, Hering called the cyclopean image (not to be confused with our usage in which we also require that the monocular image should be independent of the binocularly perceived one). Whether the perceived orientation is the mathematical mean between the left and right shifts or is weighted by the dominant eye is not our concern at this moment. If one assumes that the perceived combined image is the result of an alignment of corresponding areas followed by averaging the left and right images, even slight brightness changes (owing to reflections) can be compensated for.

However, if the two projections of an object differ in their spatial frequency distribution (as might be the case when one eye is poorly accommodated or when one image is artificially blurred) the situation changes. The random-dot stereogram of figure 3.10-3* shows such a case in which the left image is strongly blurred (by averaging nine adjacent picture elements, the center and its surround; see Julesz 1960a). The stereoscopic image pair is easy to fuse, and the binocular percept appears not only in depth but seems as detailed as the sharper image. This weighting of the "contour rich" channel in favor of the "contour deprived" channel has been system-

3.10-4*   Experiment by Levelt, showing how the richness of contour determines binocular combination.

atically studied by Levelt (1965), whose ingenious demonstration provides us with a good mnemotechnic for his claim that the contour rich channels are emphasized as illustrated in figure 3.10-4*. The three conditions in his display correspond to (a) same low and high spatial frequency information; (b) same high frequency, different low frequency information; and (c) different high and low frequency information (between the left and right corresponding targets). In case *a* the identical left and right black squares are equally averaged and appear binocularly as a black square. In case *b* the left black square and the corresponding outlined square are again equally averaged since their high frequency content is identical. The perceived binocular image is a gray square. In case *c* the left black square has to be fused with a white uniform area. There is no high frequency information in the right target. Since Levelt assumes that the high frequency information determines the weighting and the right target has no high frequency content, he therefore predicts that only the left blank square should be seen. Indeed, that is the case. For such extreme conditions where the high and low frequencies are segregated into separate bands, only those views which contain the contour rich channel are switched on.

But for realistic targets which are determined by their involved two-dimensional frequency spectrum, the situation is very complex. For instance, in figure 3.9-1* in which one image of the stereogram has no diagonal connectivities left, it is difficult to determine which of the two images is richer in contours. The diagonal-connectivity-broken image contains long, thin horizontal and vertical edges and appears richer in contours than the random-dot pattern of the other image. Since neither of the images contains low frequencies, the change in appearance is caused by the increase of some horizontal and vertical high frequency components at the expense of the medium frequencies in the disrupted image. However, in the binocularly fused image, particularly when the dominant eye views the random-dot pattern, the connectivity-broken image does not dominate, as shown in figure 3.9-1*. Rather the opposite is true. It is not the richness of contours but the complexity of the image that may underlie dominance. A uniform frequency spectrum, at least from an information theoretical point of view, is more complex (more random) than any other spectral distribution. It would require careful studies to determine numerically which spectral-distribution shape determines eye dominance and with what weights.

It is interesting to compare the experimental results of §§ 3.9 and 3.10. We were shown that binocular fusion depends on the identity of the low or high frequency spectrum in the two images. If either the low or high frequencies (or both) are identical, fusion can occur. Those frequency components that are not identical will cause binocular rivalry. In this case, whichever of the two views contains the high frequencies in a given area will dominate in the final percept.

We are now in the position to discuss inverse cyclopean stimuli from another angle. For this stimulus class, low frequencies are avoided in the two views, while the high frequency spectra are kept similar except for phase shifts. As a result, binocular fusion occurs for every picture element, and the complex monocular information with medium and high spatial frequencies appears scrambled in the fused image. Because monocular information that is rich in low frequencies (e.g., if the word YES in figure 2.4-5* were made up of uniformly white areas) cannot be suppressed after binocular fusion, we avoid it.

Thus for this case, which is neither binocular fusion nor rivalry, some new mechanisms come into play. These are in a way opposite to Levelt's high-frequency dominance. Perhaps the monocularly dominant cortical units with high-frequency extraction (narrow receptive fields) are combined into binocular units (prior to stereopsis), whereas the monocularly dominant low-frequency units stay monocularly driven even at higher stages. We can only conjecture the mechanisms. The experimental fact is, however, that under inverse cyclopean stimulation only the low-frequency information in the monocular views can escape binocular scrambling.

Recent advances in the numerical calculation of Fourier transforms made it possible to perform operations in the two-dimensional spatial frequency domain with great ease and economy (Cooley and Tukey 1965). Figure 3.10-5* shows a random-dot stereogram quantized into eight brightness levels (using eight different characters of the microfilm printer). Figure 3.10-6 shows the two dimensional Fourier transform of the left image. From the Fourier transform of the right image half of the frequency content is removed, as shown in figures 3.10-7 and 3.10-8. In figure 3.10-7 the low spatial frequencies and in figure 3.10-8 the high spatial frequencies are removed, respectively. The inverse Fourier transforms of figures 3.10-6, 3.10-7, and 3.10-8 are computed and quantized into eight levels. The left image of figure 3.10-9* shows the backtransform of figure 3.10-6 (original left image and the backtransform

3.10-5* Random-dot stereogram portraying a center square. The eight brightness levels are approximated by eight characters of the microfilm printer.

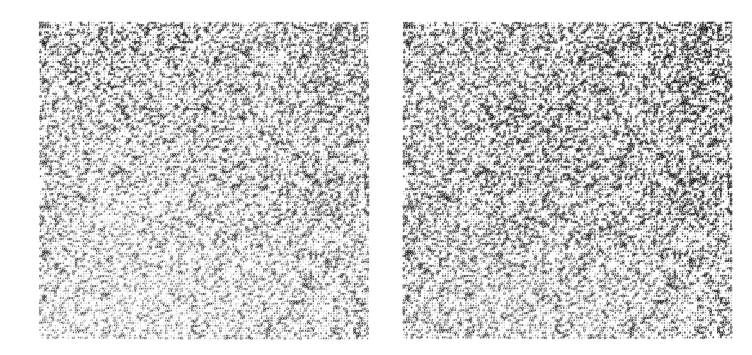

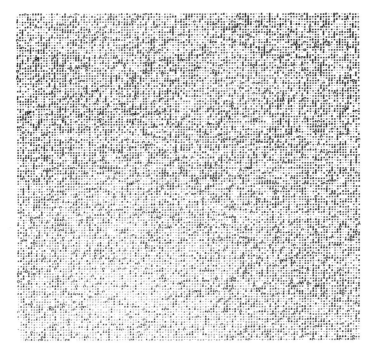

3.10-6 Two-dimensional Fourier
transform of the left image of figure
3.10-5*. The portrayed image is
quantized into eight levels.

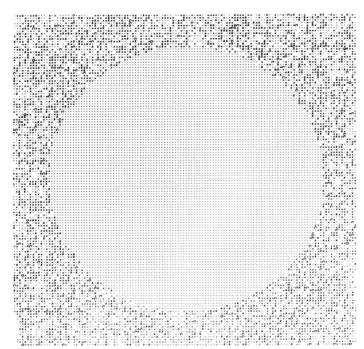

3.10-7 Two-dimensional Fourier
transform of the right image with the
low spatial frequencies removed. Only
50% of the spectrum is retained.

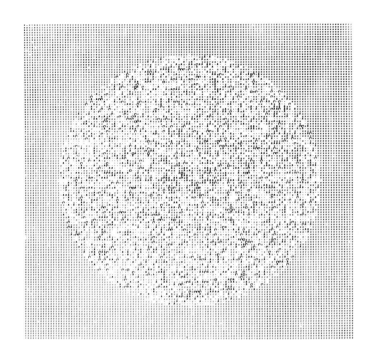

3.10-8 Two-dimensional Fourier transform of the right image with the high spatial frequencies removed. Only 50% of the spectrum is retained.

3.10-9* Stereogram in which the left image is identical to the left image of figure 3.10-5*, whereas the right image is the inverse Fourier transform of figure 3.10-7 (low frequencies missing).

3.10-10* Stereogram in which the left image is identical to the left image of figure 3.10-5*, whereas the right image is the inverse Fourier transform of figure 3.10-8 (high frequencies missing).

of figure 3.10-7; no low frequencies). Stereopsis can still be obtained. The left image of figure 3.10-10* is again the original image while the right image is the backtransform of figure 3.10-8 (no high frequencies). Stereopsis is even better than in the case of figure 3.10-9*. However, the right images of figures 3.10-9* and 3.10-10* cannot be fused, since they do not contain any common frequency spectrum. For a detailed study, see Julesz (1970b). These experiments reconfirm under controlled conditions that only those monocular processes that extract similar frequency spectra are utilized by stereopsis.

That stereopsis requires monocular stimuli of similar frequency components is equivalent of saying that spatial frequency analysis can occur before stereopsis. We have also discussed the many recent findings that indicate one-dimensional spectrum analysis in vision. Therefore, it seems reasonable to assume that a periodic grating could not be scrambled in the binocular view. Indeed, I tried inverse cyclopean techniques in order to make monocularly visible gratings disappear in the binocularly fused percept, but without any success. Even high frequency gratings with low contrast remained visible in the binocular view, provided they were monocularly visible. A typical example is shown in figure 3.10-11* (and also in figure 5.7-2*), where the right image of the stereogram is identical to the left image, except that every even row in the right image is shifted to the left by the width of the grating. Although one can easily fuse this stereogram and perceive two transparent planes, one above

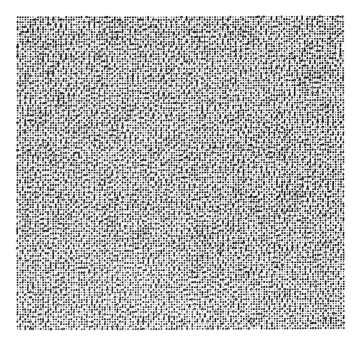

3.10-9*

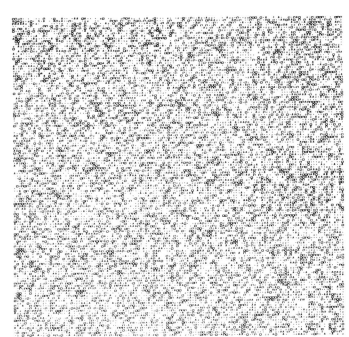

3.10-10*

the other, the grating is also seen. This means that fusion is unable to break up the grating. This experiment strongly suggests that a grating is a special entity and its processing may occur before stereopsis. However, some other evidence suggests the opposite, as discussed in § 5.7.

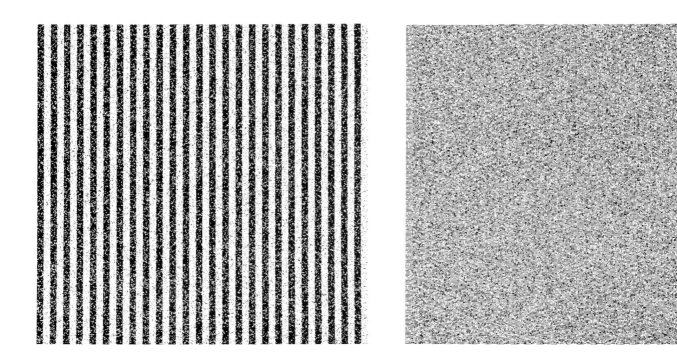

3.10-11*   Random-dot stereogram that demonstrates how a monocular grating can resist binocular scrambling.

Figure and Ground 4

The overwhelming amount of information in our visual environment is proverbial. "A picture is worth a thousand words" seems even quantitatively sound, since the ratio between the telephone bandwidth of about 4 kHz and the television bandwidth of 4 MHz is 1:1,000. But one should keep in mind that this value for hearing corresponds to the entire auditory environment—that is, for sound stimuli that originate within a 360° solid angle. On the other hand, the television screen portrays only a 10° solid angle segment from the entire surface of the visual sphere, which corresponds to an information reduction by a factor of 1,000. Thus, the visual environment that surrounds us is "worth a million words."

These television bandwidth requirements reflect some fundamental principles of vision. The retinae of most mammals, all primates, and man have a narrow center area of sharp, detailed vision called the fovea, surrounded by a large peripheral region of blurred vision which signals mainly temporal changes caused by moving or flickering targets. The diameter of the fovea depends on our criterion of visual acuity. If our criterion selects a value at which the linear acuity diminishes to 10% of its largest value, the diameter of the fovea corresponds to 10° of arc. In order to inspect our environment, complex monocular and binocular mechanisms of accommodation, eye-movement, and eye-coordination are evoked. In addition to voluntary fixations, our eyes perform about 5 involuntary saccades in a second (if we neglect the fast tremors of the eyes). How these small, voluntary, and random samples of clear vision are stored and integrated into a unified, global percept is one basic function of perception and will be discussed in detail under perceptual constancy phenomena.

This elegant solution of preselecting a moving or flickering target by a low spatial-resolution channel and then inspecting it by a tiny foveal area of high resolution must certainly serve a purpose. After all, a complex system of eye movements and eye coordination had to be developed in order to permit foveal fixations, and an even more complex storage and integration center had to be brought into existence for the reconstruction of a uniform perceptual field from a series of random samples. Whether the purpose of this system has been information reduction or selective attention by simple means, or both, is hard to conjecture. Certainly looking at an object of interest is a simple and effective preselection system that makes the work of later neural attention mechanisms much easier. On the other hand, it is hard to assess the role of foveal fixation in information reduction. The problem is to establish that the drastically reduced input information is a necessary step in order to enable the CNS to store it permanently. Unfortunately, we do not know the memory capacity of our brains. We do not even know whether a detailed texture memory exists. Except for eidetic children who might have a texture memory lasting for a few minutes (Haber and Haber 1964) some idiot savants and mnemonists (Luria 1968) we have no evidence for the existence of a texture memory. The "short-term visual memory" of Sperling (1960) is a detailed texture memory, but fades out in 0.1 sec. like the afterglow of a cathode ray tube and is merely an afterimage. However, in the spring of 1969 the existence of detailed texture memory was suggested by Stromeyer and

Psotka (1970) and Stromeyer (1970) based on cyclopean tests they performed on an eidetic woman. In the two years that have since elapsed no other investigator verified the reported amazing abilities of Stromeyer's eidetiker, nor did anyone find other eidetikers of comparable skills. While these findings are questionable, the reported cyclopean technique for finding eidetikers is so important that I devoted § 7.4 to this problem.

The question is still open as to whether both normal and eidetic persons have detailed texture memory, but only the eidetiker has access to its recall. The alternative is that only eidetikers have detailed texture memory. The lack of a detailed texture memory in normal human beings does not mean that we cannot extract and store certain stochastic parameters that specify the general appearance of a girl's skin or the grain of a walnut desk. The important distinction, however, is in the order of the statistics. A detailed texture memory would require the storage of all points of a texture, while a "bank" of texture generators may be specified by some low-order statistics. I will discuss this problem in §§ 4.4 and 5.1.

Since we know so little about memory and recall, the research reported in this book is primarily devoted to point-by-point matchings and to pattern discrimination tasks for which memory is not of importance. This limitation severely restricts the generality of our subject matter; on the other hand, it gives us some insight into the processes. This is in contrast to the vague conjectures one finds in the fields of form recognition and cognition, probably due to the enigmatic nature of memory.

In this book the only departure from this restraint to exclude memory is in § 7.4 on eidetic memory. However, it appears that eidetic images can be portrayed before stereopsis; and, therefore, this type of detailed memory perhaps requires a more peripheral storage than does common visual memory. Of course, eidetic memory could be highly central, and the eidetic brain might have access to efferent pathways that stimulate a neural site before stereopsis! Whether a detailed texture memory exists in normal people or not, the problem of pattern identification (recognition) must be solved, too. This problem is discussed next.

Even after information reduction by foveal vision, the retained information content is astronomical. If we conservatively further reduce the foveal area to a tiny central array of 100×100 receptors and assume a 10 brightness level resolution (there are about 100 discriminable brightness levels), there are $10^{10,000}$ possible textures. Needless to say, only a minute fraction of these possible patterns has some perceptual significance.

It is hard even to estimate the number of patterns that have any perceptual significance. Yet, they have something in common. They must satisfy certain criteria of the first questions of an abstract twenty-questions game. (The player of the popular twenty-questions game must guess the name of an object or entity from yes-and-no answers to his question and tries to succeed with a minimum number of questions.) Actually, these first "questions" are never consciously asked by the players of this game; they are "asked" by the perceptual processes prior to such early questions as to whether the "thing" to be guessed is a physical or an abstract object, inanimate or alive, animal or vegetable, edible or not, and so on. These first unvoiced questions establish whether a pattern or some parts of it are a thing or a not-thing. The thing is also called "figure" or "signal" while the rest, the not-thing, is also called "ground" or "noise." This separation of our visual environment into figure and ground is the fundamental process of perception as pointed out by many researchers, particularly Hebb (1949).

The Gestaltists were the first to study figure and ground perception in a systematic way and formulated a few "principles" that favor patterns to be seen as figure. Such principles, or rules of organization, which will be discussed in § 4.6, were: proximity, similarity, small area, closedness, symmetry, smooth continuation, and so on. These Gestaltist principles have great heuristic value, but when applied to concrete problems they turn out to be too abstract and too general to have any predictive value.

In this chapter I shall try to explain the problem of figure and ground in a different way from that of the Gestaltists. Instead of searching for some general central organizational rules, I shall proceed by identifying a few early perceptual processes that extract the first answers to the "twenty questions." Is the pattern or subpattern a figure because its picture elements are moving together, or because its proximate elements have similar brightness (or color), or because they have similar binocular disparity? For these three processes the Gestaltist principles of proximity, similarity, and smooth continuation gain concrete meaning; as a matter of fact, they are quite different in details for each case.

Sections 4.2, 4.3, 4.4, and 4.5 will be devoted to the grouping of movement, brightness, and binocular disparity. Section 4.2 will treat clusters in movement. Section 4.3 will discuss problems of connected brightness elements in a textureless ground, while § 4.4 discusses problems of textured grounds. Sections 4.6 and 4.7 study the role of other factors underlying figure and ground perception, particularly symmetries. Sections 4.6, 4.7, 4.8, and 4.9 are devoted to an operator-theory of perception and to some problems of selective attention. Finally, § 4.10 compares cyclopean findings on figure and ground reversals with classical results and reaffirms the importance of movement, brightness, and binocular disparity for figure and ground separation.

4.2 Perception of Moving Clusters

How would the reader construct a simple device whose survival would depend on catching a few small objects (bugs) hidden in a complex environment? If the decision logic of the device were to be limited with respect to the size of the receptor mosaic, with the bugs at a standstill and, furthermore, the irrelevant background information complex, the problem would defy even a genius. It is not by chance that such highly developed creatures as frogs would starve amidst plenty if the prey did not move. Creatures, from insects to amphibians, possess a visual system that processes primarily moving patterns falling on clusters of adjacent receptors. Even more precisely, only those proximate points having the same velocity are formed into a cluster, and this cluster has to be of a certain size and shape to be regarded as a bug. Although some recent work by Grüsser and his co-workers (1967) criticizes the "bug and snake detector" concepts of Lettvin, et al. (1959, 1961) as over-simplifications, their basic notion of early feature extractors had a revolutionary impact on current thinking. Henn and Grüsser (1969) quantified the response of the "movement gated, dark convex boundary detectors" of Lettvin, et al. (1961). If R is the average impulse frequency during the traverse of the stimulus through the excitatory receptive field, v is the angular velocity, A the size, and c the contrast of the stimulus, and $A*$ a threshold area (depending, to a limited degree, on v and c), then

$$R = kv^{0.7}c^{0.6}\log \frac{A}{A*} \text{ (impulse/sec)}. \tag{4.2.1}$$

In a treatise on human visual perception I include equation (4.2.1) describing a class of movement sensitive neurons in the frog's retina because it shows some of the stimulus parameters underlying human vision. Even if quantitatively the results may be different for human movement perception, the finding that increased stimulus size, velocity, and contrast make the detection of moving targets easier is qualitatively right (within limits) for humans as well.

The idea of transmitting only changes to the higher brain centers had such survival value that this might be the reason that even the human visual system adopted this principle. Indeed, as I have already mentioned, our peripheral visual system can detect only moving or flickering objects but is very poor in resolving stationary targets. What is more, our foveal vision cannot transmit retinally stabilized images to the cortex. If it were not for the saccadic movements of our eyes, only moving targets could be seen. This was already suspected in the last century, since entoptic images such as the shadows of the retinal capillaries, or Haidinger's brushes, could be made visible only by special manipulations. Experiments with contact lenses by Ditchburn and Ginsborg (1952), Riggs, et al. (1953), and Yarbus (1957) demonstrated the disappearance of stabilized targets of any shape, as expected from the findings with entoptic images. If the stabilization of the retinal images is not perfect because of contact lens slippage, the signals transmitted to the cortex are merely impoverished, but do not cease to exist. Under these conditions partial fadings and segmentation phenomena are reported by Pritchard (1961). These signal impoverishments can also be obtained by faint illumination, afterimages, and tachistoscopic presentations, and the fragmentation units obtained are very similar as shown by Evans (1965). Many of the perceptual elements such as parallel lines, corners, and alphanumeric characters, fade out or reappear as a unit.

Since these fragmentation phenomena are one of the few direct perceptual evidences for feature extraction, it seems surprising that after the pioneering work of Pritchard (1961) and Evans (1965), which was largely qualitative, so little thorough research has been carried out. Some auditory analog of retinal stabilization and fragmentation might be the semantic satiation experiments, which originated in Titchner's laboratory (Severence and Washburn 1907, Bassett and Warne 1919) and recently met renewed interest (Lambert and Jakobovits 1960). These experiments in their up-to-date form present an auditory pattern (e.g., the word "doberman") exactly repeated for minutes. The listener's perception will segment and regroup the original stimulus in percepts such as "overman," "oman," and so on. The only reason this auditory example is mentioned is to point out to the reader that stationary patterns are not interesting for the CNS regardless of modalities.

In this context of movement perception and retinal stabilization let me review a pilot study of mine (Julesz 1963b). I used the Pritchard technique of a small microscope mounted on a contact lens. The stimulus used was a pair of parallel vertical lines in a dark field with a letter "C" between the lines. One of the lines was vertically polarized and the other line was horizontally polarized while the letter was unpolarized. By rotating a polarizer before a light source, one could alternately illuminate the two lines, and at a certain speed best apparent movement could be perceived. The unpolarized letter was flashed twice as rapidly as either line. Because of retinal stabilization this rate of flicker for the letter "C" was above critical fusion, and the letter disappeared while the apparently moving line swept over the area of the "C" without causing it to reappear. This stroboscopic movement is thus different from real movement, since a physically moving line when sweeping over a retinally stabilized, disappeared figure causes the figure to reappear. This difference between stroboscopic and real movement has been studied by Kolers (1963) using detection tasks during normal vision. He also found it to be significant. During my experiments an interesting effect was noticed. It is known that in normal stroboscopic movement perception between best apparent movement and simultaneity regions the observer either perceives a single line moving or sees two lines simultaneously at a standstill. These two conditions quickly alternate according to the observer's state of mind. Under retinal stabilization, however, only the best apparent movement condition can be experienced. This is the case even when the rate of stroboscopic movement approaches the simultaneity condition, since the simultaneous lines at a standstill fade out. This experiment indicates that there are two neural populations: one has a time constant that yields best apparent movement, the other has a shorter time constant, and the two lines are simultaneously resolved. When one or the other neural population is predominantly firing, the observer has no voluntary control in shifting his attention and the dominant neural organization prevails. However, when the two neural organizations are about equal, the observer can shift his attention to either organization at will. Under retinal stabilization only one neural organization prevails and only best apparent movement can be perceived.

As I discussed, clusters of dots sharing some common properties are the basis of object separation. One of the most important properties underlying the grouping of many dots is movement of the same velocity and orientation. Even if the dots are not adjacent, their common motion will be perceived as a *rigid* transparent object in motion. In 1966 Julesz and Bosche prepared a computer-generated movie (Bosche 1967) in which a certain proportion of random black and white dots moved to the right, while the rest moved to the left. Even these nonadjacent thousands of random dots grouped together into two transparent oppositely moving surfaces. Of course if the dots moving together are spatially adjacent, the cluster formation becomes even more pronounced. After all, the definition of an object in its first approximation is a set of proximate dots moving together. Most of the feature extractors in the visual cortex of a cat and monkey, as discovered by Hubel and Wiesel (1962, 1968) and in the retina of the rabbit, as found by Barlow and Hill (1963), extract some adjacent clusters of dots which move together. The simplest such neural units in the visual cortex become activated when an edge (or slit) of a given orientation moves in a direction perpendicular to the edge. The complex and hypercomplex units of Hubel and Wiesel also function as moving-cluster detectors, although the properties of the clusters and their movements become increasingly more complex. For instance, a cortical unit will only respond to an end of a line at a given position provided the line is vertical and moves from left to right.

Most studies of movement have used rigid transformations. Even the first computer-generated movies by Green (1961) used rigid transformations, as did some later work by Braunstein (1966), who studied movement parallax and the kinetic depth effect. There are many important questions in depth perception for which rigid transformations of textures give good insight as suggested by the theories of Gibson (1957). I will spend some time on movement parallax in § 5.2 where the most powerful monocular depth cues will be discussed.

However, the most interesting and least explored problems are the perception of nonrigid transformations, particularly nontopological ones. My movie with Bosche in which an area of static noise expanded or contracted in a dynamic noise surround, or another scene in which every even row of a random texture moved to the right and every odd row to the left, yielded some interesting perceptual experiences. As each successive frame contains increasing areas of correlated points surrounded by uncorrelated noise, the percept is an expanding static sheet. The other scene of oppositely moving even and odd rows appears as two transparent rigid planes moving on top of each other (alternating in apparent depth at will). These computer-generated movies were first programmed, then executed, and then the displays on a high resolution video transducer (microfilm plotter) were photographed frame by frame. It is at the limit of today's computer technology to produce on-line movies with high resolution (above 200×200 picture elements and 20 frames/sec rate). But 100×100 resolution at 12 frames/sec rates has been achieved on-line by my co-workers and others. Since off-line computer movies are too cumbersome for research, on-line movies are a practical necessity for studying movement perception. In order to investigate the perception of point domains moving at various speeds and complex geometrical configurations, one must be able to change variables on-line and see immediately the outcome. Some of these changes in parameters can be initiated by typing in numerical values; however, the best way of changing geometrical relations is a light pen (or any similar graphic device). One could even try to modify the input stimulus by changes in subjects' electroretinogram, evoked potential, or eye movement. I hope that many researchers will become interested in complex spatiotemporal patterns and new insights will be gained. Until then, I will discuss some new findings in § 7.5 obtained off-line by random-dot stereogram movies that extend Korte's law.

But let us return to the problem of figure and ground. I have discussed how moving clusters of dots become the figure. In random-dot cinematograms containing correlated areas that do not move, the observer perceives areas of static and dynamic noise. Under these conditions either the static or the dynamic-noise-covered area can become the figure at the observer's will. However, if any of these areas start to move or expand, they become the figures (regardless of whether they were portrayed by dynamic or by static noise). For more conventional random-dot cinematograms in which static textures perform rigid movements, the problem of figure and ground is coupled with the monocular depth cue of movement parallax. The faster-moving areas appear closer to the observer. If in addition the faster-moving areas retain each of their picture elements while parts of the slower-moving areas become hidden by the faster-moving textures, then the other powerful monocular depth cue of interposition is called into play. Both cues, movement parallax, and interposition will cause the faster-moving area to be perceived as being closer to the observer and being the figure. This coupling between figure and the closest surface to the observer is so strong that in static two-dimensional displays without depth cues, the area which becomes the figure will be seen in depth *above* the ground. This was known to the earliest investigators of figure and ground, for example, Rubin (1915). If movement parallax and depth cues are juxtaposed by introducing stereopsis so that a hole in depth is moving around in a textured surface, the movement can make the hole (an area behind the surround) appear to be the figure. However, if the area behind the surround is at a standstill, it is very difficult to perceive it as the figure. In § 4.10 we will investigate the role of depth in figure and ground perception under ideal conditions uncontaminated by monocular cues.

Thus we have learned that moving clusters of dots are effortlessly perceived as the figure. If we want to devise schemes that separate stationary objects from a stationary surrounding, the difficulties increase by several orders of magnitude. The basic phenomenon which accomplishes this separation is the perception of figure and ground. I will discuss figure and ground perception in detail somewhat later.

I have discussed how global patterns could be portrayed by local motion; findings showing that neural units would discriminate between a certain direction or its opposite have also been reviewed. Indeed, in real life situations there are many particles that perform similar movements, such as exploding materials or a snowstorm. However, there are complex movements which usually do not occur in real life, and the question arises whether or not they can be perceived. In order to test this, Julesz and Hesse (1970) generated a movie on the computer containing thousands of tiny line segments (needles) in 48 orientations (any integral multiple of 7.5°) such that each needle appeared to rotate around its center in the plane of the projection screen. Such an array of needles is shown in either image of figure 5.3-1*. In successive frames the angle of each needle was pivoted on its centerpoint by 7.5° or a multiple thereof. Adjacent areas containing needles moving at the same speed but with the opposite rotational direction (one clockwise, the other counterclockwise) could not be discriminated from each other. Only areas of different speeds of rotation (50% speed difference or more) yielded vivid discrimination, regardless of rotational direction. This finding is different from that of translational movement of the needles or from rotational movement of parallel needles, for which direction of movement is readily perceived.

This finding shows that movement extractors are stimulated by edges which assume positions close to or parallel to their preferred orientation. They are equally stimulated by clockwise or counterclockwise rotation of the same speed regardless of directional information. Only increased speed of rotation yields global perception of areas that contain faster rotating needles than the surround.

These experiments also indicate a basic difference between local and global movement perception. Although the local movement of each needle in rotation can be readily perceived, the global movement of many adjacent needles rotating in the same direction is not perceivable. This difference between local and global movement processing is very important. While neurophysiology has revealed many mechanisms for local movement extraction, at the moment no such evidence exists for global movement extraction. Furthermore, the expense and complexities of computer-generated movies are still an obstacle in studying global movement perception. Because of this, most of the experiments in the book are on local and global stereopsis. However, I am confident that real-time studies with third-generation computers will soon reveal many new phenomena of global movement perception.

4.3 Perception of Connected Clusters under Rigid Transformations

Let us turn again to our previous question of trying to construct a device that separates objects at a standstill. If an object and its surrounding are similar, both are at a standstill and only one retinal mosaic is given (monocular vision), this task is very difficult, if not impossible. The success of animal camouflage is based on this difficulty, and even for the highly developed human visual system the task of monocularly detecting a camouflaged object is not easy.

The simplest case for determining an object at rest is when the object is formed by connected dots of the same brightness and color as exemplified by two-dimensional outlined drawings. Although it is relatively easy to define such outlined figures when they are presented alone, it is very difficult to separate two or more of them from each other when they overlap. In a way such overlapping outlined figures are camouflaged. Without being familiar with at least one of them, it is impossible to tell them apart from the rest. If a device were to be built which would separate all these outlined figures from each other, it would have to store all possible objects of interest. This raises the question of how many objects can be handled by a retina of a given size. As an example, we take a ridiculously small "retina" of a 4×4 receptor mosaic. We also assume that the background is black, while the object is uniformly white. Even this tiny retina can be confronted with 2^{16} or about 6×10^4 possible patterns. If we increase the size of the retina to a 10×10 array, which is still much smaller than the most primitive eye of existing organisms, the number of possible patterns increases to 2^{100}, which is more than the assumed number of atoms in our universe. Let us continue with the 4×4 retina. If we restrict ourselves to objects which stimulate only 4 receptors, the number of possible objects is reduced from 6×10^4 to $C_4^{16} = [(16 \times 15 \times 14 \times 13)/(1 \times 2 \times 3 \times 4)] = 1,820$. This is a considerable reduction but still a hopelessly large number when the retina approaches practical sizes. A further drastic reduction can be obtained if we restrict our objects to stimulate only connected (adjacent) receptors and regard objects under translation as being equivalent. Connectivity in a Cartesian tesselation will be in the horizontal, vertical, and diagonal directions, and we assume that the retinal mosaic of our device is a Cartesian grid. How many objects exist which stimulate (cover) n connected cells? For $n = 4$ one may readily draw all possible connected patterns $(M(n))$ and find by exhaustion $M(4) = 110$. Further reduction can be achieved if one regards as equivalent an object not only under translation, but also under rotation. Figure 4.3-1 lists all the equivalent objects of 4 connected cells and their number $(N(n))$ is $N(4) = 22$. In parenthesis $s_{n,i}$ is the number of connected figures of n cells which are equivalent to

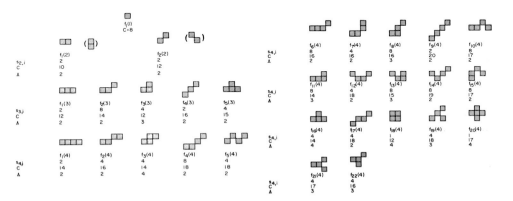

$f_i(n)$, where $f_i(n)$ is an individual connected figure consisting of n cells. It is interesting to note that rotational equivalence (or rotational invariance) divides the object number by only 5. Löfgren (1960) computed first for a given n the values of $M(n)$ and $N(n)$ using some lower and upper bounds. For large n he found that the ratio $M(n)/N(n) \to 8$. This is the more interesting since $s_{n,i} = 8$ for two-dimensional objects without symmetry axes. Thus for large n values an overwhelming majority of objects has no symmetries. One might be astonished that rotational invariancy reduces redundancy by only 8, but this is the result of the finite cell size. Because of this, 4 cells along a horizontal (or vertical) direction are different from 4 cells which lie on a diagonal; yet with finer resolution and larger n values there will be an increasing number of objects that are equivalent under rotation. In the limit, an object can be rotated by any angle without changing its shape, and indeed, we are unaware of shape changes as we look at a rotating object in spite of our finite receptor mosaic.

Löfgren's estimate of $N(n)$ is given in equation (4.3.1):

$$N(n) = \left(\frac{1}{8}\right) 6.88^{n-1} n^{-0.781} \tag{4.3.1}$$

which becomes increasingly accurate for large n values. For its derivation see Löfgren (1960). From equation (4.3.1), $N(10) = 711{,}000$, which is a minute fraction of

2^{100} possible patterns. The greatest saving is achieved by regarding only connected patterns that have translational invariance. Instead of considering patterns of retinal sizes which result in n^2 or more cells, it is adequate to restrict the computation to some function of n, the number of connected cells. This dependency on n alone is the most important fact in equation (4.3.1). In addition to translational and rotational invariance one could incorporate size invariancy in the device. For a finite cell array in two dimensions the dilation-contraction operation (or zooming) depends to a great extent on the algorithm used. Nevertheless, zooming, as well as rotation, further reduces the number of possible nonequivalent patterns.

These theoretical considerations show that even the number of nonequivalent patterns becomes astronomical when the size of the human retina is approached. In real-life situations we cannot separate unfamiliar objects from each other when some other cues such as differences in texture, movement or binocular disparity are not given. Sometimes the familiarity of objects does not have to be specific; it is adequate to know some general constraints that characterize them. For instance, one can separate convex objects from a background of concave objects or symmetrical objects from asymmetrical ones without having to be familiar with the given object. Thus it is sometimes possible to separate an unfamiliar outlined figure from the rest if it is characterized by some familiar constraints. (A computer algorithm that can separate outlined polyhedra from each other uses this principle and will be reviewed in § 5.6).

In the next two sections I skip the enigmatic problems of semantics, and turn to simpler problems of object separation which, however, are typical of real-life situations. We will see how differences in surface texture and binocular disparity permit us to separate objects in relatively easy ways. Since most objects in our environment are textured and are binocularly viewed, object separation by these cues is a most natural process, and the separation of outlined figures and symbols is a rather artificial later development that the visual system had to cope with.

4.4 Clusters in Visual Texture Discrimination

In the previous section I have estimated the number of clusters of a few points that were adjacent. In this section I turn to human psychology, to the discrimination of two side-by-side textures that are composed of many points. It will turn out that texture discrimination is strongly dependent on clusters that are formed by chance. Furthermore, texture discrimination is a special case of figure and ground perception. The textures used are usually random with controlled statistical and topological properties.

The familiarity deprivation permits the study of the primitive syntactic structure, which is ordinarily hidden by the many complex familiarity cues. The use of random textures, on the other hand, permits the stimulation of certain features many times (as they occur in the texture with some prescribed density), which usually leads to a stronger percept than the one experienced from a usual stimulus of a single feature. Originally I tried to describe the textures and predict their perception by specifying their statistics by a certain k-order (Julesz 1962a). My original efforts were orientated towards textures which had the same k-order statistics (and, of course, all lower than k-order marginal distributions were identical) but differed in their $(k + 1)$-order probability distributions. Such stochastic processes of the Markov type can always be constructed if the event space has three or more members (e.g., black, gray, and white brightness levels), and this conjecture of mine has been proven by Rosenblatt and Slepian (1962). The problem was to determine the maximum k-order for which the simple task of discriminating between two textures (presented side-by-side) was still possible and to describe the perceptual quality which remained attached to it. Figure 3.1-2, for instance, shows two textures having identical first-order statistics (same frequency of black, white, and gray levels) but different second order statistics (the frequencies of two dots have certain colors). Discrimination is effortless, and the perceived difference is in granularity (Julesz 1962a). Figure 4.4-1 shows two textures of

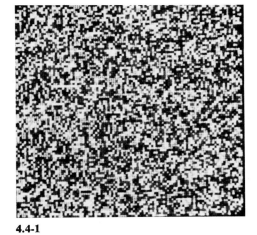 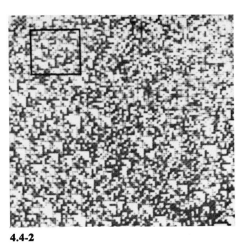

4.4-1   4.4-2

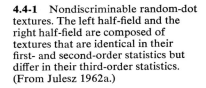

4.4-1 Nondiscriminable random-dot textures. The left half-field and the right half-field are composed of textures that are identical in their first- and second-order statistics but differ in their third-order statistics. (From Julesz 1962a.)

4.4-2 Discriminable texture based on the presence or absence of clusters of dark triangular shapes. The equally probable triangular shapes are shown in figure 4.4-3. (From Julesz 1962a.)

four brightness levels (having uniform and identical first-and second-order statistics) that differ in their third-order statistics. (Of the 64 possible trigrams the left field contains 32, while the right field contains the complementary 32 trigrams.) Discrimination of the left and right half fields into two separate entities is impossible (Julesz, 1962a). The same paradigm has been applied to study musical texture discrimination (using random melodies of given statistics by Julesz and Guttman 1965). Recent studies in audition using this paradigm have been continued by Pollack (1968, 1969) and West (1968).

Unfortunately, the next experiment of figure 4.4-2 demonstrates that n-order statistics are poor descriptors of perceptual performance (Julesz 1962a). Here the quadrant of the upper right corner is random, while the rest is made up of the six equally probable triangular patterns shown in figure 4.4-3. In spite of the fact that each of these micropatterns has the same frequency of occurrence, only the all-white triangular patterns are perceived, while the rest pass unnoticed (Julesz 1962). Thus

4.4-3 Enlarged area of figure 4.4-2 showing, besides the dark triangles, the equally probable five other triangles. (From Julesz 1962a.)

4.4-4 An illustration showing that the visual system works like a slicer. (From Julesz 1962a.)

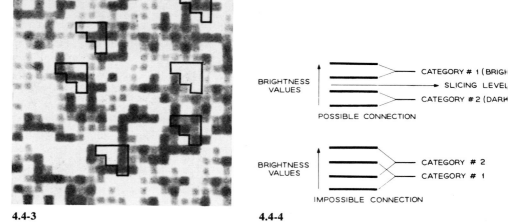

4.4-3   4.4-4

only clusters of proximate points of similar brightness values lead to discrimination. Here proximate means closeness in a Euclidean distance sense, while similar means adjacent brightness levels. Furthermore, it was shown that the visual system works like a slicer (figure. 4.4-4); that is, white and light-gray dots form bright clusters and dark-gray and black dots form dark clusters. Similarly red and yellow proximate dots (of equal brightness) form orange clusters and blue and green dots form blue-green clusters. However, it is impossible to organize nonadjacent brightness or hue values into clusters as shown by Julesz (1965a). Thus, red and green will not form a cluster nor will blue and yellow. (However, yellow and green do not group together, although they are adjacent hues.) This finding indicates again that conscious effort cannot override some of the early processings by the visual system.

The summary of these first studies is as follows: There is a cluster-detector operating in the visual system, probably at the earliest levels, and only those statistical descriptors that describe cluster formations can have some perceptual significance. Thus, the usual joint-probability distributions are inadequate descriptors in perception. There are at least two ways to handle this difficulty. One solution is to define some single cluster-formation rule and give probability distributions of the clusters and of some cluster parameters (orientation, compactness, etc.). The trouble with this solution is the arbitrariness of the cluster parameters selected. The second solution is the use of constructs of random geometry. Novikoff (1962) was the first to suggest such a solution. Frisch and Julesz (1966) applied the generalized Buffon-needle problem to perceptual studies. If we throw n-gons over a two-tone image according to a well-specified random-throwing procedure and collect the statistics of all the vertices falling on a black (or white) domain, we obtain a most useful n-order descriptor (43). For $n = 1$ (points being thrown) the area fraction of white-black domains is obtained. The $n = 2$ (line throwing) case determines the perimeter-area fraction. The $n = 3$ case (triangles being thrown) extracts the notion of convexity, and so on. The advantage of this technique is that the n-gon statistics (and the lower-order terms in their Taylor-series expansion) often have excellent geometrical and perceptual interpretations.

Frisch and Julesz (1966) used the power of this technique on a simple perceptual task. The stimuli used were computer-generated random-dot textures with various area fractions between black and white dots. Subjects were asked to judge the stimulus as being a black tabletop with white specks on it or vice versa. In this form the perceptual task was figure-ground reversal, a topic to be discussed in the next section. With very small or very large area fraction between black and white (e.g., 10% or 90% densities), subjects unanimously perceived the ground (tabletop) as black or white, respectively. At intermediate area fractions, many of the subjects could stick to the original question and were able to see a tabletop of black or white color with oppositely colored specks on top of it. The ground might have fluctuated but could be seen only one way at a given time. This perceptual weighting of either the figure or the ground (but not both) at a given instant is characteristic of figure-ground perception. On the other hand, there might have been subjects who based their decision in the intermediate region on some other criterion, such as the judged area-fraction being under or above a judged 50% threshold.

In figure 4.4-5 three area fractions of 10%, 50%, and 90% are presented. A checkerboard frame of 50% area fraction separates the stimulus from the peripheral areas in order to minimize their effects. The reader can judge for himself whether he can perceive the 50% range as figure-ground with fluctuations or is only guessing the area fraction. Whatever the subject's criterion is, the problem to be investigated was whether the perceptual response as a function of area fraction would differ for various tesselations. Thus, the black and white dots were randomly selected with various area fractions (densities) on a Cartesian grid, a triangular grid, and so on. Subjects were asked to judge the stimulus as being a black tabletop with white specks on it or vice versa. Since the n-gon statistics always differ for various tesselations used, except for the first-order throwing, we wanted to know whether figure-ground perception depends on the tesselations used. According to these findings only the area fraction between black and white dots affects figure-ground perception (or judged area fraction) and is independent of the tesselations used. Figure 4.4-6 shows averaged subjects' responses for three different tesselations, and the differences are statistically nonsignificant. It is interesting that 50% perceptual reversal (or judged 50% area fraction) occurs at a 42% black/white area fraction. Thus for figure-ground perception (or area fraction perception) of unfamiliar images, the visual system extracts only first-order statistics. This result is quite revealing in the light of recent neurophysiological findings. The concentric Kuffler-units in the retinae of the cat and monkey are probably used for figure-ground perception of random-dot textures. Of course, a human face or other familiar pattern will resist attempts to be perceived as ground and probably very high-order n-gon statistics will still be used by the visual system. This experiment with random textures served only to show that with proper stimulus control and restricted perceptual task the visual system can be greatly simplified and studied.

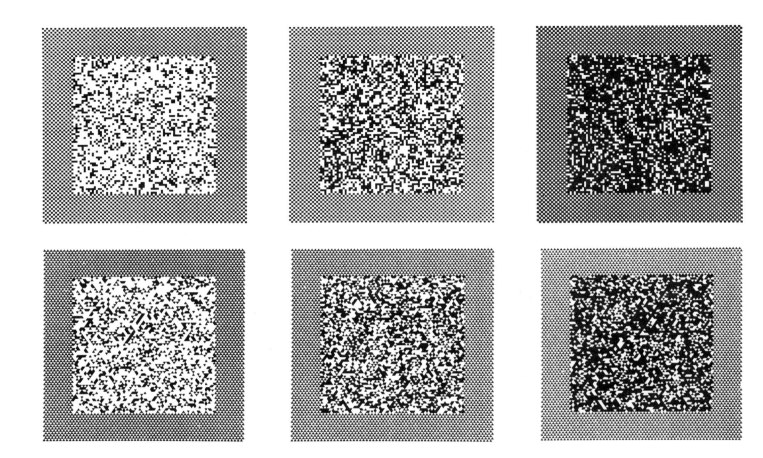

4.4-5 Random-dot textures (Cartesian [top row] and triangular [bottom row] tesselations) with 10%, 50%, and 90% area fractions, respectively. (From Frisch and Julesz 1966.)

It would be very instructive to continue with this program and use random geometrical constructs to study texture discrimination. The same problem the author tried unsuccessfully with k-order statistics could be repeated with k-gon statistics. We might generate two textures side-by-side which are identical in their one-gon and two-gon statistics and different in their three-gon statistics. By the way, this example is equivalent to two textures which have the same area fraction and perimeter-over-area fraction; but, for instance, one texture contains only convex clusters while the other texture might have concave clusters as well. One can assume that this display might still yield discrimination although no known algorithms exist which would generate textures with such n-gon statistics. Since the throwing of n-gons is inherently linked with cluster properties, the perfection of such display generating techniques with prescribed n-gon statistics would probably be powerful tools in perceptual studies.

Unfortunately for $n > 6$ the geometrical meaning of the n-gon statistics (and the lower-order terms of their Taylor-series expansion) is not known. Furthermore, some apparently simple notions such as connectivity of two points require infinite-order

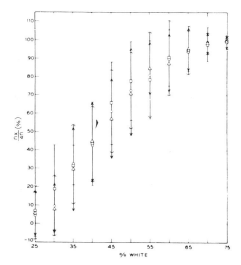

4.4-6 Subjects' responses to figure-ground reversal for three different tesselations as a function of area fraction. (From Frisch and Julesz 1966.)

throwing. This and other surprising proofs were given by Minsky and Papert (1968), who studied a simple class of perceptual models called perceptrons. Although they did not use random geometrical techniques, many of their proofs can be translated into these considerations. Such a translation might benefit both methods. The finding that infinite-order throwing is required for detecting whether two complex figures are connected or not agrees with our perceptual inability to do so. This is shown by figure 4.4-7 in which one display contains two figures, while the other contains one connected figure. Only by sequentially tracing the figures can we extract this information; when viewing the displays we cannot decide which of them is connected.

In this section I have shown how clusters of similar elements were the basis of texture discrimination. I discussed only the grouping of dots having the same brightness and color. However, multidimensional scaling of texture discrimination by Julesz (1964a) using simple 2×2 arrays hidden in random textures revealed that the most important factor was brightness and the next was orientation. Beck (1967) studied grouping and texture discrimination of line segments as well as T-shaped patterns of various tilts. He also found that similarity of brightness and orientation

4.4-7 A nonconnected and connected complex figure after Minsky and Papert. Connectivity cannot be spontaneously perceived. (From Minsky and Papert 1969.)

was essential for grouping (cluster formation), while more complex properties such as the rated similarity or familiarity of figures were irrelevant.

The method of multidimensional scaling was developed by Shepard (1962) and Kruskal (1964), and together with various cluster-seeking computer algorithms, as those devised by Johnson (1967), might be very useful for pattern-discrimination studies. I will return to these methods in § 9.5

Until now I have discussed abstract, random textures. However, most textures in our environment are not entirely random. Skin, hair, fur, wood, and similar epithelium-like structures are macroscopic aggregates of densely-packed biological cells of similar shapes. Most solids, such as minerals, metals, fibers, and so on, are macroscopic aggregates of crystals. The textured surfaces of these familiar materials and organisms are strictly determined by the microscopic structure of their elements. For instance, a salt crystal can have an endless variety of shapes, but each macroscopic representation is a finite aggregate of densely-packed NaCl molecules and differs from other kinds of crystals.

The realization that most textures around us are not random is important. I have discussed how unlikely the existence of a detailed texture memory might be if it were not for the recent eidetic experiments to be reviewed in § 7.4. However, in order to recall the texture of a walnut panel or a silk dress, one does not have to store each point of the texture. It might suffice to store a few rules that would generate some representation of these textures. In principle, one could store the shape of a single molecule or biological cell and build up adjacent macroaggregates from thousands or millions of these building blocks, according to some random rules. These rules would determine the average macroaggregate size, preferred orientation, length-width ratio, and so on. Needless to say, such a principle of texture generation is most unlikely and serves only to illustrate that a relatively limited memory can store a program capable of generating detailed textures.

Another aspect of texture memory is illustrated by this model. We are surely not able to memorize every detail of a girl's face, but we can imagine (or dream about) the same girl, recalling that her face has a certain skin texture. A relatively small number of statistical parameters of typical stored clusters such as a few higher-order moments might describe textures. The difference between this assumed model and the previous one lies in the basic elements to be stored.

Such a bank of texture generators is much easier to imagine in audition in which the timbre of various musical instruments and noise sources are the auditory textures. These auditory textures can easily be characterized by a few parameters, for example, their pitch, the temporal attack and decay functions, some amplitude and frequency modulation constants, mean power and bandwidth of noise generators, and so on. The interested reader should turn to the literature of computer music, particularly to a book by Mathews, et al. (1969).

Unfortunately, visual textures are much more complex than auditory textures. Visual textures are two-dimensional; the exact phase information has to be provided; and as we have seen, not even second-order statistics can describe visual texture discrimination, much less a two-dimensional spatial frequency spectrum (which is a particular average of the second-order statistics). In the experiments reported here we will not use familiar textures, but will contend with random textures. However, to solve the problem of visual-texture generation of familiar surfaces is important for both theoretical and practical uses. I can only hope that in the near future scientists will begin to clarify the enigmatic problem of familiar textures.

I will return to the problem of textures in § 5.1. In the light of a recent report that eidetikers may possess detailed texture memory much of this section if not our entire formulation of perceptual problems may be reevaluated. However, the fundamental question remains, whether *only* eidetikers have texture memory, or do the rest of us have it too (only we do not have access to recall)? If only eidetikers have a texture memory, the basic problems of this section remain valid.

4.5 Perception of Clusters Having the Same Binocular Disparity

We have seen how clusters of proximate points of similar velocity, brightness, or color are perceived as figure and the rest is ignored, being regarded as ground. Strangely enough it took more than a century after Wheatstone's discovery of the stereoscope in 1838 for it to become clear that clusters of proximate dots of similar binocular disparity underlie stereopsis (Julesz 1960a). Without such a simple global organization principle, the locally fused corresponding points in the two retinal projections lead to ambiguities. This is illustrated by figure 4.5-1 where each of the four points

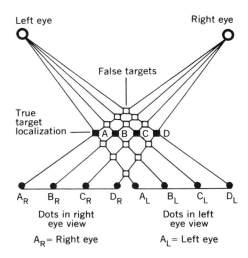

4.5-1 Illustration of how false localization of targets occurs in binocular vision. (From Julesz 1968d.)

in one eye's view can belong to any of the four projections in the other eye's view giving rise to 16 possible localizations. Of these 16 localizations, only 4 are right while the remaining 12 are phantom targets. As we increase the number of point targets that fall on the same horizontal line, the ratio of false localizations to all localizations rapidly approaches unity since P (phantom targets/all targets) $= (N^2 - N)/N^2$.

In this example we assumed that the targets to be localized were all alike. This is usually the case in localizing airplanes by two-beam stereo radar systems. If the targets are individually labelled, for instance, with the airplanes having responders with individual codes or with the points to be localized having different brightnesses, colors, or other local features, one can avoid the ambiguity problem. However, in vision this individual labelling of corresponding points is never carried out satisfactorily. After all there are hundreds of points of objects having complex textures that have the same brightness, color, or orientation. For example, figure 4.5-2* shows such a random-line stereogram in which the 100×100 cells contain a line segment of $+45°$ or $-45°$ orientation. Here each horizontal line of the two images contains on the average 50 line segments of the same orientation that could be localized in $50^2 = 2,500$ different ways. Both in figures 2.4-1* and 4.5-2* where the local information has only two values (black or white; $+45°$ or $-45°$ orientation) the simplest false global organization would be to see 50% of the cells as the background with zero disparity. Indeed, by chance alone, any two uncorrelated arrays of two values (0 and 1) will have 50% of (0,0) and (1,1) pairs. Therefore it would be a possible strategy for the stereopsis mechanism to select 50% of the dots which matched at zero disparity, then to take the remaining dots and select those that have one unit disparity and so on, until all the dots are localized at the smallest possible disparity values. These consecutive steps would lead to lacelike depth planes in close proximity, and in a few such depth planes with zero and small disparities all dots (or picture elements) would be localized. However, as stereoscopic fusion of figures 2.4-1* or 4.5-2*

4.5-2* Random-line stereogram containing line segments of ± 45° orientation.

reveals, instead of such transparent depth planes around zero disparity, one perceives a center square at a four picture element disparity (or in other displays with higher resolution such as in figure 4.5-3*, the maximum disparity becomes 80-100 picture elements) above (or behind) the background. Thus the visual system tries to find the densest possible solution regardless of disparity values, as long as it is within fusional limits.

This search for a dense surface is very strongly demonstrated in the random-dot stereogram of figure 4.5-3* where of an array of 1,000 × 1,000 dots, 10% are black and 90% are white. After stereoscopic fusion, a complex saddle surface (hyperbolic paraboloid) and a torus start to emerge and within a few seconds increase in solidity and depth. Another complex surface (a spiral staircase) is portrayed by figure 4.5-4*

The assumption that the visual system tries to find a global solution in the form of dense surfaces instead of localizing individual points in depth is proven in figure 4.5-5*. Here only 2.5% of the 100 × 100 Cartesian grid are selected as dark cells. Yet, after stereoscopic fusion, a center area emerges in depth as though it were cut out of white paper. The boundary of this white surface is spanned by the sporadic dark picture elements. The dark dots are not seen individually floating above the white surface of the printed page but are perceived at the same two depth planes as part of a speckled surface. Stereograms like figure 4.5-5* were first explored by White (1962) in a study determining the effects of texture density on stereopsis.

This search for dense surfaces can be fitted into the previous cluster-detection prin-

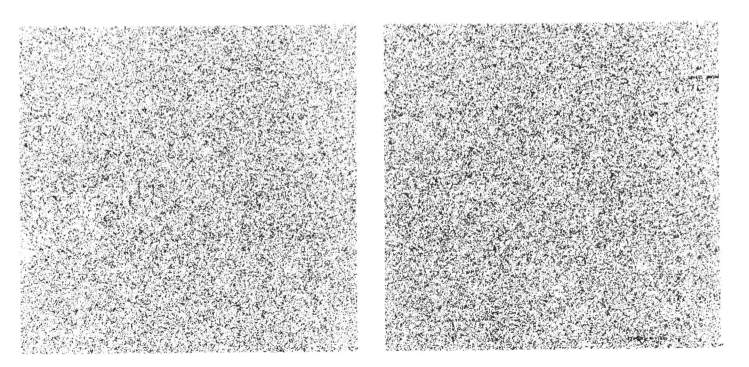

4.5-3* Hyperbolic paraboloid with torus portrayed by a random-dot stereogram.

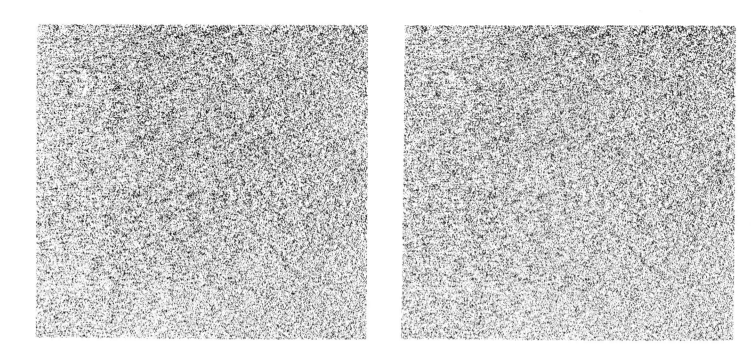

4.5-4* Spiral surface portrayed by a random-dot stereogram.

4.5-5* Random-dot stereogram with 5% black dots. The square in depth is vividly seen as forming a uniformly textured sheet.

ciple if we generalize the notion of similarity. In the previous sections we discussed how proximate points of similar brightness values are grouped into a connected cluster. This notion was generalized by Julesz (1960a,b) in his difference field model and further improved (Julesz 1962b). According to this model the left and right images of a stereogram are subtracted point-by-point from each other, and the absolute value is taken. Prior to subtraction the two fields are horizontally shifted with

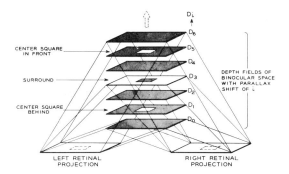

4.5-6 Difference field model of stereopsis.

respect to each other in increasing steps. Each of these shifts corresponds to a disparity unit (which may be the limit of stereo acuity corresponding to 2 sec of arc disparity), and the obtained difference fields are imagined to be stacked above each other according to increasing amounts of disparities. Such an arrangement is shown in figure 4.5-6. Each of these difference fields, D_i, consists of picture elements of various values randomly distributed except for corresponding areas of the same disparity, i, which corresponds to the disparity of the difference field in question. These are the cross sections of an object that has been sliced through by the very difference field in question, and they are composed of a cluster of dots having minimum values. For instance, in the example of figure 2.4-1*, the zero difference field D_0 and the four unit disparity difference field D_4 contain only points forming clusters of minimum values forming $(1-T)$ and T-shaped areas respectively. All other difference fields for this stereogram are devoid of clusters of minimum points.

Thus it is possible to formulate as a first approximation a simple model for stereopsis in which proximate points of zero or minimum values are grouped into clusters. These points are contained in a combined binocular field, and it is assumed that the combining operation is subtraction. The important assumption is that the operation is subtraction and not addition. It could well be division, but not multiplication. I have given several justifications as to why only an asymmetric operation (subtraction or division) can qualify (Julesz 1960a). A simple explanation will be given here: First, we assume that in the first approximation any operation can be substituted with the first two members of its Taylor expansion, which yield addition or subtraction. For stereograms that are composed of two types of picture elements, both subtraction and addition would result in similar models (except that for the latter case we would have to group points having maximum values). However, if the stereogram is composed of three or more types of picture elements, the situation changes. Take for instance a random-dot stereogram composed of three brightness values: $B(\text{black}) = -1$, $W(\text{white}) = 1$, and $G(\text{gray}) = 0$. For the difference fields the three correspondences that can occur by chance alone $|B-B| = 0$, $|W-W| = 0$, and $|G-G| = 0$ will yield differences that are less than all the differences of the other nonmatching combinations $|B-W| = |W-B| = 2$, and $|B-G| = |W-G| = |G-W| = |G-B| = 1$. However, for a summation field the corresponding points will yield differences $|W+W| = 2$, $|B+B| = 2$, and $|G+G| = 0$, which are not all more than the noncorresponding points, which yield $|W+B| = |B+W| = 0$, $|W+G| = |B+G| = |G+B| = |G+W| = 1$. Thus, for the summation field it is not possible to use a simple threshold device (a slicer) for more than two brightness levels. However, for difference fields the threshold device works for any brightness levels, and as a matter of fact, with increased effectiveness. While for two brightness levels 50% of the dots are identical in the two retinal projections by chance alone, for three levels this "noise" reduces to 33%, and for n levels it reduces to $1/n$ probability. With reduced "matching noise" it is possible to perceive much smaller areas in depth. For instance, the top half of figure 4.5-7* is made up of equally probable black and white picture elements and contains in the center a 10×10 picture-element array. The lower part of figure 4.5-7* has the same geometry, except the randomly selected picture elements are composed of eight brightness levels of equal probabilities. After stereoscopic fusion of figure 4.5-7* the small array in the top part is different from the surrounding but cannot be seen in depth. On the other hand, the same size array in the bottom part stands out in vivid depth.

Of course the difference field model is a first approximation. We have not yet included the various spatial frequency preprocessors and orientation-sensitive units. But after such preprocessors and some other normalizing processes are introduced the idea of a point-by-point difference field appears basically sound. Probably the

4.5-7* Random-dot stereogram showing a 10×10 square of two and eight brightness levels and eight picture-element disparity. Only the square having eight brightness levels yields stereopsis.

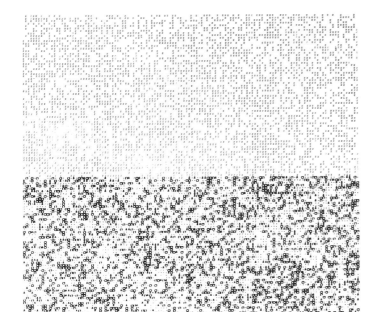
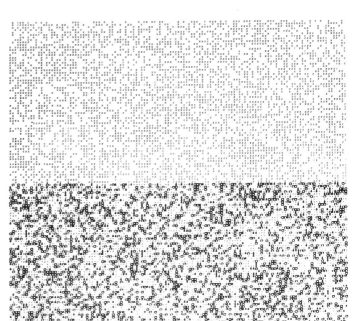

difference operator is not homogeneous throughout the entire binocular field but has to be weighted. All these additions will be added in the chapter on binocular depth perception. At the moment suffice it to say that it is possible to localize objects without recognizing them. The stereopsis mechanism regards as "figure," clusters of minimum values in the various difference fields. These figures correspond to the cross-sections of objects stacked above each other. All these figures, when perceived together, give the three-dimensional percept of an object while the rest becomes ground and is ignored. Often the grouping of minimum points requires, in addition to a given difference field D_i, the inclusion of neighboring difference fields D_{i-1} and D_{i+1} as well. After all, surfaces of physical objects usually are continuous in their coordinates. Years ago I tried some of these basic ideas by computer simulation, using a program called AUTOMAP-1 (Julesz 1962a). The interested reader is referred to this article and to § 8.4 for details. There is only one important fact to note; after a certain area is localized (i.e., found as a minimum cluster in a D_i depth plane) the corresponding points in the left and right fields are removed. In this model point removal is an explicit operation. In an improved model derived in § 8.4 point removal belongs to the inherent structure. This improves the localization of the remaining points. Probably as we try to fuse stereograms, say in figure 4.5-7*, we start with the largest corresponding areas, in this case with the background. After this is successfully accomplished, we try to fuse the remaining small area in the center. Only after all points are localized does the "urge" for fusion cease. It is interesting to compare the top with the bottom part of figure 4.5-7* from this point of view. While the bottom part gives a feeling of stability, the top continues to give a feeling of stress and results in repeated attempts for fusion.

In § 6.5 an improved model of stereopsis will be described that can cope with expansions and hysteretic phenomena that are beyond the capabilities of this simple model. However, for this chapter the difference-field model will suffice. In a way, cluster formation on which the model is based gives us good insight into many other perceptual phenomena. Besides didactic considerations, there is a historical reason for reviewing the difference-field model. This model was proposed in 1960, before binocular-disparity-sensitive cortical units were found by the neurophysiologists and suggested a way these units may interact. Since neurophysiologists still do not know this interaction, any possible model can be of help.

Are clusters of minimum values in the various difference fields all regarded equally as figures, or is there a bias toward those that are closer to the observer? This question will be discussed in § 4.10.

4.6 Perception into Figure and Ground

I have already touched upon the discrimination of the environment into figure and ground as a basic process of selective attention. The most complex form of figure and ground perception occurs when both figure and ground are stationary and at the same depth level. Yet certain patterns tend to be seen as the thing, while the rest is perceived as the background. In essence, we divide the world into two categories; into something we shift our attention to, and into the remaining items that form the not-thing and become ignored. In more modern terminology we separate the visual environment (E) into signal (S) and noise (N), as given in equation (4.6.1):

$$E = S + N = S + (E - S). \tag{4.6.1}$$

It should be noted that here, figure (signal), is a misleading term. The pattern need not be a familiar one. Figure (signal) refers only to the fact that we selectively attend to this organization while other organizations pass unnoticed.

The act of perception is to assign an operator P that acts on the stimulus:

$$PE = PS + P(E - S) = PS + PN. \tag{4.6.2}$$

Now every operator (Q) has at least one eigenfunction (R) and eigenvalue (λ),[1] such that $QR = \lambda R$. We assume that the perceptual operator P has two eigenfunctions S and N with the corresponding eigenvalues λ_1 and λ_2 respectively, thus

$$PE = \lambda_1 S + \lambda_2 N. \tag{4.6.3}$$

The peculiarity of figure-ground perception is that the perceptual process assigns the eigenvalues $\lambda_1 = 1$ and $\lambda_2 = 0$ to equation (4.6.3). In other words S is attended to, is vivid, and can be memorized, while $(E - S)$ is ignored, pale, and unmemorizable.

There are some instances when S and $(E - S)$ can alternately become the figure; thus either λ_1 or λ_2 becomes 1, but at a given instant

$$\lambda_1 + \lambda_2 = 1. \tag{4.6.4}$$

Thus the roles of figure and ground reverse. Such ambiguous stimuli are very important in the study of figure-ground since they facilitate the study of parameters that promote a given pattern organization to become the figure. The Gestaltist school, and particularly Rubin (1915), studied the perceptual factors underlying the perception of a figure. Let us review a few of these (the interested reader can find a good elementary summary in Hochberg 1964):

1. Proximity and similarity. Spatially adjacent units of similar brightness, color, and shape are grouped into clusters. These cluster-forming operations were discussed previously with the demonstration of figure 4.4-2.

2. Area. The smaller a closed region (within the limits of visual acuity) the more

---

1. For the reader unfamiliar with eigenvectors and eigenvalues the following examples might serve as an illustration. When an operator acts on a vector this will be transformed into a vector that in general has a different length, direction, and orientation from the original one. Yet for any operator there is at least one vector whose transform is identical to the original vector except for a scalar multiplier (expansion or contraction). This vector is called the eigenvector (or characteristic vector), and the scalar multiplier is the eigenvalue. For instance, the eigenvectors for the operator of expansion-contraction form an image of radial lines. Indeed, the only vectors (drawn on a rubber sheet) that remain invariant under dilation are a set of radial lines, except for a scalar multiplier. Similarly, a pattern (function) that is left invariant by an operator, except for a scalar multiplier, is called an eigenpattern (eigenfunction). For instance, eigenpatterns for the operator of rotation are concentric circles whose center is the center of rotation, while the eigenpatterns of translation are grids of parallel lines.

it tends to be seen as figure; and conversely, the larger the area of a region is the more it appears to be the ground. This tendency is clearly demonstrated in figure 4.6-1. (An extension of area to area fraction was discussed in § 4.4.)

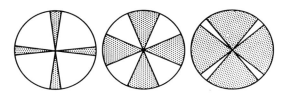

3. Closedness. Areas with closed contours are more likely to be seen as figures than are areas with open contours. This is shown in figure 4.6-2. Here it should be

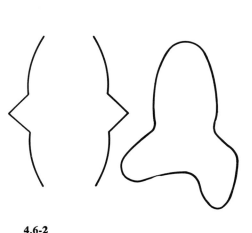

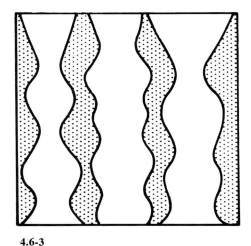

4.6-2   **4.6-3**

mentioned that the Gestaltists have been preoccupied with outlined drawings which are abstractions seldom, if ever, occurring in real life. Closedness really emphasizes the simple fact that no real objects can exist with open boundaries. (The historical fact that modern man is surrounded by artificial signs that can be "open," and are drawn on flat surfaces is probably so recent that the perceptual system is still basically a detector of objects with closed boundaries.)

4. Symmetry. The more symmetrical a closed region is, the more likely it is to appear as the figure. The greater the number of symmetries the region possesses, the stronger this tendency. This "rule" is illustrated by figure 4.6-3. The soundness of this principle seems evident if we realize that, in our environment, usually only living creatures and some man-made objects possess symmetries and natural objects usually do not appear symmetrical. (This consideration would be the opposite for a microscopic world in which the natural objects of crystals and the particles would be symmetrical while the living protozoa and bacteria would have amorphic shapes.) I will discuss symmetry perception at great length somewhat later.

5. Smooth continuation. Among many possible perceptual organizations, those that will minimize changes or interruptions in the contours (or some other features) of the constituents will be perceived as figures. This principle is shown in figure 4.6-4.

4.6-4   Demonstration of the principle of good continuation. (From Hochberg 1964.)

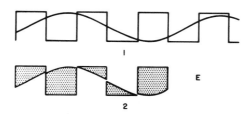

This is also a basic principle of physics whereby a moving object will continue to move along the tangent of its trajectory as long as there is no outside force to change direction.

These perceptual factors can be added or juxtaposed in ambiguous figures, and their weights for figure-ground reversal can be established. As a result of such experiments a hierarchy of figure formation exists: First, proximate points of similar brightness or color will group into clusters or lines. In the case of lines, their closure forms outlined clusters. Smooth continuation resolves ambiguities at places where outlines cross or where several perceptual organizations might occur. Both clusters and outlined clusters determine an area. The smaller the area is, the more likely it is to be perceived as figure. What is small or big depends on some complex "perceptual zooming" operation. When zooming to human face sizes, a small blemish on a face will become figure. But when zoomed on a crowd of people a blemish on a face is ignored, and probably a face of unusual complexion will be the small area to be attended to as the figure. Other properties of the clusters such as orientation, perimeter-to-area fraction, convexity, and so on, are on the next level of the hierarchy. Finally, the symmetries are observed.

4.7 Perception of Symmetries

This section can be omitted at the first reading of the book. It contains some detailed material on symmetry perception which gives good insight into the subject matter of § 4.8.

In § 4.6 I mentioned symmetries as being one of the weaker cues in figure and ground perception. In a way symmetries are one of the simplest invariances of patterns that preserve their identities under certain specific transformations (such as rotations and inversions). Of course we are not interested in the geometrical aspects of symmetry but in the question of which of these symmetries can be perceived. The other invariances, particularly perceived form (that stays perceptually constant under all rigid and many topological transformations of an object), however, strongly influence figure and ground perception. For instance, if the drawing of figure 4.6-1

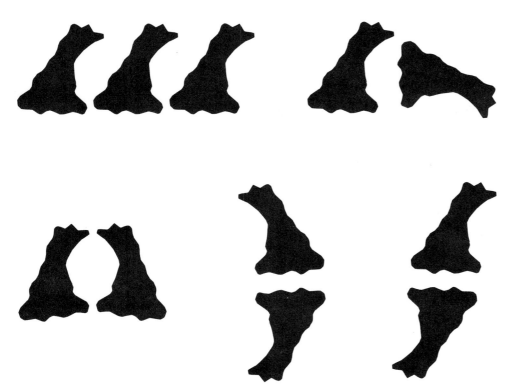

4.7-1 Mach's studies of repetition, similarity, bilateral (one-fold), vertical and horizontal, and centric symmetries using simple amorphic shapes.

contained form cues that would make one organization appear as the spokes of a wheel, the perception of this organization into figure would prevail. Unfortunately, most perceptual invariances, particularly recognition of form, are processes very hard to understand to be clarified, while symmetry perception is more amenable for treatment. Therefore, I decided to treat symmetry perception (of patterns having symmetries) separately from the other perceptual invariances.

In § 3.2 it was mentioned that Ernst Mach (1886) studied the perception of various symmetries using simple amorphic shapes. In figure 4.7-1 we show four of Mach's cases: repetition, similarity, bilateral, and centric symmetries—all of which can be easily perceived for amorphic shapes. However, for complex patterns (particularly when accidental cluster formation is prevented), only one of Mach's shown cases, bilateral symmetry, can be perceived. This has been shown in figures 3.2-4 and 3.2-5 (Julesz 1966a, 1967a). In figure 3.2-1 I showed that repetition of complex textures cannot be effortlessly perceived. Figure 3.2-6 demonstrated that centric symmetry is not spontaneously perceivable but requires scrutiny. However an n-fold symmetry (mirrored across n orthogonal axes) can be easily perceived, even when clusters are broken up as shown for two-fold symmetry in figure 4.7-2.

4.7-2 Two-fold symmetry with clusters broken.

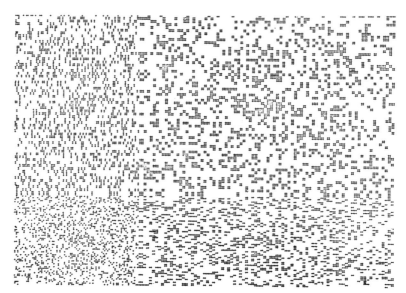

4.7-3 Two-fold symmetry in which the right and top half-fields are dilated by a factor of two with respect to the left and bottom ones.

The perception of *n*-fold symmetries for complex patterns is basically different from that for simple patterns. If a symmetrical pattern like the one shown in figure 4.7-2 is shown tachistoscopically such that the center of symmetry coincides with the fixation point of the eyes, one can perceive symmetries in 50 msec or less. (In one of my computer movies a random texture with octal symmetry is shown at 20 frames/ sec, in which each frame is statistically independent. The successive uncorrelated symmetric patterns erase the afterimages of the previous frames; nevertheless, the observer can perceive in each frame the octal symmetry.) However, if the tachisto-scopic image is viewed such that the fixation point of the eyes is a few degrees from the center of the symmetry axes, the observer cannot tell which of the presented patterns contain symmetries or are entirely random textures. The dependence on centering the patterns on the eyes is not critical for simple amorphic shapes.

This indicates that symmetry perception operates at two levels. For patterns having low spatial frequencies the symmetric relations might be extracted, and at some stage, invariant forms might be encoded by a central process. For patterns with high spatial resolution it seems that there is a point-by-point comparison process based on neural anatomy that has a symmetrical organization around the center of the fovea. The pattern of figure 4.7-3 in which the right and top half fields are dilated by a factor of two with respect to the left and bottom ones (corresponding to bilateral or quadrilateral similarity) cannot be perceived as symmetrical.

This point-by-point symmetrical representation is strongly weighted in favor of areas close to the axes of symmetry. This is shown in figures 4.7-4 and 4.7-5. Figure 4.7-4 shows a complex line pattern of 100×100 cells in which each cell contains

4.7-4 Two-fold symmetry of pattern composed of $\pm$ 45° line segments.

4.7-5 Similar to figure 4.7-4 except that the symmetry is removed in two stripes of $\pm$ 2 cells of width along the horizontal and vertical axes.

4.7-4   **4.7-5**

a $+45°$ or $-45°$ diagonal line segment. The two-line orientations are chosen at random but are "folded over" across the horizontal and vertical axes resulting in a two-fold symmetry. Figure 4.7-5 is identical to figure 4.7-4 except that the symmetry is removed in two stripes of ±2 cells of width along the horizontal and vertical symmetry axes. When figure 4.7-5 is inspected, the two-fold symmetry is not apparent. In figure 4.7-6 the dual operation is performed. In a random-line texture a horizontal and vertical stripe, each 8 cells wide, are inserted such that these are elements of two-fold symmetry across the horizontal and vertical axes. The appearance of figure 4.7-6 does not differ much from that of figure 4.7-4 which had a two-fold symmetry for each cell.

These complex random-line patterns differ in an important aspect from the random-dot textures. For the latter it is possible to change cluster formation by perceptually

4.7-6   The dual of figure 4.7-5. The
pattern is random, except for a
horizontal and vertical stripe of ± 4
cells wide along the symmetry axes,
which contain bilaterally symmetrical
cells.

4.7-7   Similar to figure 4.7-5; except
for diagonal line segments, the local
elements are square-shaped cells.

4.7-6   4.7-7

expanding and contracting the pattern to some degree. For instance the exact random-dot counterpart of figure 4.7-5 is shown in figure 4.7-7. The two-fold symmetry is not immediately apparent. But when we view figure 4.7-7 from a further distance some large symmetric clusters appear that dominate the perturbation close to the axis. As long as the average diameter of the clusters formed due to perceptual grouping is less than the width of asymmetric stripes across the symmetry axes, perception of symmetries is difficult. However, as the cluster size increases and outweighs the perturbation size, symmetry perception improves. This zooming operation does not work as efficiently for line segments. Regardless of the distance one views figure 4.7-5 the symmetries remain concealed. The counterpart of figure 4.7-6 is figure 4.7-8, in which the line segments are changed to square-shaped cells.

In a way random-dot textures are in between the simple amorphic shapes used by

4.7-8   Similar to figure 4.7-6; except
for diagonal line segments, the local
elements are square-shaped cells.

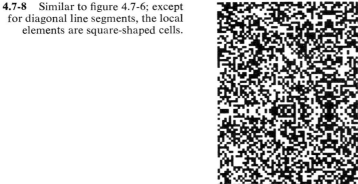

Mach and the random-line textures shown in figure 4.7-4. In order to encode a pattern such that the higher processes would be able to cope with it, one has to be able to find clusters. Without clusters the local details dominate, and only limited perceptual tasks can be performed.

The finding that centered symmetrical random-line patterns yield symmetry perception is of great interest. It might suggest that the cytoarchitecture of the cortex (after edge detectors) exhibits a certain amount of symmetrical organization. While the perception of simple clusters of bilateral symmetry might have been important for recognizing living organisms that usually exhibit such a symmetry, the perception of symmetries in complex patterns might have a different origin. Perhaps it is the remainder of the front-back movement symmetries that animals with nonoverlapping visual fields experience. Whatever the origin of such a complex pattern symmetry extractor might be, it is interesting that the CNS may have a built-in symmetric structure of its own.

In this context the report of Brindley and Lewin (1968), who evoked visual phosphenes by directly stimulating the visual cortex of a blind woman, may be relevant. Their patient reported the appearance of a new phosphene spot at a bilateral symmetry position with respect to an earlier one, as the stimulation exceeded a certain value. The few instances of such symmetric phosphene pairing in a single patient are inadequate for drawing conclusions, yet they do have heuristic value.

4.8 Perceptual Operators of Two Kinds

In § 4.6 we have seen how the perceptual system emphasizes a subset of the stimulus—the figure—almost to the exclusion of the rest, which is the ground. Another example is now given in which only *one* state can be perceived out of four. This occurs in symmetry perception. Simple arithmetic shows that any $f(x,y)$ pattern can be expressed as the sum of four patterns having two-fold symmetry (HV), one-fold symmetry across the horizontal (H), and vertical (V) axis and central symmetry (C):

$$f(x,y) = f_{HV}(x,y) + f_H(x,y) + f_V(x,y) + f_C(x,y). \tag{4.8.1}$$

Here H and V indices refer to even and odd coordinates,

$$f_{HV}(x,y) = \tfrac{1}{4}[f(x,y) + f(-x,y) + f(x,-y) + f(-x,-y)]$$
$$f_H(x,y) = \tfrac{1}{4}[f(x,y) - f(-x,y) + f(x,-y) - f(-x,-y)]$$
$$f_V(x,y) = \tfrac{1}{4}[f(x,y) + f(-x,y) - f(x,-y) - f(-x,-y)] \tag{4.8.2}$$
$$f_C(x,y) = \tfrac{1}{4}[f(x,y) - f(-x,y) - f(x,-y) + f(-x,-y)].$$

Some of these decompositions can become negative and, therefore, cannot be realized as physical stimuli, since there is no negative light energy. However, if $f(x,y)$ denotes some neural firing pattern at a retinal level, then the positive patterns refer to the ensemble of "on" units, while the negative patterns refer to the "off" units. We define the perceptual operator of symmetry, S, with its eigenvalues, λ_{HV}, λ_H, λ_V, and λ_C, by its effect on the stimulus, $f(x,y)$:

4.8-1 Addition of vertical and bilateral symmetries of equal weights. It is perceived as a random pattern.

4.8-2 Two-fold symmetry with weight 1 to which a pattern with horizontal symmetry is added with 0.25 weight. This perturbation almost eliminates the perception of two-fold symmetry.

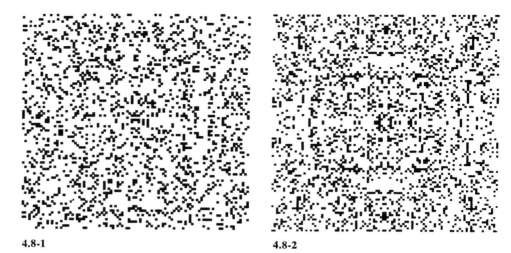

4.8-1   4.8-2

$$Sf(x,y) = \lambda_{HV}f_{HV}(x,y) + \lambda_H f_H(x,y) + \lambda_V f_V(x,y) + \lambda_C f_C(x,y). \qquad (4.8.3)$$

The four eigenfunctions remain invariant under the transformations $x \to -x$ or $y \to -y$; $y \to -y$; $x \to -x$; and $x \to -x$ and $y \to -y$, respectively. (These functions are not affine combinations of $f(\pm x, \pm y)$ since they involve subtractions.) Experiments show that except for minor perturbations, stimulus patterns composed of two or more eigenfunctions of equation (4.8.3) cannot be perceived as being symmetrical. In figure 4.8-1, f_H and f_V composed of 10% of black dots (and 90% white dots) were combined into $f = f_H + f_V$ and again quantized into two brightness levels. Since only 10% of the dots were black, only 1% of the dots, where two blacks were added, suffered a nonlinear distortion due to quantization. As figure 4.8-1 demonstrates $f_H + f_V$ is perceived as a random pattern. Only if one of the patterns were weak, say $f_V = 10\%$ and $f_H = 2\%$, could the symmetry f_V in $f_V + f_H$ be perceived and f_H be regarded as just a slight perturbation. These perceptual operators are similar to the figure-ground one and equation (4.6.4) can be generalized as

$$\lambda_1 + \lambda_2 + \ldots + \lambda_n = 1, \qquad (4.8.4)$$

where only one eigenvalue $\lambda_i = 1$, and the rest must be zero.

The result that two-fold symmetry perception is independent of the vertical and horizontal one-fold symmetry perception is not as trivial as it might seem. It is an interesting problem that horizontal bilateral symmetry (fig. 3.2-4)—the much weaker percept compared to vertical bilateral symmetry (fig. 3.2-5)—becomes equally strong for two-fold symmetries (fig. 3.2-3). (The strength of symmetry can be measured by the time required to perceive it.) One might conjecture that the presence of a vertical-symmetry axis might strengthen the perception of horizontal bilateral symmetry. This hypothesis has been disproved by the previous experiments.

A basically different class of perceptual operators are those where each decomposition of the stimulus can be perceived simultaneously:

$$Pf(x,y) = \sum_i \lambda_i f_i(x,y). \tag{4.8.5}$$

Here each perceptual feature (the eigenvectors f_i) can be perceived according to some perceptual weight (the eigenvalues λ_i). In a way this corresponds to the spectral decomposition of a function in mathematics.

These operators can be further subdivided into two classes. There are perceptual operators where each decomposition can be separately perceived; however, their sum is more than their parts. On the other hand, there are perceptual operators that are based on each decomposition but act only on their sum. An example of the first type is face recognition or listening to a musical chord. It is possible to perceive the face or chord as such, but one can shift one's attention to perceptual subunits, such as the complexion or the roundness of the face or the individual notes in the chord. The second example is illustrated by color perception in which any color percept can be produced from a proper combination of three basic colors (e.g., red, green, and blue). Yet, when experiencing a saturated yellow color, one cannot say whether it is composed of red and green or some other constituent colors. Similarly, when we listen to the timbre of a piano string, the constituent harmonics cannot be separately perceived although they act simultaneously.

The two types of perceptual operators can be bridged by assuming both parallel and sequential perceptual processes. Maybe some of the simultaneously perceived eigenfunctions are also perceived one at a time. From his reaction-time studies Egeth (1966) concluded that for the identification of multidimensional visual stimuli the dimensions (features) are compared serially, and the order in which dimensions are compared varies from trial to trial. Subjects were asked to indicate whether two simultaneously presented multidimensional stimuli were identical or different. The stimulus dimensions were color, shape, and tilt; and the reaction time decreased with increasing numbers of dimensions in which the stimuli differed. Since parallel processes would require the same processing time regardless of the number of features, the results indicate a serial process. One might argue that color is a more primitive feature than tilt or shape; and if simpler dimensions such as color, brightness, and proximity were used, the results might have revealed a parallel process. The fact remains that for more complex eigenfunctions the perceptual system is able to selectively attend to one eigenvalue f_i (by choosing $\lambda_i \rightarrow 1$) and ignore the rest at a given moment. I will discuss this process further in the next section on selective attention.

It seems instructive that many perceptual processes differ from the spectral decomposition type process, which is another name for parallel processing. The example of color perception or the parallel perception of all Fourier components of a pattern (in which it is impossible to switch attention to a single feature) represent spectral decomposition. It appears that these parallel processes occur infrequently, and it remains to be seen how complex neural feature extractors will be found. The hypercomplex units of Hubel and Wiesel (1965a) are still very "simple," when compared to the rectangular and circular shapes used by Egeth (1966). It is also most probable that the Kuffler-units (dot detectors) or the simple edge detectors synapse directly to the higher centers in addition to being connected to the next hierarchical levels.

This central level may then select the various feature extractors of color, brightness, position, tilt, and so on, serially, but perhaps in various order at each presentation.

4.9 Perception and Selective Attention: Parallel versus Serial

Figure-ground perception is the simplest case of selective attention and is already very complicated. The general problem of selective attention is an even more difficult problem, although fundamental both in perceptual and cognitive studies. The complexity of the problem is such that this section is very speculative and contains a mixture of incoherent facts. I was tempted to delete it from the book entirely, and the reader is urged to skip this section at first reading. Nevertheless, to stop at figure-ground perception without raising problems of selective attention might have given the impression that problems of perception are better understood than they really are. Of course, there is no excuse for presenting an idiosyncratic review to the expert, and I can only hope that my readers are mature enough to form their own opinion by consulting the vast literature on attention.

The serial (one-dimensional nature of language and reading tasks is very well matched to the presumed *sequential (serial)* workings of the cognitive processes. On the other hand, visual patterns are inherently two-dimensional, and one may have to assume that some of the basic perceptual processors have to be *parallel*. I already discussed in §3.3 how the early feature extractors are parallel processes and mentioned my belief that some higher perceptual processes might be sequential, a sort of serial listing of features. I discussed one of my findings (Julesz 1964a) with pentagons (or hexagons), the sides of which were presented in sequential or nonsequential order. Subjects were asked to change the presentation rate of the sides and report the slowest rate at which the polygons were perceived first as such (ignoring the flicker). It turned out that for a great variety of stimulus conditions shown in figure 4.9-1, cases *A, B, C, D,* and *E,* the mean presentation rate (in Hz) is 50% higher for the nonsequential case than for the sequential one as listed in table 2.

4.9-1   Stimulus conditions in the study of sequential and nonsequential presentation of line drawings. (The nonsequential case is specified.)

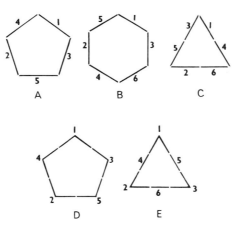

TABLE 2. *Mean Frequencies of Presenting Sides of Various Cases Given in Fig. 4.9-1 at the Threshold of Perceiving All Sides Simultaneously in Hz*

Case	Nonsequential	Sequential
A	15.8	10.2
B	25.7	18.5
C	26.2	19.8
D	17.6	11.4
E	25.3	17.8

Similar findings were obtained in a follow-up study by Bell, et al. (1969). Those who believe in a parallel pentagon extractor might have been overjoyed if the sequential or nonsequential presentation of features had caused similar effects. That this better performance for sequential scanning of features is inherent in the perceptual process and is not the result of eye tracking movements is indicated by the fact that the same results hold when the figures are zoomed down to a diameter of less than 2° of arc.

The possibility of parallel extraction of early features followed by serial listing of these features by the next stages seems to be indicated by the segmentation phenomena by Pritchard (1961) and Evans (1965) (already discussed). In these impoverished stimuli certain features such as parallel lines, corners, parallel faces of perspective cubes, and so on, fade out as units. This segmentation in retinally stabilized images, very weakly illuminated images, and afterimages seems to occur according to some random sequence. However, Mayzner, et al. (1967) showed that tachistoscopically presented letters in words are not equally readable and according to sequencing and presentation rate, certain letter positions in a word cannot be seen. This dependence of visibility on temporal sequencing is very important, and one can hope that the availability of medium-sized, very fast digital computers operating in real time will open up this field for research.

At the moment we are not examining the question whether feature extractors discovered by neurophysiological methods are identical to the feature extractors postulated by the psychologists. We assume here that feature extractors exist and develop some notions of how to specify them by perceptual techniques.

The parallel, two-dimensional nature of feature extractors makes the discovery of the relations among them very difficult. Even linear grammars are in their infancy, but the development of a pictorial syntax is in an even more primitive state. Some encouraging results were obtained for some very simple subsets of objects, such as microscopic pictures of chromosomes (Ledley 1964) or bubble chamber traces of particles (Narasimhan and Mayoh 1963). Yet the general problem of a pictorial grammar is plagued by many fundamental difficulties. Obviously, it has to be a transformational grammar. If we assume the Hubel and Wiesel notions of dot, edge, parallel edge, and so on, detectors of increasing complexity, then each level removes certain information. For instance, the absolute brightness depends on the luminance and extent of its inhibitory surrounding. Similarly, the complex edge detectors signal only the orientation of parallel edges without preserving the exact position. However, the final percept of a line contains all these data with some weighting. Suppose we perceive a dark line in a white surrounding at a position and orientation. Then all the extracted features must exist simultaneously, and their proper weighting is the basis of a syntax. Such complex levels of syntax as the recognition of familiar faces or the knowledge that a pair of gloves belong together will not even be considered at the moment.

I have already reviewed in § 4.4 the attempts to develop a primitive pictorial grammar for unfamiliar random textures. In the absence of familiarity cues, constructs of random geometry seem to give a good description of perceptual performance. In § 9.5 we will later review some recent studies in a section on multidimensional scaling in which the random polygonal figures by Attneave and Arnoult (1956) were scaled. The most important perceptual factors were compactness, jaggedness, skewness, rotation, each being highly correlated with low-order moments of stimulus parameters, such as interior angles, side lengths, perimeter, and area measures. Which of these parameters is actually used by humans is hard to determine primarily because of shifts in human attention. These parameter shifts and the large number of them greatly hamper the development of a psychophysics of form.

In the previous section I have discussed the reaction time study by Egeth (1966), who concluded that in multidimensional stimuli the attributes (features, dimensions) are compared serially but in idiosyncratic order. This finding is one of the latest in a series of experimental studies on selective attention which were initiated by Külpe (1904). Külpe demonstrated that after a perceiver is instructed to attend to only one of several attributes of a stimulus, he is able to report his perception of that attribute more accurately than he would be able to otherwise. Whether the instruction as to which attribute he should attend is explicitly given or is conveyed by implicit instructions emanating from the subject himself (such as would occur when a hungry man is looking for edible objects) is immaterial. The only important consideration is that setting the attention cannot be obtained by peripheral mechanisms such as eye movements or scanning of the stimulus to search for relevant features, otherwise the phenomena would be trivial.

An excellent review on attention, "The nature of the effect of set on perception," has been published by Haber (1966). He was particularly concerned with two hypotheses: that set (attention) enhances the percept directly, or set facilitates responses or memory organization of the perceptual experience that itself was unaffected by the set. He reviewed extensive research, yet at the present it is not possible to decide which of the two hypotheses is correct. Perhaps both may be correct, occurring together. Garner (1966) emphasized the importance of knowing a priori all the alternatives (the entire set of features) that characterize the stimulus. This is the essence of Shannon's notion of information and one reason why information theory is not applicable to perception is that we do *not* know a priori the entire set of features the observer will extract from the stimulus and put to use.

Imai and Garner (1965) showed that after the subject's attention had been set to a given attribute the presence of other competing attributes had no effect on the sorting times for decks of cards, some of which contained the criterion attribute. These results indicate that some central processor might be responsible for weighting the various feature extractors. A typical theory based on weighting has been developed by Sutherland (1964), who trained a group of rats to approach a white horizontal bar and avoid a black vertical bar. He then tested them to determine whether they were attending to orientation or color. He found that the better they performed on one attribute the worse they scored on the other. Either the brightness analyzer or the orientation analyzer had been switched out. This visual example shows striking similarity to auditory attention. According to the psychological findings of Broadbent (1958), Cherry (1953), and Treisman (1964), at least two stages of neural analysis must occur in the CNS. Simple features such as position, male or female voice, loudness, in both the left and right channel can be attended to simultaneously; but the verbal content of only one channel can be attended to at one time. Exceptions to this observation have been found: Moray (1959) showed that subjects sometimes hear their own names in the unattended channel; Treisman (1964) found that even certain words could be attended to if these were highly probable in the context; Cherry (1953) showed that if the unattended message was identical to the selected one and the time delay between the same words in the two ears was gradually reduced to 2–6 sec, the identity of the two messages was noticed; Treisman (1964) extended Cherry's findings by showing that the delayed identity between the attended and unattended channel was noted by bilingual subjects when the two messages were in different languages.

Perhaps the primitive attention mechanism in vision that uses fixation and accommodation movements of the eyes takes much of the burden away from the central attention mechanism, which for audition has to cope with the incoming information in its entirety. Furthermore, in vision there is no importance attached to shifting attention to one or the other eye, since under ordinary conditions the information is similar for the two channels. Under the artificial situation of binocular rivalry, there exist some similar reports that certain images of high interest might become seen more often as eye dominance alternates (perhaps because of increased eye-movements; Breese 1899). It remains to be seen whether this selective attention for vision (at least for the complex features) is a perceptual process or belongs to cognition.

Before we conclude the topics on classical figure-and-ground perception or selective attention in general, let us note an interesting result of Minsky and Papert (1968) in their book *Perceptrons*. In § 4.4 I reviewed one of their findings that connectivity of a figure requires infinite-order perceptions (or more precisely, if the figure falls on a finite retina of N^2 receptors, then the linear-decision function necessary to decide on the connectivity of a figure must be N^2-order). An even more surprising result of theirs is that to decide on even the simplest properties of a figure in the context of another figure requires infinite-order. For instance, the decision as to whether a figure is a line or a triangle requires only a second- or third-order perceptron, respectively. However, to decide whether a triangular area contains a line figure or not, requires an infinite-order perceptron. These two proofs are very instructive. With respect to connectivity as demonstrated in figure 4.4-7 we humans behave like perceptrons and with our finite-order brains cannot decide whether a figure is connected or not by using only perceptual means. On the other hand, we have no difficulty at all in perceiving properties of a figure in the context of another figure. For instance, to find the smaller of two circles that fall on the intersection of two rectangles is a spontaneously resolved task.

The difference between perceptrons and perceptual processes of humans and higher animals is due to the lack of selective attention mechanisms in the former. Without some feedback mechanism that would differentially weight the input in several possible ways, linear decision functions by themselves cannot separate figure from ground. In a sense the "figural properties in the context of other figures" correspond to figure and ground perception. After the "context of other figures" has been found, it becomes the ground, and the "figural properties" become the figure. This insight is very valuable. It shows that the most complex automata that are without some selective attention mechanism are simpletons compared to the lowliest creatures whose visual systems have developed such mechanisms.

4.10 Ambiguous Figure and Ground Reversal of Cyclopean Stimuli

Let us discuss in this section how factors that affect classical figure-ground perception (reviewed in § 4.6) affect cyclopean figure-ground perception. We have seen how the area fractions of black and white dots affects figure and ground reversals (Frisch and Julesz 1966); the smaller the area of a certain brightness domain, the more it tends to become the figure. In a way the dot, edge, and slit detectors in the retina and the input layers of the cortex can be regarded as primitive figure-and-ground operators that attend only to small areas of brightness changes. For random-dot stereograms, however, in which the area fraction of black and white dots is kept at 50%, this figure-and-ground operator based on brightness cannot act. One can ask whether the principles of figure-and-ground organization might be the same for cyclopean stimulation (by depth gradients). The first principle, namely the smaller the area the more it is seen as the figure, does not hold for random-dot stereograms or cinematograms. As a matter of fact, rather the opposite is true. The larger the area of a certain disparity (binocular or movement parallax), the easier it is to perceive it as being different from its surroundings. This finding indicates that processes of stereopsis and stroboscopic movement perception might occur prior to a more central selective attention mechanism.

Therefore, one can assume that many cues which are of secondary importance for the classical figure-and-ground perception become of primary importance if they can be used in stereopsis or movement perception. For instance, it is a common observation that after ambiguous figure-and-ground reversals of two-dimensional outlined drawings, the figure appears to be in depth, in front of the ground. One might assume that this weak depth sensation is just a side effect, perhaps based on the observation that objects closest to the observer are potentially dangerous and should be attended to. However, as figure 2.8-2 demonstrated, this nearness of an object when produced by stereopsis overrides every other organizational principle. Whichever area is seen nearest to the observer becomes the figure, and reversal is difficult if not impossible.

The same organization (figure 2.8-1) can be portrayed by random-dot cinematograms in which the vase (or the two faces) can be portrayed as static random texture and the two faces (or the vase) are portrayed as dynamic noise. Whether figure 2.8-1 is portrayed by a cinematogram in which the figure is dynamic noise in a static noise surround, or vice versa, figure-ground reversal is equally easy.

The essence of these experiments is that some basic parameters such as brightness, depth, and velocity precede or dominate the classical factors of figure and ground. Some other factors such as "good continuation" are valid for random-dot stereograms as well, but become enriched since good continuation can proceed in three-dimensions. Complex surfaces that are smooth will be perceived sooner than those which have abrupt changes. A particularly relevant demonstration of good continuation in the third dimension is given by figure 6.2-9* and 6.2-10 in a section on mental holography.

As mentioned in § 2.4, it is possible to scramble the percept of monocularly perceivable symmetry in the binocular view rendering the symmetry unperceivable and thus demonstrate that symmetry perception occurs after binocular combination or stereopsis dominates the perception of symmetry (Julesz 1967a,b). Figure 3.10-1* shows such a random-dot stereogram in which the monocularly perceivable one-fold symmetry across the horizontal axis had been suppressed in the fused image. Because stereopsis affects symmetry perception to this extent, the role of symmetries is even less pronounced in figure perception of random-dot stereograms than it is in the case of classical line drawings.

Finally we come to closedness. This parameter is very artificial since only line drawings in two dimensions can be open forms. Surfaces of spatial objects are always closed. Therefore, cyclopean stimulation in essence portrays areas of finite extent. Even when line drawings are portrayed, the thickness of the cyclopean lines has to be somewhat greater than the lines of classical drawings. Because of this, problems of closedness are regarded as a very special case of images conveyed by sharp brightness gradients and as such, are of secondary importance to us.

5

Binocular Depth Perception

5.1 Random-Dot Stereograms and Binocular Depth Perception

This chapter is concerned with binocular depth perception. It is not my intention to cover this field in its entirety. However, the technique of random-dot stereograms contributed some new insights into binocular depth perception that necessitate a reevaluation of the field as a whole. The many classical problems of stereopsis, such as the horopter or perceived orientation, will be avoided or just briefly commented on. A coherent model of binocular depth perception is given that was formulated as a result of new evidence collected in the last decade.

I emphasize binocular depth perception primarily to show the extent that cyclopean techniques can benefit a field. It is only a historic happenstance that the overwhelming majority of cyclopean experiments used random-dot stereograms. I hope to convince the reader that similar progress could ensue in movement perception if experiments with random-dot cinematograms were conducted on the same scale.

This chapter is more ambitious than the foregoing ones. The goal of the book was to trace the information flow in the CNS. But we will not be restricted to placing a phenomenon prior to or after stereopsis. We will also try to derive a model of stereopsis and binocular depth perception. (Some of the evidence on which this model rests has been reviewed in §§ 3.9 and 3.10. Additional evidence is provided in the forthcoming sections.)

There are several reasons why random-dot stereograms revealed a new side of binocular depth perception that stayed hidden under classical stimulation. These are summarized in table 3.

TABLE 3. *Differences between Classical and Random-Dot Stereograms*

1. Only textured objects can lead to unambiguous localization in depth. A stereogram of outlined circles could be the retinal projection of a disk, a sphere, or even a cone.

2. Only random-dot stereograms contain disparity information for every dot. Common photographs and especially line drawings have large uniform areas without disparity information. Because of this, random-dot stereograms are a much more efficient class of stimulation, and several classical limits of fusion can be extended by their use.

3. Random-dot stereograms are computer-generated. In their complexities they can simulate or even surpass natural scenes, yet with precisely controlled statistical and geometrical properties. Many perceptual phenomena can be optimally studied if the stimulus has such hypercomplexity.

4. Random-dot stereograms do not contain monocular familiarity or depth cues. In the absence of these complex cues binocular depth perception can be studied in its purest form.

5. Cyclopean techniques permit the complete separation or juxtaposition of monocular and binocular shapes and contours. The interaction between monocular and binocular processes can be studied.

6. In random-dot stereograms one of the images can be perturbed in complex, well-specified ways. This permits us to gain insight in the binocular pattern-matching process.

7. In random-dot stereograms the global (binocular) information is monocularly concealed. This permits an objective test for stereopsis.

8. For random-dot stereograms the local and global information can be independently manipulated.

9. Local stereopsis phenomena obtained by using a few dots or lines are qualitatively different from the global stereopsis phenomena obtained by using complex stereograms containing thousands of lines and dots. Global stereopsis occurs at a more complex hierarchical level than local stereopsis.

These stimuli reveal a new structure of the binocular depth mechanism which differs both quantitatively and qualitatively from those obtained under classical stimulation. Of course this does not mean that these new results are contrary to classical findings. Rather, they are complementary to those obtained by line targets. This dependency of revealed structure on the experimental methodology is a well-known fact in quantum physics. Whether matter and energy behave like waves or particles depends on the experimental conditions used. Nowadays no one would ask which of the two structures is "true"; it is accepted that both are and are different "projections" of a hyperstructure. It is time psychologists also realize that the most complex organ in the known universe—the human brain—should show different structures as a function of the input stimulation.

In § 5.2 a short review of the basic evidence concerning binocular depth perception is given. The following sections discuss the difference between the perceptual findings obtained under classical and cyclopean conditions. These sections organize the main results according to the points given in table 3.

5.2 Some Anatomical, Physiological, and Psychological Facts about Binocular Depth Perception

Although this is not a textbook, binocular depth perception and especially stereoscopic depth perception are so fundamental for cyclopean perception that some of the necessary background material is briefly reviewed here.

During the course of evolution, binocular depth perception emerged from binocular vision after some favorable bodily changes occurred. Originally binocular vision served to provide panoramic vision by evaluating the two separate views that were cast on the retinae, which point sideways. The development of head movements with the accompanying complex movements of eye-coordination gave rise to physiological mechanisms that could convey certain depth information. Such a mechanism is the accommodation of the eye-lenses, and according to Walls (1942) it evolved first and most universally among the animal kingdom, yet failed to develop into a depth sensing organ. Indeed, accommodation is a very insensitive depth cue, since the evaluation of blur is dependent on the contour properties of the stimulus; furthermore, in bright daylight the reduced pupil diameter results in increased depth of field. On the other hand, monocular parallax as a result of head movements led to the development of a very sensitive depth mechanism. Two successive images on one eye are two projections of an object from different positions. For monocular parallax the displacement between two successive eye positions is a vector, and the exact head and eye positions have to be known in order to evaluate depth. For animals that capture food with their heads monocular parallax led to satisfactory depth evaluation.

The recession of a protruding snout or beak in some animals who capture food by paws, hands, or pouncing, with the accompanying complex vergence movements of the eyes permitted a coarse registration between the two monocular fields. A finer registration between overlapping areas of correspondence was achieved by a neural mechanism that evolved and yielded a new sensation of stereoscopic depth. The advantages of stereoscopic depth perception probably outweighed the loss of panoramic vision for most animals. Indeed, with stereopsis, spatial localization of objects is perceived vividly as an independent sensation (similar to the sensation of color and brightness), and as such, helps to form an internal model of the outside world.

Although there are several other depth cues (both monocular and binocular), stereopsis is the strongest single depth cue. The monocular depth cues (as we will discuss later) require very complex image processing and can only be utilized under favorable circumstances. The binocular depth cues besides stereopsis are vergence and correlative accommodation (differential focusing of the two eyes). Both are weak depth cues since we cannot sense the absolute positional state of the muscles that are responsible for accommodation or convergence-divergence movements. The role of vergence and accommodation is primarily to provide sharp and coarsely aligned retinal images for the stereopsis mechanism. Although accommodation and vergence are very weak depth cues by themselves, their strong linkage with stereopsis can influence stereoscopic depth perception in several ways.

Stereopsis is based on the geometrical fact that two-dimensional projections of a three-dimensional object on the left and right retinae differ in their horizontal positions. This horizontal shift between corresponding points in the two retinal images is called retinal disparity, or simply disparity. The terms binocular parallax or parallax shift are also used. Wheatstone's discovery of the stereoscope (1838) first demonstrated that disparity gives rise to stereopsis. That disparity alone is adequate for stereopsis was demonstrated by Dove (1841), using exposures too brief for any convergence movement of the eyes to be initiated.

Stereopsis itself gives us the experience of relative depth only. It enables us to rank-order the nearness-farness of objects within a region of space around a fixation point. If we fixate on some point in space—that is, its retinal projections coincide with the centers of the foveae—then the projections from all features lying on the circle through this point will, by definition, have zero horizontal disparity. This is called the Vieth-Müller circle. Projections from a point lying inside this circle will have what is termed convergent (or crossed) disparity; points further away will have divergent (or uncrossed) disparity. This terminology refers to retinal disparities; for describing stimulus disparities, convergent disparity corresponds to a nasal shift, while divergent disparity is caused by a temporal shift. Vertical disparities will occur because of slight differences in elevation of the eyes and also because the magnification differences between the two eyes can be significant for close objects. However, the range of vertical disparities for points lying at reasonable viewing distances is much less than the range of horizontal disparities. As long as the horizontal and vertical dis-

5.2 Some Anatomical, Physio-   145
logical, and Psychological Facts
about Binocular Depth Perception

parities are within Panum's fusional area (which in the center of the fovea is a disk of 6 min of arc diameter and increases to 30 min of arc at the peripheral angle of 8° of arc), the disparate points are seen as a single fused image in depth. It is even possible to extend the fusional limits by several times and obtain a vivid depth sensation, although, the fused image breaks into two. Often one of the double images is suppressed, but the sensation of depth is obligatory. This is the region of patent stereopsis, which has been studied intensively by Ogle (1952). The quantification of the relation between perceived depth and disparity was the subject matter of extensive psychophysical investigations summarized by Ogle (1962) and led to several theoretical problems. The generalization of the Vieth-Müller circle in three dimensions resulted in variously defined spatial curves, called the horopter; attempts to establish a metric of three-dimensional perceived space by Luneburg (1950) and his followers brought some theoretical refinements.

In order to obtain retinal disparity, however, the stereopsis mechanism has to establish correspondence with those parts of the images that are projections of the same object on the left and right retinae. Strangely enough this basic problem of stereopsis was completely ignored until the introduction of random-dot stereograms. This mechanism establishing binocular correspondence in the simplest cases can occur prior to stereopsis by assuming that the targets have a few contours of unique shapes and orientations. However, for complex targets containing many contours that have the same shapes and orientations, ambiguities will arise. This can only be resolved by a global evaluating mechanism that weights the results of all possible locally established correspondence and selects the right global organization.

One implication of random-dot stereograms provides us with a better understanding of why stereopsis evolved in higher animal species. Since stereopsis yields only a relative sense of depth, it is only one of the many other depth cues that enable us to make an absolute depth localization. As a matter of fact, before random-dot stereograms several psychologists questioned the importance of binocular disparity and stereopsis for depth localization. For instance, Gibson (1950) and Ittelson (1960) regarded disparity as being of minor importance. They assumed that the retinal gradient of texture or movement parallax was more powerful.

The demonstration that binocular disparity alone can give powerful depth sensations reestablished the importance of stereopsis as an important depth cue. Furthermore, when anaglyphs are viewed during horizontal head movements the usual movement parallax appears to reverse. The objects above the background do not move opposite to the observer's movement but together with the observer. Thus movement parallax can depend on stereopsis and can be modified by stereopsis.

Even if binocular disparity were a weak depth cue, it is a powerful asset. Since time immemorial animals developed camouflage and easily blended with the environmental background. It is possible to hide rather successfully this way when the predator has only monocular vision. In order to see a camouflaged moth on a tree branch requires very complex form recognition capabilities. However, as random-dot stereopsis demonstrates, even under ideal monocular camouflage, the hidden objects jump out in depth when stereoscopically fused. What is more, this object separation

146

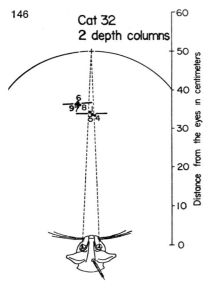

5.2-1 Neurophysiologically determined horopter curve after Blakemore. The marked units respond to the same binocular disparity values. (From Blakemore 1969.)

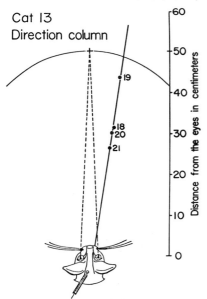

5.2-2 Locale of stimulation of cortical units that respond to contralateral eye orientation after Blakemore. (From Blakemore 1969.)

does not necessitate any familiarity with the stimulus and therefore stereopsis could evolve at a relatively early stage in the processing chain of vision.

Knowledge of the anatomical and physiological bases for stereopsis go back to work by Descartes, but it was Newton who first proposed a partial decussation of the optic nerve fibres. (For a review of early investigators, see Polyak 1957.) The optic tract of one side receives information from the same half-fields of both eyes. This decussation of the optic tract occurs at the chiasma, and the corresponding neurons from the two retinae project to two nearby layers in the LGN. There is apparently little interaction in the LGN between the two visual pathways (Bishop, et al. 1959), whereas the majority of neurons in the striate cortex can be binocularly driven (Hubel and Wiesel 1962). Sanderson, Darian-Smith, and Bishop (1969) reconfirmed that no binocular facilitations occur in the LGN of the cat; however, they found binocular inhibition. I have summarized in great detail these findings in § 3.3. A recent histological result of Hubel and Wiesel (1969) that should be mentioned was obtained by using degenerating neurons stained by Nauta's technique. It appears that the columnar organization of the cortex with respect to common orientation of receptive fields has another slab-shaped organization with respect to eye-dominance. The fourth layer of Area 17 in the cat is alternately segregated into bands that are driven by left or right retinal stimulation respectively, and these bands or slabs run parallel into the cortical surface. Neural units outside the fourth layer receive their inputs from neighboring bands and belong to various binocular dominance classes.

I also discussed the findings by Barlow, Blakemore, and Pettigrew (1967), who established that a class of the binocularly driven cortical units in Area 17 of the cat responded optimally for specified disparity values. This is illustrated by one of the figures by Blakemore (1968) shown as figure 5.2-1. Here the depth distribution is given for optimal stimulation of cortical neurons. Each numbered dot shows the horizontal position and depth in space an object would have to occupy in order for its retinal projections to optimally trigger that disparity sensitive unit. Blakemore (1968) found another class of cortical units that responded to contralateral eye orientation. These units fired at targets in space that were lying on a single line of gaze. This is illustrated in figure 5.2-2. The two classes of binocular neurons complement each other: one gives the depth information, the other the positional information. Here again one must assume that only a few objects cast their images on the retinae, otherwise it would be ambiguous as to which of the disparity sensitive units belongs with the proper orientation sensitive units. These physiological findings are a mere beginning to explaining stereopsis. There is still no physiological evidence for the next hierarchical level, which resolves ambiguities for complex stereograms and will be termed global stereopsis. Evidence for global stereopsis comes from psychological experiments with complex stereograms; the next step in physiology is the elucidation of this mechanism. Probably this global stereopsis mechanism will be found only in visually sophisticated animals and certainly in the monkey.

Just before the work on this book was closed, Hubel and Wiesel (1970) reported binocular depth cells in Area 18 of the macaque monkey cortex. About half of the units in Area 18 were ordinary binocular units responding about equally to the stimu-

5.2 Some Anatomical, Physio-   147
logical, and Psychological Facts
about Binocular Depth Perception

lation of anatomically corresponding parts of either retina. For the other half of the cells (termed binocular depth cells), however, stimulation of either eye separately gives no response, whereas appropriate stimulation of the two eyes together results in strong responses. For best responses there is a well-defined disparity in the positions of the left and right receptive fields, and this disparity is usually perpendicular to receptive field orientations. Maximum response as a function of this disparity is very steep; the region of rise and fall being a fraction of receptive field dimensions. Responses of cells with obliquely oriented fields also peak sharply for purely horizontal displacement. No such binocular depth cells were found in Area 17 in the monkey, though several hundred binocular cells were studied. Thus, at last we have neurophysiological evidence that the cyclopean retina must be at least in Area 18 or even more centrally located.

Besides the pathways that combine homonymous half-fields of the retinae after decussation in the chiasma, there exists another pathway that combines heteronymous half-fields of a small region around the vertical midline. After all, the psychological fact that we can fixate at a point lying on the midline and perceive points closer or farther along the midline in depth indicates some neural interaction between areas in the two retinae that fall on opposite half-fields. This observation, although suggestive, is not conclusive evidence for a second binocular pathway since subjects might fixate slightly off-axis. This objection can be partly repudiated by using tachistoscopic presentation for the near and distant test targets while trying to maintain fixation on a midline point. Conclusive evidence can be obtained by testing humans whose chiasma has been cut (by accident). Blakemore (1969) reported stereopsis in a chiasma cut human subject within a $\pm 1.5°$ strip across the midline. He also proposed, based on physiological evidence, that the heteronymous hemispheres interact through the corpus callosum. Hubel and Wiesel (1967) also found in the vicinity of the cortical representation of the midline (in the cat) some neurons that responded to stimulation of the opposite hemiretinae. Whether the binocular interaction of opposite hemiretinae in the vicinity of the midline is the result of an imprecise mapping of the two hemiretinae on the two cortical hemispheres or is due to corpus callosum fibers, the main result is that two binocular pathways exist. The anatomical flow chart of stereopsis is schematically given by figure 5.2-3. The solid lines indicate the main binocular interaction between the same hemiretinae, while the darkened lines show the binocular interaction between the opposite hemiretinae. Figure 5.2-3 is after Blakemore, based on the findings of Choudhury, Whitteridge, and Wilson (1965). However, there is also evidence that the nasotemporal overlap projects via the optic tract of the same side, since Leicester (1968) reported that the overlap in the cortex survives the section of the corpus callosum. Furthermore, Kinston, Vadas, and Bishop (1970) found a similar overlap in the LGN. Hubel and Wiesel (1967) found that units they recorded from the corpus callosum were all binocularly driven, in contrast to the hypothesis of a solely transcallosal projection being responsible for nasotemporal overlap (Bishop 1969a). The latter hypothesis would predict monocular units in the corpus callosum. Bishop (1969a) assumes that the transcallosal projection reinforces a nasotemporal overlap which exists independently of it.

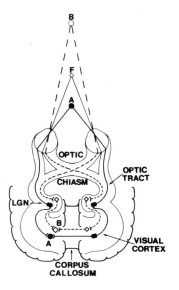

5.2-3 Anatomical flow chart of stereopsis after Blakemore. (From Blakemore 1969.)

Most of these findings on stereopsis were obtained in cats. It is interesting that in Siamese cats some of the retinogeniculate fibers have been misrouted, and the usual laminar organization of the LGN is abnormal (Guillery 1969). Because of this, in the Siamese cats binocular interaction can occur for only limited parts of the visual field. Whether other animals also have "scotomas" for stereopsis remains to be seen.

Stereopsis is innate. Hess (1956) showed that newly hatched chicks wearing distorting prisms would peck too close and might starve in the midst of plenty, without relearning. Bower (1968) showing random-dot stereograms to six-day-old human infants could elicit a stable eye fixation if a cyclopean bar appeared in depth; otherwise the eye movements occurred at random. This is most remarkable since the eyes of these babies were covered by silver nitrate (required by law) prior to these experiments (and were cleaned by the experimenter), and the first encounter with patterned stimulation resulted in stereopsis. Only about a quarter of the babies exhibited this behavior, and Bower assumes that the rest had problems with monocular vision. In the same report Bower introduced a conflict between movement parallax cues and stereopsis. The conflict was only apparent after four days of being exposed to patterned vision. This might be the calibration time necessary for the stereopsis mechanism, assuming that movement parallax is absolute. The reason why stereopsis requires calibration is because at birth the centers of the foveae are shifted by about 10° of arc temporalward and drift to their final positions at an age of three years. This drift of the eye centers changes the geometries underlying depth localization. Although stereopsis gives only the sensation of relative depth with respect to the fixation point, a shift in retinal coordinates changes the relation between vergence and stereopsis, thus the metric of perceived three-dimensionality changes as well. Bower (1968) also reports a grasping reflex in neonatal babies whenever an object comes nearer than 20 cm.

The primary function of the vergence mechanism is to bring the corresponding retinal projections within Panum's fusional areas. However, it serves another purpose. With increased convergence, the cortical representation of a retinal image decreases in size. This is the well-known Emmert's law, which states that a retinal afterimage decreases in perceived size when the eyes are fixated at a nearby surface and increases in size when a remote surface is viewed. The result of this phenomenon is to provide a certain amount of size constancy even if this compensation is not perfect. Nearby objects cast larger projections on the retinae than do distant objects, and this perceptual zooming mechanism tries to stabilize the size of the percept. The largest perceptual zooming occurs in the range within 100 cm which corresponds to distances of objects that we can reach for with our hands. I will discuss this binocular size constancy mechanism in chapter 9.

While the stereopsis mechanism is cortical through the LGN pathway, the vergence mechanism is located in the tectum and resides in the superior colliculus. This structure gets input from the optic nerves and from the cortex as well. This is evident from psychological observations with random-dot stereo movies that portray a cyclopean square moving back and forth in depth. The vergence (and pursuit) movements of the eyes follow this monocularly invisible target as if they were triggered by the usual monocularly-seen edges. The only difference is in the time delay between stimulus movement and the onset of vergence movements, which for cyclopean stimulation is assumedly much longer than for monocularly perceivable stimuli. This is expected since the monocularly perceivable contours can trigger the superior colliculus directly, while the cyclopean targets have to be processed first by the cortex, thereby affecting the superior colliculus through the cortical pathways. That an interocular transfer exists between the oculomotor systems of the two eyes during tracking tasks has been established by Fender and St. -Cyr (1969). They showed that if one eye is occluded it still performs motions that are fairly well correlated with the movements of the seeing eye. Recent neurophysiological work by Wickelgren and Sterling (1969) and Sterling and Wickelgren (1969) using microelectrode probings reveals complex feature extraction capabilities, such as moving-edge detection of specified orientation and direction, by the superior colliculus. The same researchers also established that this complex feature extraction is under cortical control and can be abolished if the cortical interconnections are severed.

In addition to the LGN-cortex system, there is a second pathway to the cortex through the pulvinar. While the LGN pathway connects to Area 17 of the striate cortex, the pulvinar fibers synapse on Area 18, that is one hierarchical level higher. For a comparative anatomical review of the development of the two visual systems, see Diamond and Hall (1969). Also the Ph.D. dissertation of Abplanalp (1968) is most informative. The function of this second visual system is unknown. My guess is that the many perceptual constancies that keep the percepts of outside objects invariant during eye, head, and body movements are perhaps obtained by this second visual system. It is not by chance that this system is linked to Area 18, which is the site of complex and hypercomplex units. Indeed, in order to obtain, say, translational invariancy one must report an occurred physical shift in eye position to units that combine many parallel-displaced edge detectors. Of course, this is only a conjecture and remains to be verified.

This concludes the necessary background material. Some recent neurophysiological findings in stereopsis will be discussed in § 6.7. We now turn our attention to the psychological findings of global stereopsis.

5.3 Local versus Global Stereopsis

The basic difference between findings in stereopsis and fusion of simple line targets and stereopsis and fusion of complex stereograms is in the levels of hierarchy. Molecules behave differently from atoms even if their constituent atoms obey the rules of atom physics. Organisms have a very different structure from their building blocks, the biological cells. Phenomena of social psychology differ from the psychology of individuals. The set of real numbers is a different entity from single real numbers. Similarly the binocular fusion of a single dot or line segment is not only quantitatively but qualitatively different from the fusion of a set of dots or line segments.

Consider the notion that stereopsis or fusion of a single line or dot is qualitatively different from the stereopsis or fusion of a set of such lines or dots. When a random-dot stereogram of a low density texture is stereoscopically viewed (such as fig. 4.5-5*), the sporadic black dots in the surrounding white field are not individually perceived in depth above a common white background. Rather, the entire texture is seen in depth, as though a center area were cut out of white velvet paper and seen in front of the background. Of course, a single stereoscopic pair of white dots in a black surround appears in depth as a single target above the black background. A small number of such dots appears likewise. However, there is a minimal set of dots which appears not as individual dots in depth, but carries with it the black area between the dots as well. In the 100×100 dot array a 1% density of white (or black) dots—that is, 100 dots—is adequate to be seen as a random-dot texture and not as individual dots in depth. The role of density in random-dot stereograms has been intensively studied by White (1962).

This qualitative change in stereopsis can be attributed to the *ambiguities* arising with increased numbers of dots. Whenever more than one dot has the same ordinate in the two retinal projections, we cannot be sure which of the dots in one view belongs to a given dot in the other view. I discussed this problem of phantom targets in § 4.5 and refer the reader to figure 4.5-1. As long as the density of targets (dots) is low, the probability that nearby dots will fall on the same horizontal line is also small, and the problem of false localization does not arise. This is the stimulus condition when the dots can be individually resolved and are perceived as such. With increased dot density, the visual system cannot find uniquely the corresponding points, and a new process has to be evoked which can resolve ambiguities by global considerations. Several examples were given indicating that this global requirement is to localize points which yield the densest textured surface. This is probably the reason why the few hundred dots in figure 4.5-5* appear as a textured surface.

Thus, to obtain local stereopsis of a few edges or dots, one can visualize how the binocular dsparity units will maximally fire for similar receptive fields in the two retinae of the same shapes, orientations, and retinal ordinates. On the other hand, for complex textured surfaces, another level of neural processing has to be evoked that evaluates the possible local solutions and selects the densest firing units of the same disparity. This processing I will call global stereopsis, and it is on another level of complexity from the commonly quoted local stereopsis of the textbooks.

This difference between local and global stereopsis has several other ramifications. For instance, it is well established that local stereopsis and local fusion are different experiences. If a target is locally fused, it also yields local stereoscopic depth perception. However, the converse is not true. Targets can be too far apart to be fused and are perceived as double, yet are in vivid depth. This patent stereopsis can occur at disparity values that are several times the disparity limit for fusion (called Panum's area, which is only 6 min of arc for foveal vision). Global stereopsis can be produced from local elements that are not fused but are within the stimulus conditions for local patent stereopsis. Marlowe (1969) showed such a case with random-line stereograms that contained line segments rotated from alignment by at least $\pm 15°$. The line segments appeared to be tilted in depth and the horizontal ones were often seen as double, yet portrayed the global organization of a center square above the surround with great clarity. Such random-line stereograms are demonstrated in figures 5.3-1* and 5.3-2*.

5.3-1* Random-line stereogram, such as used by Marlowe; corresponding line segments in the left and right fields differ by $\pm$ 7.5°. Global stereopsis can easily be obtained.

5.3-2* Similar to figure 5.3-1* except that the corresponding line segments in the two fields differ by $\pm$ 15°. Global stereopsis can still be obtained.

Marlowe's experiments showing global stereopsis owing to local patent stereopsis are very different from the interesting experiments of Kaufman (1964) and Kaufman and Pitblado (1965). The latter showed that during global stereopsis a small proportion of the array can be locally suppressed owing to binocular rivalry. I pointed out earlier that for global fusion at least some spectral components must be similar in the two fields. However, in Marlowe's case the local suppression is due to exceeding the limit of local fusion. Yet all the suppressed elements are attracted to the corresponding elements because of patent stereopsis. This explains why in figure 5.3-2* practically all elements are locally suppressed but global stereopsis still can be experienced, while in Kaufman's case only a few percent of the elements can be suppressed owing to rivalry in order to experience global stereopsis.

    Can the global percept of a center square above the surround, when conveyed by local elements which are perceived as double, be regarded as global fusion or only as global stereopsis? After all, the line segments are seen as doubled, yet the square in depth is a clear single percept. I would regard this case clearly as an example of global fusion conveyed by locally unfused elements. Theoretically it might be possible to speak about global stereopsis without global fusion. That means that the center square would be seen double but still in depth. However, the maximum disparity limit for global fusion seems to occur where the maximum disparity limit for local patent stereopsis is reached. Outside the region of local patent stereopsis the global percept collapses, thus it is not possible to evoke global stereopsis without global fusion!

I will explore these problems in greater detail in § 5.9 wherein a cooperative phenomenon discovered by Fender and Julesz (1967a) is discussed. When thousands of local points are first simultaneously fused, then slowly pulled apart in the horizontal direction of the retinae (by preventing vergence movements), it appears that local stereopsis can be extended beyond the limit of patent stereopsis. Furthermore, this local extended stereopsis can give rise to global fusion even with a temporal shift exceeding 120 min of arc for foveal vision. Above this 2° shift, the global fusion suddenly jumps apart, and two random-dot images are perceived at the actual physical separation. The question arises whether prior to the break-away limit the global percept was due to global fusion or only to global stereopsis. The latter condition would result in seeing the center square above the background; however, the two random-dot arrays would not be fused but seen as double with the array viewed by the nondominant eye being suppressed. Since the complexity of the dominant array prevents any awareness of the suppressed array, it is very hard to devise experiments that could verify the existence of global stereopsis without global fusion. However, in experiments (Fender and Julesz 1967a), subjects reported a sudden loss of global fusion. The sensation was one of sudden jumping apart of two images from a single position. The global suppression hypothesis would predict that at the instant when suppression fails (above 2° of arc disparity), the suppressed image merely reappears without any sudden jump. The observation of this sudden jumping apart is strongly in favor of the global fusion hypothesis. It is most likely that the global stereopsis mechanism has only two states: fusion and nonfusion.

One could argue that if each local element of a complex stereogram is beyond local fusion-limits (but within stereopsis limits) with one eye's view suppressed, then the global percept must have similar suppression. But this argument has several weaknesses. First of all, dominance of eyes can vary over the retinae. There might be islands of right-eye dominance surrounded by areas of left-eye dominance. Of course, one could extend the notion of global stereopsis without global fusion to cover this case of alternate suppression. But more importantly the global pattern-matching process can evoke an entirely new global percept in depth which is portrayed by locally fused or suppressed elements. It might perhaps be better to refrain from using the words "global fusion" or "global stereopsis" altogether and call this ordered state "global pattern matching." There is just no necessity to assume that the doubling (and suppression) of each local element must result in the doubling (and suppression) of the global percept. I argue that point in such detail because Sperling (1970) in his otherwise stimulating model of binocular depth perception attributes the same properties of fusion, doubling, and suppression to the global process rather than the ones observed for the local process. When we use the term global stereopsis in this book, our use is merely an analogy and refers to the percept of global depth, without meaning global patent stereopsis with one eye's view suppressed.

A final example might further elucidate the difference between local and global phenomena. In random-dot stereogram movies in which each of the successive stereograms is portrayed by uncorrelated random-dots, the local dots seem to perform a sort of Brownian-movement. Yet the global percept is a static square in depth (or moving according to a predetermined trajectory). Although each picture element is perceived as moving, the global percept is static. Similarly, when each picture element of a complex stereogram is seen as double, it does not mean that the global percept cannot be a single one.

5.4 Textures in Visual Perception

Physical objects in our environment must be textured before we can perceive their shapes. A smooth, evenly colored, uniformly illuminated sphere, for example, cannot be distinguished from a disk or a cone by vision alone. In the past it was impossible to generate textured surfaces having exactly specified properties that could be used in psychological experiments. The psychologist who wished to study visual perception found real-life objects too complex for his purposes, therefore he was forced to use greatly simplified artificial stimuli in the form of textureless, outlined figures. These line figures were used for so many years that it was not even questioned whether the results obtained might be applicable for stimuli that had textured surfaces. Perhaps another reason why outlined figures were so frequently used in perceptual studies is the importance of contours in perception and memory.

The use of randomly patterned visual stimuli is relatively new to psychology, yet antedated computers by decades. The familiar Rorschach inkblot test falls into this stimulus class, although it is used to test complex perceptual-cognitive phenomena. Interest in texture as a variable in perception by Gibson (1950) and his followers has been focused primarily on texture gradients and their roles in slant and depth perception. Studies of psychological interest have been conducted by engineers concerned with the detectability of photographic film granularity (McBride and Reed 1952, Zweig 1956).

It was also systems engineers who first attempted to test complex, often nonlinear systems by using Gaussian noise as an input signal. The era of cybernetics and information theory is marked by the study of complex systems with stochastic input signals and the measuring of stochastic output parameters. MacKay (1961) first applied random visual noise in a cybernetic sense and evoked his "orthogonal after-image" phenomena. The interested reader is referred to his review on the use of visual noise in psychological research (MacKay 1965). This probing of the CNS with dynamic noise was started much earlier by workers in audition as we mentioned in another context; however, we shall restrict ourselves here to vision (for further references see Julesz 1968b).

A real technological breakthrough in the probing of the CNS with complex displays was the advent of fast digital computers and accurate graphical display devices under their control. For two decades it has been possible to generate complex visual patterns having precisely defined statistical and geometrical properties. Thus, for the first time in history, it is possible to devise artificial visual stimuli that approach or even surpass real-life objects in their complexities, but are devoid of all undesirable cues. It is difficult to trace the first computer-generated stimuli ever used in psychology. Green, Wolf, and White (1959) studied the detection of a statistically defined matrix of dots. But this was preceded by Graham and Kelly (1958), who developed a computer-simulation device for encoding TV pictures and studying the perceptual effects of various picture degradations. I used their device for studying the perception of fast brightness transients in pictures by replacing the convex or concave shaped transients by linear gradients (Julesz 1959). My observation was that the visual system cannot delineate the exact shape of brightness changes of contours. In the same year, 1959, I produced the first computer-generated stereograms, the basic subject matter of this chapter.

Without the computer these studies would have been extremely difficult, if not impossible. With enough patience, of course, one could turn out random-dot stereograms and cinematograms by hand. But the generation of movies that contain several hundred such arrays would be certainly a very impractical undertaking.

After this general introduction, let us discuss some concrete problems of texture generation by computers. The reader who would like to explore the many other uses of computer-generated displays in science and art is referred to a recent article by Harmon and Knowlton (1969).

The displays in this book were executed mainly by a General Dynamics (Stromberg-Carlson) 4060 microfilm plotter under digital computer control. Input and some output displays were scanned by a facsimile system of 196 lines/inch resolution (and 8 inch $\times$ 8 inch aperture), followed by a DDP24 medium-sized digital computer that performed gray-scale correction, digitalization, and wrote a magnetic tape (or produced a half-tone image from digital tape). The most elaborate programs were not handled by the DDP24 computer, but the digital tape of the input display was sent over to the computer center. Interestingly, some complex processings were prohibitively long (24 hours on the largest and fastest computers of 1 μsec cycle time) to be run on computers owned by computer centers. These we handled ourselves on our DDP24 departmental computer on weekends. From Friday afternoon until Monday morning we collected line-by-line the processed displays on tape. Such elaborate techniques were only necessary for certain complex displays discussed in chapter 6.

The usual procedure was as follows: Random-dot stereograms or cinematograms were produced off-line on the microfilm plotter. If the global information to be displayed was relatively simple, such as the T-shaped areas in figures 2.3-1 and 2.4-1*, the portrayal of the information was done by program instructions. For complex global displays such as the Müller-Lyer illusion in figure 2.6-2*, the input was given as an ordinary black-and-white drawing (e.g., fig. 2.6-1) to be scanned by the facsimile system. The resulting digital tape controlled the random-dot stereogram routine.

Various brightness values of the input stimulus were converted into different binocular disparities. The final output was displayed by the microfilm plotting routine. For certain stimuli the quality of the microfilm plotter was not sufficient because of the two-tone limitations and certain halo effects; therefore, we used our own facsimile installation (built by Walter Kropfl and myself). The possible 1,600 × 1,600 dot resolution and about 30 discriminable brightness levels surpassed our usual needs (and TV standards) by one order of magnitude. Only figure 5.4-1* in this book

(portraying some complex spatial surfaces) is portrayed by the facsimile system. It should be noted that the resolution of our facsimile system is an order of magnitude coarser than the resolution of the high quality facsimile systems used in the news media. The scanning of the input and output images takes about 8 min for the facsimile system, while the microfilm plotter portrays a 1,000 × 1,000 dot image in a minute or two (depending on image complexity). A typical display such as figure 2.6-2* would require a classical pencil drawing, eight minutes of scanning time for the input during which a digital tape is recorded, two minutes of computer processing time on a big computer with disk storage, and another minute or two for the microfilm printer to display the output stereogram. These are on-line times, while the loading of tapes, their physical transport from one computer installation to another, and the film processing add considerable time. Even so, one can convert a classical line drawing into a cyclopean stimulus in less than a day.

5.4-1*   Same as figure 5.4-2* except a facsimile system is used portraying each of the 10^6 dots in four brightness levels.

If the image (surface) to be displayed as a random-dot stereogram is mathematically defined, one can skip the facsimile system entirely and do the computation and display in the computer center. Julesz and Miller (1962) developed a system that automatically displays any function of two variables as a random-dot stereogram so that for multiple values the closest point to the observer is portrayed. Such a complex surface is shown in the form of a random-dot stereogram in figure 5.4-2*. For this

5.4-2* Truncated hyperbolic paraboloid of 1,000 × 1,000 resolution, but only 10% of the computed dots are portrayed, using two brightness levels.

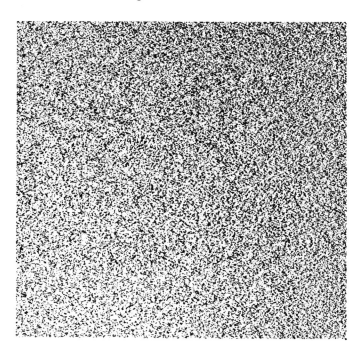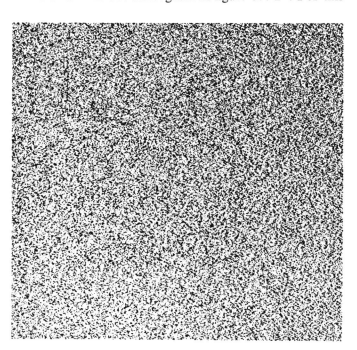

5.4-3* Random-dot stereogram of 33 × 33 resolution. The center square is only 15 × 15, yet stereopsis can be easily obtained.

surface the resolution is $1,000 \times 1,000$ dots; however, only 10% of the dots—that is, 10^5 dots—are actually portrayed at two brightness levels. In figure 5.4-1* the same random-dot stereogram has been portrayed by our facsimile system, displaying the entire million dots in four brightness levels. It is hard for present printing techniques to cope with such fine arrays, and in most of our research we have found that the quality of figure 5.4-2* is more than adequate. After all a stereogram consisting of an array of 33×33 cells having 15×15 cell center square in depth yields good stereopsis as demonstrated in figure 5.4-3*. The decrease in resolution from figure 5.4-1* to figure 5.4-3* is a thousand times!

5.5 Complex versus Simple Stereograms

These textured surfaces differ from real-life scenes in several important ways. First, the many monocular familiarity and depth cues that influence perception of common objects are entirely absent. Furthermore, photographs of real-life scenes usually contain some uniform areas without textures (cloudless sky, smooth lake without mirrored images, etc.); such areas do not carry binocular disparity information. On the other hand, random-dot stereograms (and cinematograms) are covered with texture elements over the entire display and as such are more effective stimuli than real-life stereograms, not to mention outlined drawings. I have already illustrated this increased effectiveness by comparing figures 2.8-7* and 2.8-8*. Here an aniseikonia of 15% linear size difference can be tolerated for the random-dot stereograms.

The complexity of random-dot stereograms leads to other discrepancies nonexistent in outlined drawings. Since Helmholtz (1909) presented simple line drawings stereoscopically such that one eye received the complement (negative) of the corresponding image, obtaining stereopsis, there is a common belief that monocular contours are crucial for stereopsis. Such a stereogram is shown in figure 5.5-1*, which can be easily fused. However, for random-dot stereograms a positive and negative image cannot be fused, as shown by Julesz (1960a). This is illustrated by figure

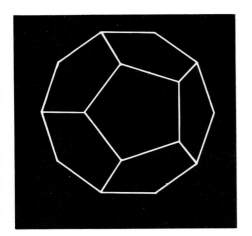
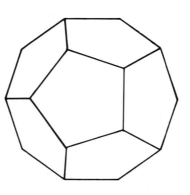

5.5-1* Simple outlined stereogram showing that a positive and negative corresponding image yield fusion. (After Helmholtz 1909.)

5.5-2* Complex (random-dot) stereogram showing that in this case
a positive and negative image pair
cannot be fused.

5.5-2*, which is identical to the basic stereogram of figure 2.4-1* except that one of
the images is complemented (black cells become white and vice versa). I tried to
make an even clearer analogy to figure 5.5-1* (Julesz 1963), taking only the outlines of figure 2.4-1* and the complemented one of the fields as shown in figure
5.5-3*. Again fusion is impossible to obtain.

This experiment clearly shows that simple extended outlines are very different from
densely packed line segments when one of the stereoscopic images is negative and the
other positive. Real-life images lie somewhere between these two extreme conditions.
Another experiment from the same publication (Julesz 1963a) shows even more
peculiar phenomena. Figure 5.5-3* is the outlined version of figure 2.4-1*, and since
both images are of the same polarity (i.e., black outlines on a white background)
fusion is strong. When the black cells (of 10 × 10 dots size) in the right image of

5.5-3* Positive and negative image of
the outlined version of figure 2.4-1*.
Again fusion is impossible. (From
Julesz 1964b.)

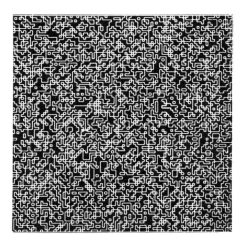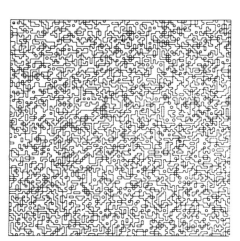

5.5-4* Random-dot stereogram similar to figure 2.4-1* except the black cells in the left image are expanded by 20% ($f_L : Bf_R$). From Julesz 1963a.)

figure 2.4-1* are expanded by two dots (20%) in two dimensions at the expense of the white dots, while the left image is kept unaltered, stereopsis can be easily obtained as shown in figure 5.5-4*. However, when figure 5.5-1* is again outlined, the fusion of the resulting stereogram depends on subtle factors. In the quoted paper I summarized the results of these experiments in a table, which will help explain the outcome of the last experiments shown in figures 5.5-2* and 5.5-3*. The *relation* between two fields of the basic stereogram of figure 2.4-1* is viewed stereoscopically and will be designated by $f_L:f_R$. Similarly, the relation $Xf_L:f_R$ is a shorthand notation for a modified left and right image viewed stereoscopically. The operator (X) that modifies the images can be of four basic transformations or any cascaded application of these: $N \equiv$ *Complementation* (Negation). Black picture elements are transformed into white and white into black; $B \equiv$ *Black Expansion*. Black picture elements are expanded in both dimensions by 20% at the expense of the white; $U \equiv$ *Uniform Expansion*. Every picture element is uniformly expanded in both dimensions by 10%: $C \equiv$ *Contour Formation*. Black outlines (of 1 dot width) are formed on a white background (or white outlines on a black background) at the boundaries between black and white clusters of picture elements. The black outlines lie entirely on the black side of the boundaries of the given field (or similarly white outlines lie on the white side). (See table 4.)

Table 4 shows how figure 5.5-5* and 5.5-6* give a different quality of stereopsis although they differ in minute detail: In figure 5.5-5* the black outlines in the right image fall on the black side of the boundaries separating the unchanged black and white clusters in the left image, while in figure 5.5-6* the black outlines in the right image fall on the white side of the boundaries in the left image. That a uniform expansion by 10% of one field preserves stereopsis, while a slight nonuniform expansion destroys it, is an interesting outcome. However, that the right fields of figures 5.5-5* and 5.5-6* yield strong stereopsis although they have the same nonuniform local expansions as figure 5.5-6* is even more revealing. These experiments suggest that if

TABLE 4. *Summary of Experiments Performed Including Quality of Stereopsis*

Description	Quality of Perceived Stereopsis	Demonstration
f_L:f_R	strong	fig. 2.4-1*
f_L:Nf_R	none	fig. 5.5-1*
f_L:Bf_R	strong	fig. 5.5-4*
f_L:Uf_R	strong	fig. 2.8-8*
f_L:UBf_R	strong	. . .
f_L:Cf_R	medium	fig. 5.5-5*
f_L:UCf_R	medium	. . .
f_L:CBf_R	none	fig. 5.5-6*
Cf_L:CBf_R	strong	right fields of figs. 5.5-5* and 5.5-6*
Cf_L:NCf_R	weak	. . .
Cf_L:$NCBf_R$	weak	. . .

5.5-5* Random-dot stereogram showing $f_L : Cf_R$. Stereopsis is medium. (From Julesz 1963a.)

 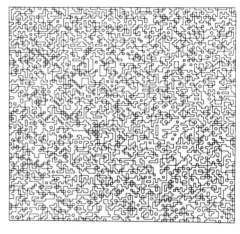

5.5-6* Random-dot stereogram showing $f_L : CBf_R$. Stereopsis cannot be obtained. (From Julesz 1963a.)

 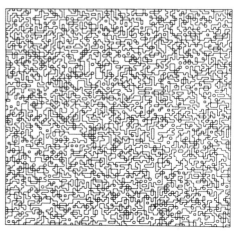

similar micropatterns occur in the two fields, fusion can be obtained locally. If the micropatterns themselves are not similar, then the observer tries to fit the entire field optimally. In this case, however, only uniform expansion or contraction is permitted.

Both uniform expansion and black expansion must be compensated for solely by the central nervous system. Therefore, it is not surprising that the results (in table 4) are identical under tachistoscopic illumination as well. The relations of table 4 suggest a complex structure for stereopsis that has to be accounted for by any theory of stereopsis. The complexity of these relations is not reflected in any neurophysiological result to date. It remains to be seen whether a class of binocular units will be found that will respond to the more complex stimuli studied here.

5.6 Removal of Monocular Depth Cues

Binocular depth perception is based on many depth cues that can be divided into two main classes: monocular and binocular depth cues. These cues interact with each other in complex ways which previously hindered the study of binocular depth perception. The use of computers permits us to manipulate these depth cues and remove or juxtapose any number of them.[1] This book is not intended to give an introduction into these depth cues, therefore the interested reader should consult the many good available textbooks on visual perception. Only recently have models been proposed that can utilize these cues under some simplified conditions.

Such a development began with the first computer algorithm invented by Guzman (1968), which can separate monocularly presented outlined drawings of complex polyhedra from each other. Guzman tried to determine the faces of polyhedra that belonged together but were partly hidden by others. The famous cue of interposition —the hiding of distant objects by close objects—is of no help when tested by the rigorous demands of a computer algorithm. Only when the notion of interposition is carefully examined do we realize that it is based on familiarity with the objects. In visual perception familiarity with objects and patterns corresponds to the semantic information processing which is as enigmatic for vision as it is for linguistics. The semantic difficulties for utilizing interposition as a depth cue are illustrated by a simple example.

Imagine that no stereopsis or texture cue is given as exemplified by two-dimensional outlined drawings. In a monocularly viewed outlined drawing we see a rabbit before a background fence. We know that the rabbit is in front of the background because it hides part of the fence. But the only reason why we know that the ears of the rabbit are also in front is because of our familiarity with rabbits; we know that the ears belong to their faces and not to the fence. On the other hand, we recognized the rabbit solely because of its long ears which we had to separate from the complex background. Thus there is a vicious circle: we can only separate textureless, monocularly viewed

---

1. The idea of manipulating and juxtaposing depth cues is not new. In the many Ames demonstrations (see a review by Ittelson 1952) false depth is created by distorting familiar shapes. However, only computer-aided stimulus generation gives us enough flexibility to explore fully the many possibilities.

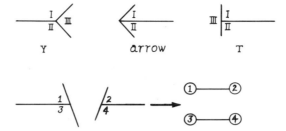

5.6-1 Vertex configuration classes in Guzman's interposition algorithm. (From Minsky and Papert 1969.)

5.6-2 A simple scene of polyhedra separated by Guzman's algorithm. (From Minsky and Papert 1969.)

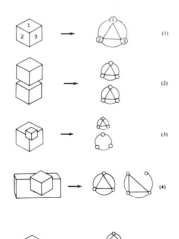

objects after their recognition, but in order to recognize them we have to first separate them.

One way out of this dilemma is the use of two eyes. As we have seen in § 4.5, stereopsis can separate objects in depth without requiring their recognition. The other way is to restrict ourselves to real-life objects with rich textures and separate these objects by texture discrimination (as discussed in § 4.4).

Canaday (1962) developed the first algorithm for monocular vision that could analyze textureless scenes composed of two-dimensional overlapping objects limited to straightsided pieces of cardboard. Roberts (1963) produced a three-dimensional automatic representation of line drawings with the restriction that the lines must be perspective projection of surface boundaries of a set of three-dimensional objects with planar surfaces. The algorithm of Guzman (1968) is the most sophisticated, yet it is restricted to polyhedra.

I shall give a brief account of Guzman's work (based on Minsky and Papert 1968), since some of his ideas might be generalized to curved boundaries. Also his mastery of treating local and global cues in order to grasp hidden semantic content is very instructive. Guzman's idea is to treat different local vertex configuration classes as providing different degrees of evidence for connecting the faces that meet there. For example, in the three different vertex configuration classes shown in figure 5.6-1 the Y provides evidence for linking faces I to II, II to III, and I to III. The "arrow" only links I to II. However, a T usually results if one object occludes parts of another, therefore it is assumed to be evidence against linking I to III or II to III (and is neutral about I and II). There are many other rules and vertex types, but using just these, we can analyze pictures and separate linked groups of faces as follows: We represent Y links by straight lines, arrow links by curves, and T links as weakening the existing links.

In figure 5.6-2 a few simple scenes are shown in which only separate objects are linked. For more complex configurations, such as those shown in figure 5.6-3, Guzman uses a hierarchy of links that form a subset of linked faces (called nuclei) which in turn compete for more weakly linked faces. He uses more global vertex configurations such as "collinear T-s" shown in figure 5.6-1 which is assumed evidence for linking faces I to II, and III to IV.

I have dwelt on this example to show that even one of the simplest monocular depth cues requires complex processes for its utilization, and for general objects we

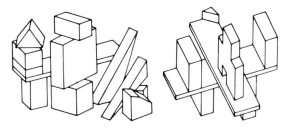

5.6-3 A complex scene of polyhedra
successfully separated by Guzman's
algorithm. (From Minsky and Papert
1969.)

cannot imagine how the visual system can utilize interposition. For other monocular
depth cues, for instance the known size of objects, the semantic problem is even more
formidable. The very moment we recognize that the assumed cat walking on a distant
hidden roof is not a cat but a tiger, its perceived distance increases.

Of course there are important monocular depth cues that require fewer semantics
than do the cues of interposition or known size. Rules of perspective, retinal gradients
of texture, and movement parallax are powerful depth cues, yet are devoid of familiarity considerations. I have discussed a few of these in §§ 4.2 and 5.2. Monocular-
movement parallax is a particularly strong depth cue. Its action is very similar to
stereopsis. As we move our eye's position, objects closer to us appear to have larger
disparities (and as a consequence appear to move faster). Similarly with stereopsis, the
changes in disparity are a consequence of geometrical rules. The real problem is not
so much the quantification of perceived depth as a function of disparity (parallax
shift), but how to establish this disparity. Since random-dot cinematograms can produce moving areas in spite of the individual frames being uniformly random, we
can conclude that no prior form recognition is necessary for establishing movement
parallax. The only difference between a random-dot stereogram and cinematogram
is that in the former the disparity is always horizontal (being parallel to the line join-
ing the two eyes) while for the latter, disparity can be in any direction.

This similarity between stereopsis and movement parallax does not mean that the
neural mechanisms are the same. While for stereopsis the constant distance between
the two eyes provides a permanent reference, for movement parallax the body, head,
and eye movements have to be exactly known to evaluate the new positions of the
eyes. This in turn requires a complex coordination between proprioceptive and visual
mechanisms. We will see that stereopsis is also affected by convergence and accommodation movements of the eyes, but to a much lesser extent.

For real-life scenes most of the monocular depth cues provide information compatible with stereopsis. However, under artificial conditions these cues can be juxtaposed with each other and rank ordered. For instance, one can show a concave
painted mask of a face that will appear normally convex when it is far from the
observer. The familiarity cues counteract the weak cue of binocular disparity. However, as the mask is brought closer to the observer the increased stereopsis counteracts the familiarity considerations and the mask appears concave. It is also known that
stereograms when exactly aligned and separated by polarizers exhibit an inverse
movement parallax. Objects in front do not move against the observer but with the

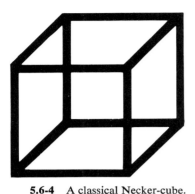

5.6-4   A classical Necker-cube.

5.6-5*   A cyclopean Necker-cube. This can be seen in perspective depth and reversed like the classical one.

observer. In spite of this, objects in front are seen vividly in depth, which suggests that stereopsis is a stronger depth cue than movement parallax.[2]

Most of these monocular depth cues are highly central and certainly occur after stereopsis. They do not affect stereopsis, and the final depth evaluation operates on these separate processes. That linear perspective operates after stereopsis and independently of it has been shown by Hochberg (1963). He used the technique of random-dot stereograms in order to portray a Necker-cube as shown in figure 5.6-5*. The classical Necker-cube is shown in figure 5.6-4. It is possible to see the cyclopean Necker-cube in depth and even to reverse it, similar to figure 5.6-4. That this cyclopean depth is not coupled with stereopsis is evident from the fact that the sides of the cube seen in different depth from the rest do not disappear. In figure 5.6-5* the random-dot bars that portray the cube have the same disparity and appear first at the same depth. If the perspective cues (that make certain sides recede in depth) were coupled with stereopsis, then the receding sides would have the wrong disparity and because of this would disappear.

2. Lee (1969) gives another explanation which, however, arrives at the same conclusion: When the observer moves in front of a stereogram there is no movement parallax for any of the objects in the stimuli. The reason why the stereoscopic percept of objects appears to move is because when the observer moves, the binocular disparity changes. This means, however, that binocular disparity changes alone can affect movement perception (similar to the way monocular movement parallax can give rise to depth perception).

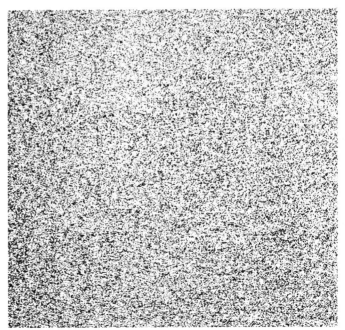

Similarly to linear perspective, the monocular depth cue of retinal gradient of texture (as introduced by Gibson 1950) is processed after stereopsis. Complex textures with changing granularity are perceived as tilted. Even if one views such a display binocularly, the tilt is apparent although disparity cues reduce this tilt. However, the display is not seen as an untilted plane, which would be expected from stereopsis alone. Retinal gradient of texture does not require familiarity cues, yet the determination of the average granularity size of a complex texture, such as an aerial view of a plowed field or a forest, can be a complex task. Even linear perspective is a complex cue that must be learned. Therefore it is not surprising that stereopsis precedes most, if not all, monocular depth cues. The most complex depth cues that necessitate familiarity with the stimulus are, of course, highly central.

An important monocular cue of shading permits the spatial reconstruction of objects to some extent provided their surface reflectivity functions are known. This problem will be briefly discussed in § 8.5.

Even if the monocular depth cues are processed after stereopsis, the final depth percept is strongly influenced by these depth cues. Since these cues interact with each other in unknown ways, the technique of familiarity deprivation, which is accomplished with random-dot stereograms and cinematograms, helps to remove all the complex depth cues from the stimuli except the one under study. The remaining sections in this chapter are concerned with findings obtained under the purest possible conditions: (1) All monocular depth and familiarity cues were removed from the stimuli (through the use of random-dot patterns). (2) The statistical and geometrical properties of the stimuli were precisely known (since they were generated according to a specific computer program). (3) Convergence movements of the eyes and proprioceptive cues were eliminated (through the use of tachistoscopic illumination). (4) The time of presentation was under control (through erasure of the afterimages).

5.7 A New Paradigm: Stereopsis and Ambiguities

Stereopsis gives us the experience of relative depth. It enables us to rank-order the nearness-farness of objects within a region of space around the fixation point. The quantification of the relationships between disparity and perceived relative depth has been the main object of research for a hundred years. The pioneering work of Panum, Hering, Helmholtz, Hillebrand, and Tschermak was aimed primarily at exploring the maximum and minimum limits of disparity. Ogle (1962) gives a comprehensive review of this extensive psychophysical literature. Some of these classical problems are related to the concept of a horopter surface. If one imagines an observer with the center of the fovea of each eye fixed at some point in space, then the images from all points that lie on a circle—which goes through the fixation point and the two anterior nodal points of the eyes—will have zero disparity. This defined curve, the Vieth-Müller circle, is an arc on the horopter surface, a surface of space having points which appear at the same depth. In addition to this surface there are two other surfaces (nearer and farther away from the observer), which are defined as being at the threshold of fusion (having maximum convergent or divergent disparities) when fixing at a given (reference) point. One can define another pair of surfaces as representing the threshold for patent stereopsis. The shape of the horopter curve is closely related to Luneburg's theory (1950) of hyperbolic metric (as will be discussed in the chapter on perceptual invariances). Until recently, these and similar questions were the preoccupation of researchers.

However, the study of the relationships between disparity and perceived depth is perhaps one of the less intriguing problems of stereopsis. The main problem is the establishment of corresponding points that are disparate. Perhaps classical research neglected this problem because it depended too heavily on the use of simple-stimulus materials consisting of a few lines or dots. Indeed, when only one dot projects on the left and right retinae (having the same ordinates) there is no ambiguity as to which picture elements belong to each other on the retinae. Yet, as shown in figure 4.5-1, with four targets there are already 16 possible localizations out of which 12 are false. As more and more points of the same color lie on a horizontal line with the same ordinate values in the two retinal projections, the ratio of false localizations to the sum of correct and false localizations rapidly approaches unity.

For complex stimuli, such as faces in a football crowd, it has been tacitly assumed that the ambiguities would be resolved by organizing the many elements into a few gestalten separately in the left and right eye views. After recognizing a certain face in the two monocular views and detecting their sameness, the amount of binocular disparity could be determined. If monocular form recognition would be necessary for removing the ambiguities in stereoscopic vision, then stereopsis would be one of the most complex processes, since problems of form recognition are still enigmatic. Another possibility was proposed by Julesz (1960a), assuming that first the two retinal images are combined in a binocular view and all further processing (that is usually simpler than form recognition) is performed on the combined binocular fields.

I have discussed this problem in § 4.5 in some detail and pointed out how this latter hypothesis can be verified by the use of random-dot stereograms. Particularly it was demonstrated that no monocular form recognition can aid the stereopsis mechanism to resolve ambiguities. However, there are several monocular cues—simpler than form recognition—that help stereopsis in reducing false localizations. For instance, if the corresponding targets are edges or slits of a given orientation and no other such orientation occurs with the same ordinate values, the ambiguity is resolved. Similarly, the identical length-width (aspect) ratio of corresponding rectangular areas might fire binocular neurons of identical left and right receptive field shapes. The color of a corresponding point or line pair might serve as a label to identify them. Thus, color, orientation, and aspect ratio of corresponding edges or slits can serve as identification. However, the experiments of figure 3.9-1* (in which the diagonal connectivities were broken) demonstrated that the stereopsis mechanism can find corresponding areas on a point-by-point basis. For further discussion the reader is referred to Julesz (1968a).

In figure 3.9-1* the fine details (high spatial frequencies) have been changed in one of the images, but the coarse-detail information remains very similar. However, it is simple to generate perturbations in which no correspondence is similar either in fine or coarse details—except for the highest spatial frequencies or a small region in the two-dimensional spectrum—yet fusion can be obtained. Such stereograms were generated by Julesz (1960a, 1968c) in which every second or third row has the same disparity. For instance figure 5.7-1* contains three global organizations; every third row has the same disparity; —2, 0, or +2 units, respectively. A transparent plane in front, a middle plane, and a third plane behind the previous two can all be perceived simultaneously with stereoscopic viewing. In these stereograms only those Fourier components remain identical that fall on the horizontal axis of the two-dimensional spatial frequency plane.

Blakemore (1970), however, presented vertical gratings of slightly different frequencies to the left and right eyes, respectively. As expected, he produced the percept of a tilted striped surface. For a certain maximum spatial frequency ratio between the left and right gratings the global percept of a tilted surface disappeared (and instead some sporadic regions were fused). This maximum frequency ratio was greatly reduced for very low and very high frequencies. Blakemore concluded from these findings that stereopsis might utilize monocular one-dimensional spatial frequency analysers in the left and right pathways that have similar frequencies. Blakemore's notion of one-dimensional frequency analysers cannot explain the fusion of Fig. 5.7-1*. For this, we have to introduce two-dimensional spatial analysis, since the information varies in the vertical dimension as well. However, the notion of some spatial analysis occurring before stereopsis is not unlikely.

We have discussed how it is possible to scramble up monocular patterns in the fused binocular percept. However, when one of the monocular images of a random-dot stereogram portrays a spatial grating, it is very difficult, if not impossible, to make this grating disappear in the binocular percept. Fig. 5.7-2* shows such a random-

5.7-1*   Random-dot stereogram containing three transparent depth planes.

dot stereogram that portrays a vertical grating in the left image. Even after fusion (when two transparent planes are seen as one being above the other) the grating is usually seen. Thus monocular edges and gratings do resist scrambling in the binocular view. This indicates that monocular edge detection takes place before stereopsis, a finding that is expected from neurophysiological evidence. However, Blakemore's conjecture, that these edge detectors are also organized into spatial analysers for various frequencies and these extracted frequencies are utilized by stereopsis, is a novel idea.

An experiment similar to figure 5.7-2* has been discussed in § 3.10. There, figure 3.10-11* contained a low-contrast grating similar to figure 5.7-2* but with an octave lower frequency, and this grating also withstood scrambling in the binocular view. However, before one would jump to the conclusion that spatial frequency analysis occurs before stereopsis, let me note a recent observation by Blakemore. He modified the experiment reviewed above. Instead of producing a tilted surface by presenting vertical gratings with slightly different frequencies to the left and right eyes, he presented identical gratings to both eyes. But prior to this he adapted one eye to a high (low) frequency grating in order to produce an aftereffect of perceived frequency shift as discovered by Blakemore and Sutton (1969). If this aftereffect (owing

5.7-2* Inverse cyclopean stimulus of a vertical grating. Monocular bars and grids resist binocular scrambling.

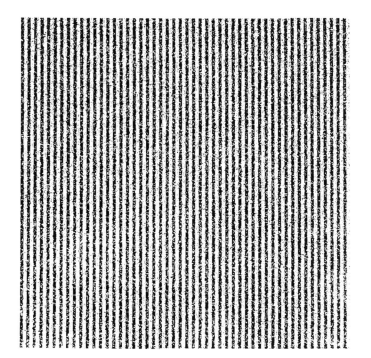

to the selective adaptation of spatial frequency analyzers) were to reside before stereopsis, one would expect a tilt in the fused percept. In fact, Blakemore was unable to produce such a tilt. Until some cyclopean version of the Blakemore and Sutton aftereffect is produced, the locale of spatial spectrum analysis will remain unknown.

The inability of producing a tilted, fused plane by Blakemore is even more enigmatic in the light of a report by Blakemore, Nachmias, and Sutton (1970). They found partial interocular transfer of the Blakemore and Sutton aftereffects. The interocular frequency shift was 1.5 times less than the shift perceived through the adapted eye. Why the strong peripheral component of this aftereffect did not yield the percept of the tilted plane in Blakemore's fusion experiment is a mystery. One might assume that this one-dimensional spectrum analysis is not utilized by stereopsis, but this assumption is difficult to reconcile with our findings that showed how stereo images of partly identical spectra would yield stereopsis.

These experiments show clearly that the stereopsis mechanism selects from the countless global organizations only those that yield the densest surface. Even if this surface is a subarea to a small extent, it will be selected if it gives a dense surface containing points with the same disparity values. Thus in figure 5.7-1* the stereopsis mechanism selects every third row as the global solution—since every picture element in these rows has the same disparity—rather than the chance solution of 50% coinciding picture elements at any disparity value, including zero disparity.

The interesting problem of how the global stereopsis mechanism might operate when the stimulus contains two or more ambiguously perceivable dense surfaces will be discussed in chapter 6.

5.8 Hysteresis in Binocular Depth Perception

One of the most significant phenomena of vergence, accommodation, and of fusion is the existence of multiple stable states. In the simplest case an obtained state can lag behind a new change, which otherwise would result in a new state; and the system can suddenly flip from the old state into the new one. This lag between cause and effect is called hysteresis.

Such a hysteretic phenomena for vergence was demonstrated by Helmholtz (1909). He presented a stereogram binocularly in alignment so that fusion was obtained. Then slowly pulling the two images apart he caused the eyes to diverge and maintain fusion. Actually the eyes can diverge by as much as 8° of arc before the fused images suddenly jump apart. Yet if the left and right images are presented suddenly at a horizontal separation which requires only a few degrees of divergence, the eyes could not diverge to fuse them. Thus when the images are separated by 8° of arc, the eyes are either correctly verged or may be in their neutral position. There are two stable states of vergence; the particular vergence state reached by the eyes depends not only on the present stimulation, but also on the history of the stimulation.

Similar hysteresis can be demonstrated for accommodation. If the reader is still capable of a few diopters of accommodation, he may easily demonstrate for himself two different stable states of accommodation to a single stimulus condition. The following is a procedure described by Sperling (1970): (1) Cover one eye. (2)

Establish the smallest distance at which a sharp image is possible—the near-point—by moving a pencil toward the open eye while fixating on its point. (3) Fixate a textured object, for example, a picture, at a few meters distance. (4) While maintaining focus on the picture, move the pencil to the near point, such that its retinal image coincides with the point of fixation. (5) Focus on the pencil keeping the eye position. By alternating between (4) and (5) the reader is sure to find two stable states: the accommodation to the picture or the accommodation to the pencil. The identical external stimulus then will have evoked two different states of accommodation. However, in order to alter states one has to fixate on one or the other target and bring the two slowly in coincidence. The actual accommodation state depends on the prior history of stimulation.

Hysteresis also occurs in stereopsis. Suppose that vergence of the eyes is artificially maintained on a particular depth plane by optically cancelling vergence motions using binocular retinal stabilization (Fender and Julesz 1967a). Because of the importance of this experiment I shall include a detailed summary of our procedure as well as the main findings; the reader is, however, advised to consult the original paper.

In this article we reported experiments in which the fused images were actually pulled apart on the retinae; nevertheless, fusion was maintained for some distance outside of the classical fusional areas. Although there are simpler ways to prevent vergence movements of the eyes during the pulling of the images than by using binocular retinal stabilization (i.e., pulling the images temporalward beyond the divergence limits of the eyes), it would be difficult to monitor accurately the positions of the eyes. The advantages of the technique of binocular retinal stabilization are twofold: it permits us to control the position of the stimuli on the retinae, and it provides a high accuracy for tracking eye motions, which cannot be matched by other methods.

The left-eye component of the apparatus which provided a stabilized image is shown in figure 5.8-1. The monocular design of this equipment was originally undertaken by Ditchburn and Ginsborg (1962) and by Riggs, et al. (1953). The actual apparatus shown in figure 5.8-1 was designed by Fender and Nye (1961). In the present work the apparatus was duplicated for the right eye; the distance between the two halves of the equipment could be adjusted to suit the interocular distance of the subject. The subject wore close-fitting scleral contact lenses over his eyes from which a stalk protruded carrying a small mirror (M). By suitable manipulation, the subject could adjust the optic axis on each side of the equipment to coincide with a comfortable fixation direction of the corresponding eye.

Several other optical components concerned the correct initial positioning of the stabilized image in the center of the fovea. In normal vision there is only a very small region of the fovea which gives the sharpest vision; evidence was advanced by Polyak (1941) that this region is as small as 10 min of arc diameter (not to be confused with Panum's fusional area). This is not the case, however, for stabilized vision; under certain circumstances the target may be moved as much as 30 min of arc with equal sharpness experienced by the subject. This finding of increased area of the sharpest vision under retinal stabilization is a phenomenon in its own right, not to be discussed here.

Two targets were used: (a) a single vertical black line subtending 60 min of arc vertically and 13 min of arc wide and (b) a random-dot stereogram (such as fig. 2.4-1*), which subtended 3.43° of arc in the visual field and each picture element was little over 2 min of arc square; the disparity of the center square was about 8 min of arc.

Results with Line Targets

All subjects could adjust the lines to fusion in normal and stabilized vision: they reported the percept as a single line, steady and unchanging. As the lines were slowly moved apart in stabilized vision, the percept did not change, and none of the four subjects was able to detect a change in perception of depth of the line. This should be contrasted with normal-vision in which disjunctive motion of the targets produced a very powerful perception of changing depth.

In either viewing condition after a certain amount of pulling the fused images "broke" apart. The break from fusion was always sudden and well-defined; subjects then perceived two lines in the visual field. The perceptual separation of the lines which had been lost after fusion could be estimated by the subjects, since each target included a few subsidiary marks at known angular spacings from the line. Under normal-viewing conditions, estimates of the separation were equal to the true separation within an error of about 15 min of arc. In stabilized vision the separation remained constant. Upon moving disparate targets towards each other, refusion of the images took place when the disparity was reduced by about 30% in either normal or stabilized vision. The results are illustrated for stabilized vision only in figure 5.8-2. In order to test the area over which fusion is possible, the targets were moved slowly and symmetrically temporalward in the visual field until the subject reported a break in fusion. The angular separation of the target was noted. The targets were then moved

5.8-1 The left-eye component of the apparatus used to obtain binocular retinal stabilization. (From Fender and Julesz 1967a.)

5.8-2 Hysteresis phenomenon of vertical line targets under binocular retinal stabilization and horizontal pulling. (From Fender and Julesz 1967a.)

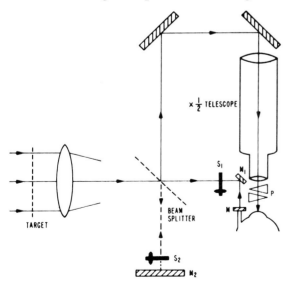

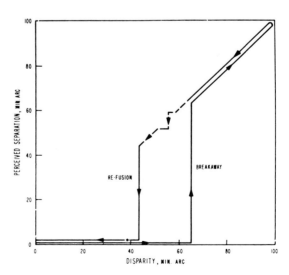

5.8-1   5.8-2

nasalward until fusion took place once more; the separation at which fusion was re-established was noted. Observations were taken around this cycle many times.

This was the only target-movement routine permissible in these experiments; the individual visual axes tried to track the target displacement, but the divergence movements were severely limited. Target motion in the nasalward direction or displacement of one target only in any direction causes extreme convergence or conjunctive motion until the contact lenses strike the fornices and are displaced on the eyes, destroying stabilization.

For the line targets, horizontal and diagonal orientation were also tried, and the effect of vertical and oblique disparity was determined. Results both for normal and stabilized vision are illustrated in figure 5.8-3, showing that fusion can be achieved over an elliptical area of the retina without the assistance of disjunctive eye movements.

Results with Random-Dot Stereograms

The experimental procedure was similar to the previous one except that random-dot stereograms were used in place of line targets. In normal vision, of course, the targets could be adjusted to fusion and stereopsis; however, depth could also be perceived in stabilized vision. This is not novel, for the tachistoscopic experiments by Julesz (1963a) indicated this. One out of the four subjects was unable to obtain more than fleeting glimpses of stereopsis. His difficulty was that during stabilized vision his area of high acuity shrank to an ellipse at most 20 min of arc wide. Most subjects can examine a considerable area of a stabilized image with high acuity even though they cannot shift their line of regard over the target. This area is usually elliptical, sub-

5.8-3 Hysteresis phenomenon of vertical-line targets under normal vision and binocular retinal stabilization with horizontal, vertical, and diagonal pulling. (From Fender and Julesz 1967a.)

5.8-4 Hysteresis phenomenon of random-dot stereograms under binocular retinal stabilization. (From Fender and Julesz 1967a.)

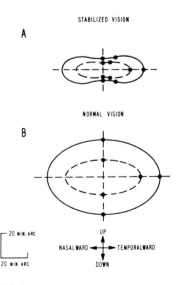

5.8-3

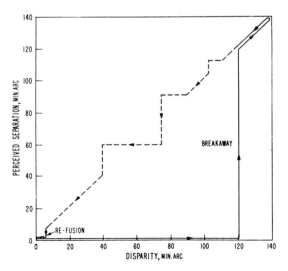

5.8-4

tending about 2° horizontally and 1° vertically; outside of this area, acuity in stabilized vision falls off rapidly. Some authors regard this as the area of attention.

Evans and Clegg (1967) reported stereopsis of random-dot stereograms in afterimages lasting for 15 sec. They were produced by very bright flashes and the characteristic fading and regeneration cycle of the afterimages was forestalled by illuminating
the closed eyes of subjects with a strobelight flashing at 5 Hz. Since these afterimages
are perfectly stabilized with respect to the retinae, the fact that 85% of their subjects
made correct depth judgments proves that stereopsis can occur in the absence of all
eye movements.

In the Fender and Julesz experiment subjects were allowed to make small changes
in the positions of the retinal images until stable stereopsis was experienced. Target
disparity was then introduced slowly by the experimenter during the periods for which
the center square was perceived in depth. With increasing disparity, the failure of
stereopsis and the fusion of the two images occurred simultaneously, and the two
images were seen to separate. The results of this experiment are shown in figure 5.8-4.
These results are in broad agreement with the results obtained for the single line, but
it is noteworthy that the targets can be pulled apart by a much larger amount, exceeding 2° before fusion is lost. On the other hand, the targets do not refuse until they are
moved very close together, within about 6 min of arc. For vertical disparity the
threshold for refusion is also about 6 min of arc, while the breakaway disparity occurs
about 20 min of arc.

The results are qualitatively similar to those obtained with line targets: targets presented with zero disparity are seen as one fused image; if they are then moved gradually into disparate positions, the images remain fused until some limiting disparity is
reached. At this limit, the two images break apart perceptually and are seen
separately in their disparate positions. Further increase of the disparity merely causes
the images to appear and move further apart. If, however, the disparity is reduced
below this limit, refusion does not occur until a much lower limit is reached. The targets then form a single fused image once more, and this percept is maintained until
the real disparity is reduced to zero. This lagging of effect (magnitude of perceived
disparity) behind cause (retinal image disparity) is another example of hysteresis in
the classical sense.

We also tested the dynamic properties of stereopsis (Fender and Julesz 1967a). In
normal vision the motions of the eyes are not perfectly correlated; this results in a
constantly varying amount of binocular disparity. It is a puzzling fact that this fluctuation of disparity does not affect stereopsis. The errors of convergence or divergence
caused by the drifting component of eye motion developed at slow rates (of the order
of 1 min of arc per sec). The previous experiments demonstrated that error rates such
as these can be compensated adequately by the hysteresis process which preserves
the cortical registration. On the other hand, the saccadic components of eye movements cause convergence or divergence errors which are much larger than those
caused by slow drifting motions. These errors develop in a very short period. The
duration of a spontaneous saccade is at most 40 msec and the vergence error exceeds
stereo acuity thresholds by orders of magnitude; nevertheless, even experienced visual

observers do not detect loss of fusion or changes in perceived depth during saccades. I will turn to these dynamic factors of binocular depth perception in the next section.

At the point of break-away the loss of fusion was abrupt and total; at a lesser disparity by a minute amount the fusion was perfect with no hint of impending disaster. However, an additional experiment clarified the conditions before the break-away state. Two random-dot stereoscopic images were set up so as to be well registered and to give good depth perception. The targets were then moved slowly by the experimenter into disparate positions at a rate of 1 min of arc per second. Both targets were then occluded for short intervals varying from about 10–600 msec; the interruptions occurred at random times with a mean exposure period of 1 sec. The subject was asked to signal when the central square could be perceived and seen in depth. The disparate motion of the targets was halted during the periods of occlusion and when the subject signalled that the square could not be perceived.

In this experimental condition, we found that once fusion was lost, it was never regained, although with longer exposure periods the result might be different. The subjects also reported that loss of the stereoscopic effect coincided with one of the periods of occlusion. The maximum duration of occlusion which could be tolerated without breaking fusion is an inverse function of image disparity as shown in figure 5.8-5. We assume that the curve climbs to about 120-min-arc disparity at zero duration of occlusion, and from figure 5.8-5, that the images would be refused at 6-min-arc disparity, after a very long occlusion.

The concept of corresponding retinal points has undergone many changes since it was introduced in 1613 by Aguilonius's notion of the horopter. The discovery by Wheatstone and Panum that fusion was possible over small corresponding "areas" of the retinae broadened the concept, since the corresponding points of the two images no longer need have the same coordinates in the left and right retinae, but merely have to be positioned within the fusional areas. The hysteresis phenomenon further changes the classical notions: According to the new idea, the recent history of the stimulation of the visual system has to be available to whatever central mechanism is responsible

5.8-5   Effect of occlusion time on the hysteresis phenomenon. (From Fender and Julesz 1967a.)

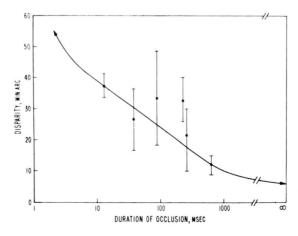

for binocular fusion before it can determine whether two points on the two retinae are corresponding points or not. Moreover, the context of the entire stimulation, for example the detail surrounding a given image point, has an effect on the determination of binocular correspondence. Our experiments show that two points falling on the perifoveal regions of the left and right retinae with a disparity of 2° may be corresponding points if they are members of a random-dot ensemble and if, prior to this amount of disparity, they were brought into alignment and then slowly pulled apart. If any one of these criteria fails, for example if the image points are not members of a dense ensemble forming a sheet of texture or have not previously been seen as fused before the onset of the disparity, the points do not function as corresponding retinal points, fusion does not occur, and stereopsis is not perceived.

These experiments reveal the dependence of the function of the visual system on the spatiotemporal characteristics of the probing stimulus. Many of the phenomena reported would have been disguised if only simple line stimuli had been used. On the other hand, some properties of binocular vision can best be examined by using line targets, since they can be perceived with one eye as well.

Two fundamental properties of random-dot stereoscopic images emerge from these experiments: First, stabilized vision reconfirmed that these images possess a fusion region only. There is no region where the images are seen as double but still perceived in depth. Second, the ratio of disparity at breakaway to disparity at refusion, is 20:1 (2°:6 min arc) for random-dot stereoscopic images and only 1.5:1 (65 min arc:42 min arc) for vertical-line targets. This dramatic difference between the two cases gives some insight concerning the fusion mechanisms. Since the random-dot stereoscopic images are devoid of monocular shapes, the binocular correlation between corresponding areas has first to be established. It seems that this labeling of corresponding points can occur only within Panum's fusional region. We believe that the correlation process assigns the proper labels to the corresponding points; these labels can then be preserved for large retinal-image shifts provided that the disparity is less than a critical value and that the velocity of pulling is also less than a certain limit. The abrupt transitions between ordered and disordered states indicate a cooperative phenomenon requiring the near-simultaneous participation of all of the constituent elements. In our case, the establishment of fusion (correlation or labeling) and its preservation correspond to the ordered state, whereas prior to fusion and after breakaway a disordered state exists.

The strong dependence of the breakaway threshold on the previous perceptual state, whether fusion occurred or not, indicates a simple memory process. We called it according to accepted usage "hysteresis" or "an hysteretic phenomenon." In our opinion this name should be assigned to all phenomena in which effect lags behind cause, and "memory" should be used for more complex information storage. For instance, the phenomenon that the break point for diplopia is always greater than that for recovery (when experimenting with prisms) is another example of hysteresis; whereas, the time necessary for aniseikonic distortions to appear after an observer puts on aniseikonic glasses might indicate a phenomenon complicated enough to be regarded as a memory process.

For a maximum hysteretic effect, a large number of elements have to cooperate simultaneously. Random-dot images are densely covered with points, each carrying information concerning image disparity, and it is reasonable to assume that stronger interactions may take place between the neural representations of ten-thousand picture elements than between the representations of a few points required to form a straight line. This might explain the small hysteresis effect obtained for lines, where only a few receptors are stimulated.

5.9 The Role of Convergence

Since Dove (1841) demonstrated stereopsis using brief flashes produced by an electric spark much too brief for any convergence motions of the eyes to be initiated, we assume that the main purpose of vergence is to bring the images within Panum's fusional area where the CNS process of stereopsis can operate. This experiment proved that stereoscopic depth sensation is not the result of vergence movements of the eyes.

This belief in the secondary role of vergence in binocular depth perception was further amplified by experiments that showed our inability to judge the positions of the eyes. Helmholtz (1866) reported one of the first observations which can easily be verified by the reader: Suppose one is looking at stationary objects. At each voluntary fixation movement of the eyes these objects appear stationary, although their retinal images have moved. However, if instead of an active voluntary fixation a passive movement is forced on the eyes (by catching hold of the eyelids at the outer canthus and pulling on them) the viewed objects appear to jump.

Helmholtz concluded that the effort of will compensates for the expected retinal image movement. He also regarded these experiments as proof that no information about the position of the eyes is derived from the sense endings in the eye muscles. Observations on afterimages confirm this conclusion. During active and passive eye-movements the sensations of position from an afterimage are the opposite of the previous case. Thus, if one has an afterimage at the fovea and views it against a featureless background, its apparent position seems to move with the fixation point during voluntary eye movements. During passive eye movements, however, the afterimages appear motionless.

Recently, Brindley and Merton (1960) and also Irvine and Ludvigh (1936) checked Helmholtz's conclusions. They turned anesthetized eyeballs physically by $20°$ or more at rates of several times a second, yet the subject was unaware of these eye movements. Therefore, it is not surprising that neither the extent of conjugate (asymmetric) eye movements nor the extent of convergence and accommodation of the eyes (disjunctive or symmetric eye movements) is a reliable cue for absolute positional or depth judgments. Perhaps one reason that only the voluntary command exists and no signal is transmitted back to acknowledge successful execution, is the frictionless movement of the eyes in their sockets in contrast to the limbs (which require such report).[3]

3. This lack of inflow signal is regarded as one basic dogma of vision. The evidence supporting an inflow theory is scant and rests mainly on Sherrington's (1898) discovery of stretch receptors in

Not only do afterimages follow voluntary eye-movements, they also change sizes. The well-known Emmert's law (1881) states that when fixating (i.e., accommodating and converting) at a near surface the afterimage shrinks, while when fixating at a far surface the afterimage expands. Since near objects cast larger projections on the retinae than objects further away, this phenomenon compensates for the changes in retinal image size which occur with changes of distance. A similar size constancy phenomenon that operates in complete darkness has been reported by Gregory, et al. (1959) for forward and backward head movements. Advancing reduces the size of afterimages; retreating increases their size.

These phenomena showed how afterimages change their position or size for instance during asymmetric or symmetric eye movements. The direction of these changes is such that for real images projected on the retinae these compensations diminish the effect of the projective transformations due to eye-movements. In addition to voluntary eye-movements and forced (passive) eye-movements the eyes can be immobilized by force (physical or chemical) during a voluntary eye-movement attempt. In this case the perceived image jumps in the same direction as the attempted voluntary eye-movement commands. This finding has been expected from another observation by Helmholtz (1909), who asked perceptual reports from patients (whose eye muscles were paralyzed) during voluntary eye-movement attempts. In order to make Helmholtz's "effort of the will" more palatable for twentieth-century tests the explanation was given in terms of afferent and efferent signals, black-boxes, and feedback mechanisms by Holst and Mittelstaedt (1950) and by Holst (1957). These notions still have no concrete neurophysiological parallels but have a strong heuristic value. According to Holst and Mittelstaedt's theory the voluntary command for eye movement sends efferent signals to the eye muscles but simultaneously is stored as an efferent copy. Because of the executed eye-movements, the retinal image changes, and this change is signalled back to the CNS as an afferent. Since this afferent signal is the result of a central command it is called reafference. If the reafference equals the efferent command stored in the efferent copy mechanism, the two signals nullify each other and the turning of the eyes stops. As long as the reafferent signal is less than the efferent command, the eyes continue to move. The difference between the efference copy and reafference is perceived visually; furthermore the voluntary afferent command is conscious. The efference copy and the reafference are unconscious processes. This model predicts very well the perceived phenomena

---

human extraocular muscle. However, recent research has shown that, in general, stimulation of stretch receptors does not lead to a conscious sensation of changes in limb position (Gelfan and Carter 1967, Rose and Mountcastle 1959). Recently, however, Skavenski and Steinman (1970) concluded that an extraretinal signal (that does not arise from retinal stimulation) must exist, because the eye, in total darkness, could be kept near the position at which a target appeared two minutes earlier. Furthermore, Skavenski (1970) in his Ph.D. thesis reported that subjects could maintain their eye position in total darkness when loads were pulling their eyes. The subjects were aware of the inflow signal and could correctly report the direction of load even when the eyelids and conjunctivae were anesthetized. Even if inflow signals were to have some role, this may not be perceptual, since Skavenski also showed that, of juxtaposed inflow and outflow (retinal) cues, only the latter affected perceived direction.

during voluntary, passive, and inhibited voluntary eye-movements. Some such mechanism is necessary for the visual system in which the muscle spindles do not convey the positions of the eyes. Similar efference copy/reafference models can be formulated for size changes owing to convergence and accommodation movements.

The most recent work on Helmholtz's paradigm concerning positional constancies during saccadic conjugate eye-movements is reported by Matin, et al. (1966) and Sperling (1965).

Size constancy during symmetric (disjunctive) eye-movement for real images cast on the retinae follows the predictions obtained with afterimages. This phenomenon might be a result of different processes depending on whether the object is within grasping reach (within 1m) or further away. As Leibowitz and Moore (1966) have shown, size constancy due to convergence and accommodation cues alone diminishes beyond 1 m. For farther distances contextual cues enter as shown by Holway and Boring (1941). In figure 5.9-1 are Holst's measurements (1957) for the perceived

5.9-1 Perceived size as a function of convergence and accommodation. (From Holst 1957.)

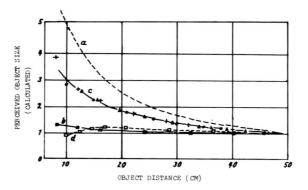

size of objects while fixating at various distances. As curve *b* shows, with free convergence and accommodation the relative perceived size approximates a constant line. For *c* in which either the convergence or accommodation is kept constant, the size constancy is reduced, yet the perceived size is much less than the size of the retinal image given in *a*. There are several studies, Leibowitz and Moore (1966) and Heinemann, et al. (1959), that confirm these results at near distances. As I have already discussed, convergence and accommodation are weak cues for distance judgments. (Distance in figure 5.9-1 means physical distance which sets the convergence-accommodation mechanisms without the need for estimating the perceived distance.)

These findings that vergence has a rather passive role in yielding a sensation of depth, but participates in a perceptual size zooming are modified by the findings of Fender and Julesz (1967a). We found several perceptual phenomena under binocular retinal stabilization that differ from normal vision. In the following viewing conditions, disjunctive motion of the targets gives rise to the following perceptions:

Normal vision. Powerful impression that the line is moving toward the observer, or weak impression that it is moving away from observer.

Stabilized vision. Stationary line, fixed in space.

RANDOM-DOT TARGETS

Normal vision. Powerful impression that the fused stereoscopic image as a whole is moving toward or away from the observer. No change of the apparent depth difference between the central square and the surround.

Stabilized vision. Surround fixed in space, but the center square advances and grows in size as the targets are returned to the central position after temporalward pulling.

In normal vision the convergence-divergence motions of the eyes are correlated with the disjunctive motions of the targets. That convergence gives a stronger perception of apparent motion than divergence is interesting by itself, although in the equipment used for this research the target is seen at optical infinity; conflict of information between the accommodation system and the disjunctive eye-movement system may therefore be responsible for this effect. Under stablized vision the line target appears to stay stationary in space, while for random-dot targets, which contain both a center square and a surround, the center square appears to advance in depth and grow in size whereas the surround stays stationary. The maximum increase of size was estimated by the subjects to be 30%, but immediately following this report the subject would estimate the size of the 40×40 central square as 40% of the 100×100 surround. These two cases are not strictly comparable; the line target lacks the reference plane which is provided by the surround of the random-dot pattern. Nevertheless, we have an indication that disturbing the correlation between disjunctive motions of the target and the convergence motions of the eyes causes this phenomenon.

These observations show that no depth sensation ensues when the disjunctive motion of the targets cannot elicit corresponding vergence movements of the eyes. Since the ensuring vergence movements of the eyes cannot change the position of the retinal images, there is no reafference change. Vergence movements are pursuit eye-movements and only partly voluntary. On the other hand, the model by Holst and Mittelstaedt applies to voluntary eye-movements. In our case the claim of their model does not hold; it is not the difference between the efference-copy and reafference which is perceived visually. During binocular retinal stabilization the reafferent signal is constant while the subject sends a voluntary command for eye-movement which is stored as an efference-copy and also causes vergence movement of the eyes. Since the perceived depth remains constant during changes in the difference between efference-copy and reafference, the model is not valid for binocular depth perception. It appears that the percept of depth is independent of vergence movements (efference-copy) and depends only on retinal disparity (reafference).

At the same time the vergence mechanism causes perceptual size scaling, regardless of whether the stimulus is normally viewed or under binocular retinal stabilization. Richards (1968) argues for a size constancy mechanism by sensory remapping and proposes the LGN as a plausible site. Prior to him Bishop (1963) pointed out the neurophysiological possibilities inherent in the LGN. There are several psychological and physiological findings that make the LGN an excellent candidate for this size rescaling operation. One is an observation by Richards (1967b) that retinal rivalry depends on the changing image size due to various fixation changes. Since the apparent size change affects binocular rivalry, it can be argued that the size scaling process occurs prior to binocular combination. I discussed this experiment in § 3.5. Furthermore, Bizzi (1966a,b) and Kawamura and Marchiafava (1966) showed that the LGN receives inputs correlated with eye movements. Quite recently Bizzi performed some pilot studies on an unanesthetized rhesus monkey as reported by Richards (1968). The main result was that the activity of "on" units in the LGN slowed down during convergence and increased during divergence. The "off" units behaved oppositely. Feldman and Cohen (1968) reported similar results.

Let me note, however, that Richards' suggestion of the size-zooming process being in the LGN is not supported by some recent psychological evidence. For instance, Harris (1970) studied the effect of convergence on size-scaling by the McCollough's orientation-specific aftereffect. This color aftereffect is strongest when the retinal width of neutral test stripes approximately matches that of the colored adaptation stripes. Harris found that this required identity between adaptation and test stripe-width for optimal aftereffect did not depend on viewing distance. Either the McCollough effect is not cortical but occurs before it and even before the alleged size-zooming site in the LGN (a most unlikely event) or the size-zooming does not occur in the LGN, but is cortical. An alternative hypothesis, that the size scaling occurs in the LGN but that increasing convergence does not enlarge the width of the receptive fields only their length, is improbable since the perceptual size-zooming yields an isotropic perceptual scaling. As a matter of fact, Richards and I studied the effect of convergence on vernier acuity thresholds. In a brief pilot study we found no change in vernier resolution with convergence, which is also contrary to the assumption that the LGN is the site of the size-zooming process, since it is known that vernier acuity depends on the length of the line segments.

The vergence mechanism can be driven both by monocularly perceivable targets and cyclopean targets. The former occurs with less delay. The vergence mechanism that is triggered by monocularly perceivable targets has an additional important dynamic function as shown by Fender and Julesz (1967a). In order to understand this dynamic function we must review some dynamic aspects of vergence. We already noted that in normal vision the symmetric motions of the eyes are not perfectly correlated; this results in a constantly varying amount of binocular disparity that does not yield fluctuations in stereopsis.

The distributions of errors of disparity caused by saccades when a pinhole or a random-dot stereoscopic pattern is viewed binocularly for a period of 2 min are shown in figure 5.9-2. These histograms were obtained as follows: The eye movements of the subject were recorded and transmitted to a computer as described previously. The computer was programmed to identify saccades by noting any displacement of the visual axis of either eye greater than or equal to 3 min arc occurring in a

5.9-2 Histograms of disparity errors caused by saccades when (*A*) a pinhole or (*B*) a random-dot stereogram is fixated for two minutes. (From Fender and Julesz 1967a.)

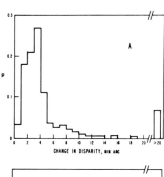

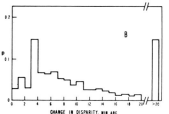

40 msec interval. The corresponding displacement of the other visual axis was then calculated; these two vectors were then combined to give the absolute value of the change of disparity caused by the saccade. This value is displayed as the abscissa in figure 5.9-2. The diagrams show that the probability that any saccade changes the disparity by 3 min arc or more in 40 msec is 0.57 when a pinhole is viewed and 0.89 for a random-dot pattern. (The actual binocular parallax in these targets was 8 min arc, so some of these saccadic changes of disparity may have been purposeful. The probability of a saccade changing the disparity by less than 5 min arc or more than 11 min arc is 0.65.) Some saccades produce changes greater than 20 min arc.

Large binocular disparity changes during saccades may be compensated by a number of processes: the cortical registration demonstrated by the previous experiments may be able to follow rates of change of disparity up to 500 min arc per sec, or the cortical projection of the retina may be rezeroed after each saccade, as suggested by Beeler (1965). The following experiment tested the first of these hypotheses.

Method

Vertical-line targets and random-dot targets were used as in the earlier studies. The targets were carried on linear-motion electromagnetic transducers, and each could be moved horizontally through angles up to 100 min arc. For these experiments the duration of the motion was 30 msec. The motion was slightly underdamped, permitting about 8% over-shoot in the step motion of the target. This is characteristic of the motion of the visual axes during a spontaneous saccade.

Initially the targets were adjusted to fusion by the subject; the transducers were then energized and pulled the targets apart in a temporalward direction through a known small angle. The subject pressed a key whenever he lost fusion. At intervals of 5 sec, the targets were returned to coincidence, remained there for 5 sec and then were moved apart again.

Results with Vertical-Line Targets in Normal Vision

All subjects reported that if only one line was moved through distances smaller than about 15 min arc there was no loss of fusion, but that the fused pattern moved in the direction of the moving line. If both lines moved, the fused line appeared to be motionless. In this case fusion was not lost for motion less than 30 min arc total disparity, but for values of disparity greater than this there was a transitory loss of fusion. The unfused period becomes longer with larger disparity until a disparity is reached at which fusion is no longer possible. These results are illustrated in figure 5.9-3. It is noted that outside of the region in which fusion is not lost there is a range of disparities for which fusion can be reestablished in 1 or 2 sec; but then a sharply defined limit is reached beyond which fusion is not possible.

Results with Vertical-Line Targets in Stablized Vision

The results of this experiment are also shown in figure 5.9-3. Qualitatively, the outcome is similar to the results in normal vision, but the permissible image movement for maintenance of fusion is much smaller. However, it is evident that refusion is always possible within the region of hysteresis reported in the first experiment. The fusion limit for subject *N* in this experiment was 35 min arc; this is consistent with the value of 42 ± 10 min arc recorded and averaged over the subjects especially since the subject was allowed only 5 sec during which the targets could refuse. In this case the limits of the range of fusion were approximately the same for subjects *G* and *N*, although their performances in normal vision were rather different.

5.9-3 Effect of fast pulling of vertical-line targets under normal vision and binocular retinal stabilization. (From Fender and Julesz 1967a.)

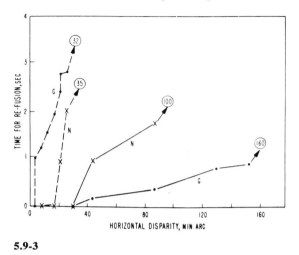

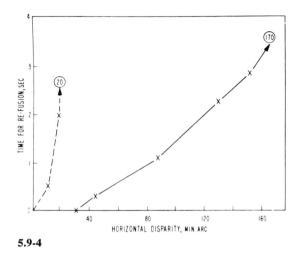

5.9-3

5.9-4

5.9-4 Effect of fast pulling of random-dot stereograms under normal vision and binocular retinal stabilization. (From Fender and Julesz 1967a.)

Results with Random-Dot Targets in Normal Vision

Phenomena strictly analogous to those reported for vertical-line targets were reported when random-dot targets were viewed; there is a transitory loss of fusion and of depth perception but both are reestablished after a short interval. The results are given in figure 5.9-4.

Results with Random-Dot Targets in Stabilized Vision

In this case it was found that pulling the targets rapidly apart, even by amounts as small as 10 min arc total disparity in 30 msec, caused transitory loss of perception of the square. After a short period, the square could once more be perceived and seen in depth; the results are also shown in figure 5.9-4. The fusional limit (20 min arc) is considerably smaller than the limit of the hysteresis effect which can be achieved with slow pulling of the targets (figure 5.8-5).

The problem of steady depth perception in spite of large and sudden disparity changes caused by vergence errors in normal vision can only partly be explained by hysteresis. I have shown that slow drifts of the order of a few min arc/sec can be adequately compensated by the labeling mechanisms. On the other hand, the fast saccadic components of eye-motion cannot all be compensated by this effect alone. The large differences between refusion for normal and stabilized vision (figures 5.8-5 and 5.9-4) show the importance of vergence motion of the eyes in compensating sudden errors. In the light of these results the fusional process depends at least on a convergence-divergence mechanism and on a cortical registration mechanism. The finding that random-dot stereo images can be refused for large shifts in normal vision (figure 5.9-4), without monocular form cues to assist convergence-divergence motions, indicates an intricate interrelationship between these two mechanisms.

5.10 Dynamic Phenomena in Stereopsis

In the previous section I discussed stereoscopic phenomena having a simple dynamic element—a slow pulling of each point by the same amount. The dynamic phenomena reported here are much more complex; they were produced by computer-generated movies in which each frame was a random-dot stereogram.

The movies were directly generated by a Stromberg Carlson (General Dynamics) 4060 microfilm printer on 16-mm film strips, were projected by an adapter that contains prisms and polarizers, and fit ordinary movie projectors. A large number of random-dot stereo movies were generated by Bosche and myself in 1966 and were demonstrated at many scientific meetings. Brief accounts of these movies are given by Julesz (1966 b) and Julesz and Payne (1968). In this section a detailed description of all the stereo movies of this period follows. Some recently generated stereo movies will be reviewed in other contexts in chapter 8. The stereo movies were generated by a computer program developed by Knowlton (1965) called BEFLIX and modified by Bosche (1967). The resolution of each stereo half-field was 180×120 dots. The animation (accompanying text) was also produced by BEFLIX.

Stereo Movie 1

Each frame contained a random-dot stereogram portraying a center square at $+4$, $+2$, 0, -2, -4 picture element (dot) disparities, respectively, in subsequent frames. The random-dot textures were kept identical in successive stereograms. *Results:* The monocularly and binocularly perceived texture appears as static noise. Monocularly a center square is perceived as horizontally moving from left to right and from right to left like a pendulum. The stopping of the movie projector, of course, makes the percept of a center square disappear. When stereoscopically viewed the center square appears to move in and out in depth from the surrounding depth plane. Since the "no-man's land" (an area seen by only one eye) appears to be localized at the farthest depth plane (see § 7.8), the receding square appears as a rectangle carrying with it the two undetermined areas on the left and right sides.

Stereo Movie 2

Each frame contained a random-dot stereogram portraying a center square at 2-picture-element disparity. The random-dot textures were uncorrelated in successive stereograms. *Results:* When monocularly viewed the entire image appears as dynamic noise wherein each dot appears to move in a random fashion. However, when stereoscopically fused, the dynamic noise segregates into two depth planes and a center square is seen in vivid depth before the surround. With reversed disparity the center square recedes behind the surround. Depth is easily perceived, even at 30 frames/sec

—the upper limit for apparent-movement perception. Above this threshold in the simultaneity region, stereopsis is still obtained since the successive frames integrate and remain binocularly correlated as long as the contrast stays above 2%. At very high frame rates when the sum of several hundred uncorrelated images results in a washed-out pattern under 2% contrast so that the patterns cannot be resolved monocularly, stereopsis ceases. I have noted in § 3.10 that under stereoscopic fusion the perceived dynamic noise never crosses the boundaries between two depth planes. Thus proximate dots in successive frames perceived as a single moving dot are proximate in three-dimensional space and not in the two-dimensional space of monocular vision. This is evidence that stereopsis occurs first, and the localized dots are then perceived in apparent movement.

Stereo Movie 3

Same as stereo movie 2 (i.e., it uses dynamic noise), except that the square in depth moved in a zig-zag trajectory. *Result:* When monocularly viewed, the image appears as a uniform array of dynamic noise. When stereoscopically viewed the dynamic noise segregates into two depth planes, and a square in depth with clear boundaries is perceived as moving along an irregular path. This experiment demonstrates that a single frame duration of 30 msec or less is adequate to yield global stereopsis. At higher frame rates the global movement of the cyclopean square disappears.

Stereo Movie 4

Same as stereo movie 2 (i.e., it uses dynamic noise), except that the center square was portrayed by a monotonic sequence of disparity values as in stereo movie 1. *Result:* When monocularly viewed the image appeared as a uniform array of dynamic noise. When stereoscopically fused, a center square appeared to be moving back and forth in depth. Thus apparent movement can be produced in depth as well. This is even more interesting since the texture elements in successive frames are uncorrelated. Each dot of the receding-penetrating square was connected to a proximate dot of the same color in successive frames giving rise to a trajectory in depth and was formed by chance alone.

Stereo Movie 5

Same as stereo movie 2 (i.e., it uses dynamic noise), except that every odd frame portrayed a center square by +2 dots disparity, while every even frame portrayed a center square by −2 dots disparity. *Result:* When monocularly viewed, the image appeared as a uniform array of dynamic noise. When stereoscopically viewed the global movement of the square was in the simultaneity region. Thus two squares were perceived in depth, one behind the other. Although the squares were 100% binocularly correlated, they appeared transparent; one could see through the square in front as well as the second square behind. Because of the alternate parallax shifts, the monocularly perceivable local movement (dynamic noise) was uncorrelated in the center of the left and right views. The occurrence of stereopsis indicates that apparent movement perception must occur after global stereopsis (see also § 3.10).

Similar to stereo movie 2 except that the left images were slowly expanded-contracted in the horizontal dimension by $\pm 20\%$, while the size of the right images was kept fixed. The expansion-contraction followed either a triangular or a sinusoidal time sequence with a period of 1 sec. Twice each period the left and right fields of the stereograms had identical sizes. *Result:* The images could be fused and fusion could be maintained even for the 20% expanded or contracted end states. The percept was either of a center square and its surround expanding-contracting, but parallel to the surface of the projection screen; or one could perceive the two depth planes of the center square and its surround slowly tilting in depth. For this latter percept the tilted planes (corresponding to the $+20\%$ and -20% dilation) appeared to be almost at right angles to each other. This experiment was intended to illustrate the essence of Fender and Julesz's hysteresis phenomenon (1967a) during binocular retinal stabilization, and it did not require contact lenses. The horizontally expanding-contracting square could not be compensated for by vergence movements; fusion could be maintained only by the cortical labeling and label preserving mechanisms.

Stereo Movie 7

Line segments (needles) of 24 possible random orientations were sequentially rotated by $2\pi/24$ angles in successive frames. A center square contained similar needles rotating at twice this speed. The left and right images portrayed this center square with a slight disparity, but the corresponding line segments were perpendicular to each other. (This experiment was performed by Merritt and myself in 1969.) *Result:* When monocularly viewed the center area (composed of the needles that rotated twice as fast as the needles in the surround) could easily be discriminated from the surround as shown by Julesz and Hesse (1970). In spite of the clear boundaries between the center square and its surround, stereoscopic fusion of the left and right images could not be obtained because of the strong binocular rivalry between the corresponding orthogonal line segments. One could have assumed that perceived rotation of the same direction in the two fields may have overcome binocular rivalry. (In the experiment by Merritt and me only one monocular movie and in front of one eye a double prism that rotated the entire display by $180°$ was used.)

In addition to these several other stereo movies were produced and will be discussed in chapter 7.

6 A Holistic Model

6.1 Ambiguous Surfaces in Three-Dimensions

Reversible figures yielding two (or more) stable perceptual organizations have been frequently used in psychology for more than a century. From the first description of an illusory figure by the geologist Necker (1832), "on an apparent change of position in a drawing or engraved figure of a crystal," several variants of the Necker-cube have been devised, such as Schröder's staircase. However, these stimuli were selected from a small repertoire and exploited certain ambiguities inherent in two-dimensional perspective drawings.

In three-dimensional examples reversible surfaces have been limited to a single case, known as the "wallpaper effect." Grid-like structures containing vertical bars of constant periodicity (such as wallpaper, bathroom tile, and old-fashioned radiators) can be fused at multiple depth levels, since the binocular disparity can be any integral multiple of the horizontal periodicity. Such a periodic random-dot stereogram generated by Julesz (1964b) could be perceived as a plane in front of or behind the real plane of the printed page. These periodic random-dot stereograms have been successfully used in studies of perception (to be reviewed in the next two sections) yet are limited in their scope, since only parallel planar surfaces can be portrayed by this method.

Whether ambiguities can be produced in three-dimensions without practical limitations on the organizations to be portrayed is a far from trivial question. Before the success of such an attempt is reviewed, let us note the history of three-dimensional representation.

There are many limitations to portraying our three-dimensional environment on a two-dimensional surface. One of the primary shortcomings is the difficulty of effectively representing the hidden surfaces of objects together with the visible ones. Perspective drawings and even stereoscopic images do not alleviate this limitation. The cubist artists tried to place the hidden surfaces of their objects side by side with the visible ones, but the results were rather confusing. The usual representation by multiple projections solves the problem geometrically, yet it is very difficult to combine these projections into a unified spatial percept. The invention of panoramagrams using lenticular screens by Ives (1928) and multiple lens arrays (fly's eye) by Lippmann (1908) portrayed the three-dimensional objects within a wide angle, but in order to inspect some hidden surfaces the observer still has to move around the panoramagram. Furthermore, certain hidden surfaces (such as the boundaries of inside cavities of opaque objects) stayed invisible. Since holograms (invented by Gabor 1948; and improved by Leith and Upatnieks 1962) are similar to panoramagrams, only more simply made, some change in the geometry of the optical rays has to be initiated in order to obtain various stored organizations. This is usually achieved by the observer when inspecting the hologram from various angles.

This section describes a method of generating ambiguously perceivable stereograms. These stereograms contain several predetermined surfaces, out of which only one can be perceived at a time. Moreover, some surfaces might be internal—that is, hidden from every angle. In order to inspect the various surfaces the viewer can sit still; it is his mind that wanders around the object.

Random-dot stereograms have shifted interest to the problem of how the visual system resolves ambiguities. In several sections we have seen that in random-dot stereograms, hundreds of dots are presented on a horizontal line in the left and right eye's views, and the observer is confronted with ambiguities as to which of the many dots in the left retinal projection correspond to a given point in the right retinal projection. This problem is further amplified for random-dot stereograms, since no monocular familiarity or depth cues are provided that would aid perception. Findings with random-dot stereograms have shown that the visual system selects that organization which yields dense surfaces, and all other possible organizations pass unnoticed.

Would it be possible to create random-dot stereograms which portray simultaneously more than one dense surface? If possible, which organization would be preferentially perceived? The next section will describe a general algorithm which generates a single stereogram portraying two (or more) specified surfaces. That such stereograms can exist is based on the fact that certain areas are seen by one eye only and thus can be freely selected for one surface. Also segments in which the two (or more) surfaces coincide add to the degrees of freedom, since these surfaces can be covered by any random texture at will. In general, there is no restriction on the surfaces to be portrayed simultaneously. Provided that the number of surfaces is restricted to two and the resolution is fine (the number of samples is large) there is enough freedom in choosing the texture elements that the formation of monocularly perceivable short periodicities can be prevented.

6.2 Ambiguously Perceivable Random-Dot Stereograms

The first ambiguously perceivable random-dot stereograms were based on the wallpaper effect (Julesz 1964b). As figure 6.2-1 indicates, two bars, each containing identical random textures (A and B) can be periodically alternated as $ABAB$ . . . in the left view and $BABA$ . . . in the right view. If the width of A and B is less than the maximum limit of disparity (i.e., within Panum's area) then stereoscopic viewing of this display can produce a multiplicity of organizations to be perceived at various depth levels. Such an ambiguous center square is demonstrated in figure 6.2-2* flanked on the top and bottom edges by two unambiguous areas, one in front and one behind the depth of the surround. These unambiguous areas facilitate the depth reversals for the center. It should be noted that these ambiguous stereograms yield quite a stable organization once one is obtained. They are unlike Necker-cubes for which a perceived organization satiates after prolonged viewing and causes the other organization to become dominant. Rather they are comparable to Boring's young-wife-mother-in-law figure for which the first perceived organization can last indefinitely. These two kinds of ambiguous percepts were first recognized by Leeper (1935), and it is probable that the spontaneously reversible one is the less central process. However, the fact that periodic random-dot stereograms yield one stable organization is not so much the result of a highly central process but rather because of vergence. As we will see, stereopsis becomes increasingly difficult with larger disparities. Converging at one depth plane can bias that organization to such an extent that the other organization can never be perceived as long as the eyes remain fixated. With periodic

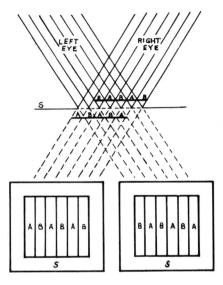

6.2-1 Illustration of how an ambiguously perceivable random-dot stereogram is generated based on the "wallpaper effect."

6.2-2* Stereogram of ambiguously perceivable center square flanked with unambiguous areas in front of and behind the surround.

random-dot stereograms there is an inherent depth difference between the ambiguous planes, therefore the role of vergence cannot be assessed. However, with general random-dot stereograms discovered by Julesz and Johnson (1968a,b), it is possible to portray any surface shapes, even those that stay within the same depth range, and the bias due to vergence can be eliminated.

6.2-3 Illustration of the holistic constraints imposed on the textures of ambiguously perceivable random-dot stereograms. (From Julesz and Johnson 1968b.)

Before the general algorithm of Julesz and Johnson is discussed in detail, a brief illustration is given of the basic idea in figure 6.2-3. Figure 6.2-3 (solid line) shows

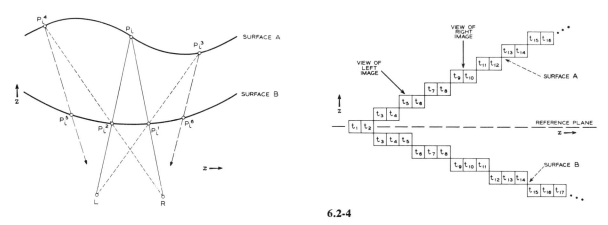

6.2-4

6.2-4 A simple example of two cylindrical surfaces for illustrating the workings of the algorithm. (The cross-section is indicated in the x-z plane.) (From Julesz and Johnson 1968b.)

how points P_i^1 and P_i^2 (belonging to surface B) have to be colored identically to P_i (the original point of fixation, belonging to surface A), in order to obtain the same retinal projections for both surfaces. Figure 6.2-3 (dotted line) shows how this requirement forces P_i^3 and P_i^4 to be colored the same. Figure 6.2-3 (dashed line) shows the next step P_i^5 and P_i^6. As this procedure is continued the algorithm assigns the color of P_i to a gamut of points belonging to the two surfaces.

For a new point of fixation on either surface A or B, then there are two possibilities. Either this point of fixation already has been assigned a color by the previous algorithm or it can be assigned an arbitrary brightness value. In the latter case the above described algorithm is continued which assigns this chosen color to another set of points belonging to the two surfaces. This procedure continues until each point on either surface has been colored. It is interesting to notice that the color of any point depends on some global relationship between surfaces A and B.

In this algorithm, as long as only two surfaces have to be portrayed, each iterative step assigns one new constraint. In the case of three in surfaces, each iterative step generates two new constraints which in turn generate 4, 8, 16, . . . new constraints. This exponential proliferation of constraints drastically reduces the degree of freedom for three or more surfaces. Therefore, in the forthcoming examples we restrict ourselves to two surfaces to be portrayed.

The study of how such ambiguous stereograms are perceived can be of considerable scientific value but also permits some useful applications. It is now possible to portray hidden surfaces of objects together with the visible ones. Since in these ambiguous stereograms only one surface can be seen at a time, multivalued functions of two variables can be portrayed. For instance, if one surface is chosen to be the front view of an object (such as a statue or a machine part) and the other surface is chosen to be the rear view, one can often obtain an entire 360° impression of an object, since the perception of the two surfaces may be alternated at will.

This algorithm thus produces a holistic structure. Any local point influences and is influenced by a gamut of other points. Such holistic structures were first postulated in the monad theory of Leibnitz and should not be confused with holography in the limited sense referring to diffraction patterns produced by coherent light. Holistic structures have this connectivity (dependency) among a large number of elements, forcing identical values on them. Whether these constraints are imposed by the boundary conditions of coherent optics or by cell divisions of embryogenesis or by artificial means such as a computer algorithm is immaterial. They all refer to the same redundancy principle of "equidistributed information mapping" or "equidistribution" (a word Julesz and Pennington [1965] coined in order to draw analogies between holograms and human memory). I dwell on this holistic aspect of the algorithm in order to point out that calling this technique "mental holography" is quite reasonable. It does not refer to the similarity between holograms and ambiguous stereograms in that both can portray visible and hidden surfaces of objects, rather that both have multiple identical points forced upon them, which depend on the whole geometry to be portrayed.

I will now describe the general algorithm. The algorithm is an extension of the technique of random-dot stereograms. The following simple example will give an insight into the workings of the general algorithm. The two surfaces, A and B, to be portrayed by the stereogram are given in the x-z plane in figure 6.2-4 and for simplicity are selected as cylindrical; that is, $z = f_A(x)$ and $z = f_B(x)$, independent of y. Since stereoscopic vision operates on corresponding single rows in the two views, the algorithm given here applies to any surface (not only to cylindrical ones).

In order to construct the stereogram we must specify the textures $T_L(x)$ and $T_R(x)$ for the left and right images, respectively. The right image of the stereogram is selected as the perpendicular projection of figure 6.2-4, while the left image is viewed from an angle of $45°$.

Examine for a moment the case where we have just one figure, $z = f(x)$. We may pick the texture $T_R(x)$ at random; the texture $T_L(x)$ is now basically chosen as follows:

$$T_L[x + f(x)] = T_R(x). \tag{6.2.1}$$

This just expresses the fact that a point x seen in the left image is displaced horizontally by a distance $f(x)$ when viewed in the right image.

There are two necessary qualifications to the above rule:

1. If $x + f(x)$ and $x' + f(x')$ are equal for x unequal to x', in fact only the point corresponding to min (x,x') is seen by the left eye, the other being "in the shadow." The constraint (6.2.1) thus does not hold for the larger of x and x'; we say the larger is obscured in this case.

2. In the event that after applying all the constraints there are some values of $T_L(x)$ not determined by T_R, these values may be chosen at random. (Julesz and Miller (1962) developed these ideas extensively.)

Now to color two figures, $z = f_A(x)$ and $z = f_B(x)$, we apply a similar method; the main difference is that there are more constraints, so that the right image can no longer be chosen at random. The images must represent both A and B; thus, if T_L and T_B represent the left and right textures as above we have basically

$$\begin{aligned} T_L[x + f_A(x)] &= T_R(x) \\ T_L[x + f_B(x)] &= T_R(x) \end{aligned} \tag{6.2.2}$$

for all x. Qualification 1 is still valid, and may serve to eliminate one or both of the above constraints for certain values of x. From this it is also seen that if x and x' are distinct, and neither x nor x' is obscured, then

$$f_A(x) + x = f_B(x') + x' \text{ implies } T_R(x) = T_R(x') = T_L[x + f_A(x)]. \tag{6.2.3}$$

These are also easily seen to be the only constraints on T_R.

TABLE 5. *Projections*

x	1	2	3	4	5	6	7	8	9	10	11	12	13	14
$f_A(x)$	0	0	1	1	2	2	3	3	4	4	5	5	6	6
$f_B(x)$	0	0	-1	-1	-1	-2	-2	-2	-3	-3	-3	-4	-4	-4
$L_A(x)$	1	2	*	3	4	*	5	6	*	7	8	*	9	10
$L_B(x)$	1	2	4	5	7	8	10	11	13	14	16	17	19	20
$T_L(x)$	t_1	t_2	t_4	t_3	t_4	t_6	t_3	t_5	t_9	t_4	t_6	t_{10}	t_7	t_3
$T_R(x)$	t_1	t_2	t_3	t_4	t_3	t_5	t_4	t_6	t_7	t_3	t_5	t_8	t_9	t_4

NOTE: These are the left and right projections of the surfaces given in figure 6.2-4 before and after the constraints.
* Indicates a position that has been "uncovered" by the shifting process—the texture here may be chosen at random if not otherwise constrained.

Once again, qualification 2 is valid; anything not explicitly constrained may be chosen at random.

A simple example of the computational case of this algorithm is given in table 5. Notice that x, $f_A(x)$, and $f_B(x)$ are given in the first three rows. The rows labeled L_A and L_B are formed by putting the integer x into positions $x + f_A(x)$ and $x + f_B(x)$, respectively, subject to qualification 1.

We may read off our constraints directly from table 5: if $L_A(x) \neq L_B(x)$, and neither $L_A(x)$ or $L_B(x)$ is an asterisk, then

$$T_R[L_A(x)] = T_R[L_B(x)]. \tag{6.2.4}$$

After all such constraints have been applied to the right image, the left image can be generated by reference to equation (6.2.2) and qualification 2. Table 5 gives final values for T_R and T_L in terms of randomly chosen texture values $t_1, t_2, t_3, \ldots$.

Let us give a few concrete examples. The interested reader may consult the original papers by Julesz and Johnson (1968a,b) for details.

The demonstration will be quite general and the $1,000 \times 1,000$ dot resolution permits the portrayal of surfaces having complex shapes. The only restriction on the surfaces will be the use of cylindrical shapes. For this case, the algorithm determines the same constraints for each row, but of course within the degrees of freedom each row is independently colored by a random process. For general surfaces each row would have to be computed separately, which would increase the computation time (now about two minutes on a GE 645 computer) nearly a thousand times. The cylindrical surfaces have another advantage; they permit us to use two unambiguous surfaces at the top and bottom margins of the stereogram respectively, to facilitate perceptual reversals for the inexperienced observer.

Besides the $1,000 \times 1,000$ dot resolution there are a few stereograms composed of 128×128 picture elements. In the $1,000 \times 1,000$ dot array each dot (picture element) can take three different brightness values; for the coarser array, each picture element can take eight brightness values. This is done by using three (eight) characters (blank, period sign, degree sign, asterisk, etc.) of the General Dynamics (Stromberg Carlson) 4060 microfilm printer. For the 128×128 array the probability of each of the eight characters is $\frac{1}{8}$. For the $1,000 \times 1,000$ array the probability of using the light and heavy period signs is 0.05, respectively, while the probability for the blank is 0.9. Thus the average number of portrayed dots in these stereograms is 10^5, which is within the resolution capabilities of the printing process.

Figure 6.2-5* shows an unambiguous pyramidal staircase in front of the real plane of the printed page; the picture element resolution is 128×128. In figure 6.2-6, the left and right images of figure 6.2-5* have been interchanged, and the pyramidal staircase is descending behind the printed plane. To view these illustrations use the anaglyphoscopes fastened inside the back cover of this book. Put the red filter over your right or left eye, respectively. If you have good stereopsis, the depth should become apparent to you within several seconds.

Figure 6.2-7* demonstrates an ambiguous stereogram that contains both the pyramidal staircases above and below the printed plane as given in figures 6.2-5* and

6.2-5*

6.2-6

6.2-7*

6.2-6. Both of the organizations can be obtained when stereoscopically viewing figure 6.2-7*, but only one at a time. In a brief study figure 6.2-5* or figure 6.2-6 has been shown to 21 subjects who have never seen these stimuli before. After viewing one of these unambiguous stimuli for a minute, the ambiguous stimulus of figure 6.2-7* has been shown. Ten subjects perceived that organization in the ambiguous stereogram which corresponded to the previous unambiguous organization. Nine subjects perceived the ambiguous stereogram always as the descending staircase, while two subjects always saw the ascending staircase. There was no attempt on our part to train these subjects to learn to reverse the organization.

On the other hand, the reader can easily learn the reversal if he alternates between figures 6.2-5* or 6.2-6 prior to viewing figure 6.2-7*. In figure 6.2-7* the maximum disparity is 8 picture elements. The degrees of freedom are 46 (out of 128). Convergence movements of the eyes can influence the reversals, but it might require careful studies to determine whether reversal could be obtained while the eyes are immobilized.

Particularly interesting is figure 6.2-8*, which has $1,000 \times 1,000$ resolution. Here the upper margin contains the unambiguous surface A, which is a plane with 40 picture element disparity, while the lower margin contains the unambiguous surface B, which portrays a descending wedge having a maximum disparity of -60 picture elements. In this example there is no gap between the ambiguous and unambiguous surfaces. Organization A is very strong, and our everyday experience would suggest

6.2-5* Random-dot stereogram portraying an unambiguous ascending pyramidal staircase in front of the plane of the printed page. (From Julesz and Johnson 1968b.)

6.2-6 Random-dot stereogram portraying an unambiguous descending pyramidal staircase behind the plane of the printed page. View the anaglyph version of 6.2-5* with red filter over left eye. (From Julesz and Johnson 1968b.)

6.2-7* Random-dot stereogram portraying an ambiguous pyramidal staircase either ascending or descending at the will of the observer. (From Julesz and Johnson 1968b.)

193

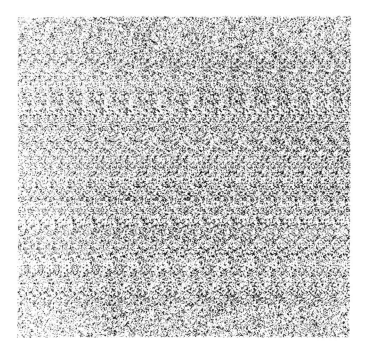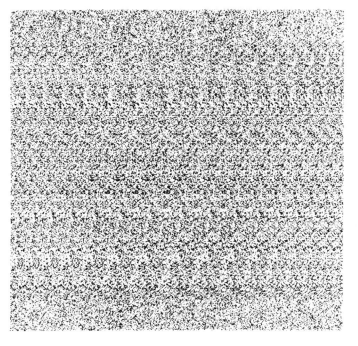

6.2-8*

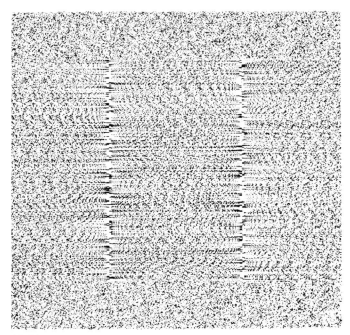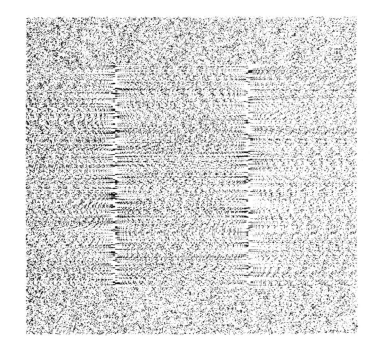

6.2-9*

that when each dot of a front plane is seen by both eyes without any hidden areas present, then this plane should be the only percept. Yet, as figure 6.2-8* demonstrates, it is relatively easy to obtain the other organization, too. Here the degrees of freedom are only 99 (out of 1,000); yet in spite of this relatively small degree of freedom, the image quality is very good.

When we try to portray more than two surfaces, the degrees of freedom rapidly diminish. On the other hand, some of the stereograms with two surfaces yield some additional percepts. For instance in figure 6.2-8* after the front plane is perceived, sometimes the percept of an ascending wedge above the plane can be obtained, too.

Figure 6.2-9* shows a stereogram that can be perceived in many different ways as illustrated in figure 6.2-10. The unambiguous margins correspond to the two shapes

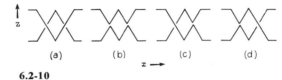

6.2-10

in figure 6.2-10, but the reader might obtain the other percepts as well. Obviously the strongest constraints are obtained for parallel surfaces A and B with a few dots separation in depth. This occurs near the intersection of the two surfaces, yielding visible clusters of dots having the same brightness values.

Figure 6.2-11* portrays a cosine function and a cosine function of lesser ampli-

6.2-8* Ambiguous stereogram with unambiguous margins in the upper and lower areas. Surface A is a horizontal plane in front of the printed page, while surface B is a wedge behind the printed page. (From Julesz and Johnson 1968b.)
6.2-9* Ambiguous stereogram with unambiguous margins in the upper and lower areas. Two slanted planes (above the plane of the printed page) that intersect each other are portrayed. Figure 6.2-10 shows the various percepts which can be obtained. (From Julesz and Johnson 1968b.)
6.2-10 Schematic illustration of the various ways figure 6.2-9* can be perceived. The cross-sections in the x-z plane are indicated. (From Julesz and Johnson 1968b.)
6.2-11* Ambiguous stereogram with unambiguous margins in the upper and lower areas. It portrays a cosine function and a cosine function of lesser amplitude (height) and half periodicity. Both surfaces appear in front of the printed plane. (From Julesz and Johnson 1968b.)

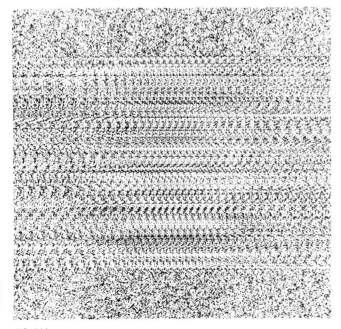

6.2-11*

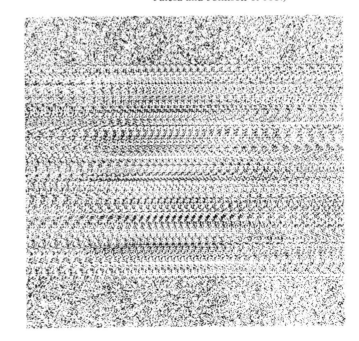

tude and half periodicity. Since both surfaces are in front of the surround and close together, an interesting perceptual phenomenon can be experienced. For the previous demonstrations only one organization could be perceived at a time, and considerable convergence was required to get rid of the prevailing organization and to bias the other organization. In figure 6.2-11* it is possible to retain one organization and, meanwhile, start to perceive the other organization. The perceptual effect is that of a transparent surface behind which another transparent surface is seen. However, this double perception is not a stable state, and it is easier to perceive only one organization at a time. The degrees of freedom are 119 (out of 1,000).

Figure 6.2-12* shows a case in which three surfaces are portrayed. Besides a cosine function in front and behind the surround there is a plane with zero disparity. Interestingly enough for this special case the degrees of freedom are not additionally reduced. As long as the surfaces *A* and *B* are each other's mirror images and the third surface is a plane with zero disparity, it is possible to portray three surfaces without additional constraints. The degrees of freedom are 78 (out of 1,000). An unambiguous rectangle at the top and bottom margin is presented at ±70 picture-element disparities in order to aid perceptual reversal.

From these demonstrations it is possible to see some of the limitations of the ambiguous random-dot stereograms. In order to avoid short periodicities the surfaces to be portrayed should not intersect or come closer than a few picture elements in the *z* direction. The worst case is if the two surfaces are one picture element apart.

6.2-12* Ambiguous stereogram with two unambiguous rectangles in the upper and lower areas. It portrays three surfaces. A cosine function in front of the plane of the printed page, the same function behind the plane, and the printed plane itself (a plane with zero disparity). (From Julesz and Johnson 1968b.)

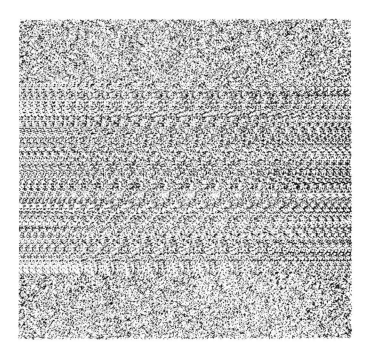
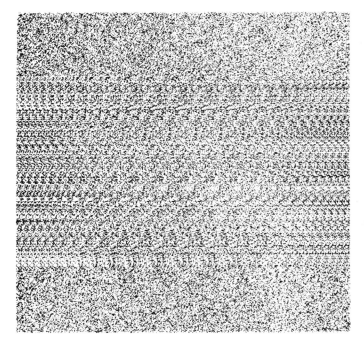

In this event the periodicity is one, and the two surfaces are formed of horizontal lines having the same color. As we discussed, the two surfaces can coincide, but after initial separation they must be spaced in a discontinuous fashion having a jump in depth of several picture elements. In figure 6.2-13*, special care has been taken to separate the two surfaces in depth. Therefore a cosine function (to be seen in front of the plane of the printed page) has been placed on a 10-picture-element-high pedestal. The other surface is a wedge (behind the printed plane). The degrees of freedom are 128 (out of 1,000). Because of this pedestal the shortest periodicity is limited to 10, and the resulting stereogram can be easily fused and reversed.

Another limitation is the rapid decrease in the degrees of freedom as the number of surfaces is three or more. By increasing the resolution of the images, the absolute degree of freedom increases as well, so that three or more surfaces could be adequately portrayed. Unfortunately, the resolving acuities of the eyes limit the size of individual picture elements to about 1 min of arc. With finer image resolution more than one picture element (of different brightness levels) falls on a single receptor of the retina, and the image contrast rapidly decreases.

A third limitation of ambiguous stereograms is the unavoidable fact that because of the constraints there will be some other dense surfaces perceivable besides the selected two surfaces (as pointed out in the demonstration). Since most of these phantom surfaces are perceived at greater depth than the desired ones, there is a way to eliminate them. If the desired surfaces span the depth limits for fusion, then

6.2-13* Ambiguous stereogram with wide unambiguous margins. It portrays a cosine function in front of the printed page and a wedge behind the plane. (From Julesz and Johnson 1968b.)

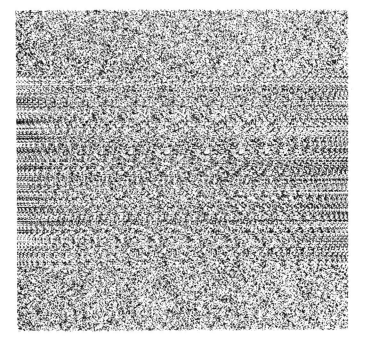
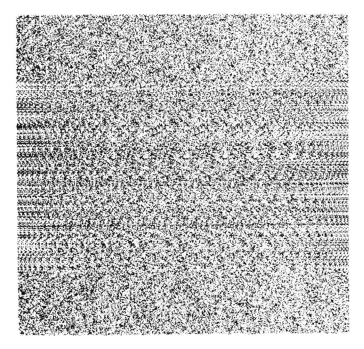

the phantom surfaces will be outside the region of maximum disparity for stereoscopic fusion.

As long as the surfaces to be portrayed are placed such that they stay separate in distance and the number of surfaces is two, the above technique gives satisfactory results. The obtained results are analogous to holography, but only superficially. After all, holograms contain a vast number of stereoscopic views, while ambiguous stereograms contain only a single one. Holography is based on the diffraction properties of coherent wave optics, while our technique uses plain geometrical optics. For holography the observer has to move around the hologram in order to inspect it from various angles, while for ambiguous stereograms the viewer can stand still. It is his mind that wanders around the object. Furthermore, ambiguous stereograms can portray any mathematical surface, including completely hidden ones, from any view; this could be obtained by computer-aided or computer-generated holograms as well, but would require more effort.

Particularly interesting is the example of figure 6.2-11*. Here the two organizations are almost in the same disparity range, and the vergence movements of the eyes have little effect on perceptual bias. It is possible to retain one organization and simultaneously perceive the other organization with both surfaces appearing transparent. This is similar to stereo movie 5 in § 5.10 in which two depth planes were presented alternatingly. It seems that the simultaneous perception of two dense surfaces produces two lacelike organizations. One of the perceived organizations is not stable enough to inhibit every point of the other organization and vice versa. In the stereo movie case, the alternating two stimuli prevent the adaptation to any one of the two percepts, and this labile percept owing to inadequate suppression of the other perceptual state can last indefinitely. However, for ambiguous stereograms this labile state is only transitory; and as the older state becomes satiated, the new state becomes dominant and rapidly suppresses every point in the previous percept. Thus in the absence of vergence the ambiguous stereograms are in the same category as Necker-cubes and other spontaneously reversible figures. Yet the perceptual bias of vergence can permanently suppress one organization if the two surfaces have adequately different disparities.

6.3 Perception Time of Stereopsis

While the perception of random-dot stereograms portraying complex surfaces can take several seconds, the perception of simple stereograms containing only two planar depth planes is on the order of 50 msec (Julesz 1964b). This perception time can be measured by tachistoscopically presenting a random-dot stereogram (such as figure 2.4-2) and erasing (masking) the ensuing afterimage by a second uncorrelated complex stimulus. The quoted 50 msec time is the minimum time between the onset of the stimulus and the onset of the masking field that yields the percept of two depth planes of clearly perceivable surfaces. Obviously this time is not only the perception time necessary for depth but also for the entire image to be seen as a textured surface.

In order to separate the perception time for stereopsis itself from the other processes, the following technique was devised as shown in figure 6.3-1 (Julesz 1964b, 1965a):

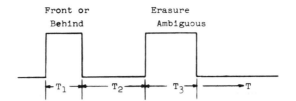

A. T_1 = ?; T_2 = 0-5 msec; T_3 = 50-100 msec

     $T_{1\ min}$ = Perception Time

B. T_1 = Constant; T_2 = ?; T_3 = 50-100 msec

     $T_{2\ max}$ = Attention Time

6.3-1 Stimulus conditions for measuring perception time of stereopsis.

An unambiguous random-dot stereogram (having a center square with either a temporal or nasal disparity) was tachistoscopically flashed and was followed by the presentation of an ambiguous stereogram (having the same disparity but in both directions). The picture elements in the second stimulus differed from those in the first. Thus, the second stimulus erased the afterimages of the first stimulus; therefore the real presentation time for the first stimulus was known. It was found that when presentation time was adequate, the second (ambiguous) stimulus was consistently perceived at the same depth as the first (unambiguous) stimulus. (The unambiguous stereogram was presented with temporal or nasal disparity in mixed order.) Perception of the ambiguous stimulus was influenced by perception of the unambiguous stimulus even when the first stimulus was not consciously perceived. When the first stimulus was presented for a time shorter than this "perception time for stereopsis," or when the onset of the second stimulus was delayed by an interval longer than the "attention time," the second stimulus became independent of the first and could be perceived as having depth opposite the first. This finding and the fact that perception and attention times were typically under 50 msec make it most unlikely that vergence motions might have been initiated. The subjects were unaware that the second stimulus was ambiguous. These facts imply that the first stimulus served as a depth marker and determined which of the possible depth organizations was attended to. Under favorable conditions perception time for stereopsis (depth marker) was as short as 10 msec. This value was obtained for a picture size of 6° of arc (image size, 100 × 100 picture elements; size of center square 48 × 48 picture elements); disparity, 7 min of arc; stimulus brightness, 5 foot lamberts (55 lumens/m²).

The ambiguous stereograms, when presented by themselves for a brief period, were usually perceived in only one way. When a slight (10%) bias was introduced (to counteract the subject's natural bias), it proved to be excessive and reversed the perceived depth.[1] Were the preferred depth level the first given attention and the

---

1. Such an ambiguous stereogram without and with a 10% bias is shown in figures 6.5-5* and 6.5-6*.

other depth levels were searched sequentially only afterwards, then the 90% match at the preferred depth level would be more than adequate for stopping the search. That, instead, the 100% match is perceived at the unpreferred depth level implies certain models for the attention mechanism in stereopsis. One model in agreement with this finding assumes that stereopsis is a parallel process in which each depth plane is simultaneously processed, and the one which contains the most activity is attended to. Such a parallel model was proposed by Dodwell and Engel (1963), who used random-dot stereograms and a technique by Efron (1963) for presenting alternate left and right stereo images with various time delays.

Another model primarily emphasizes vergence. It assumes that the natural bias of subjects is simply due to their habits of converging or diverging first, prior to fusion. Until some stimulus bias is introduced which changes the subjects' preferred vergence direction, the ambiguous stereogram is always perceived at that depth plane which is first reached by the idiosyncratic vergence motions. According to this model the vergence mechanism is triggered by the cross-correlation between the left and right images. For ambiguous stereograms the cross-correlation contains two equal peaks at opposite disparities if the subject fixates at the depth level of the surround. Therefore the stimulus does not determine the direction of vergence, but only the subject's natural bias. However, a slight stimulus bias can increase one of the peaks, and the vergence mechanism tries to bring the two images in alignment around this disparity value. This second model does not account for the fact that the tachistoscopic exposure "freezes" the afterimages on the retinae, which cannot be affected afterward by vergence movements; yet most subjects have a natural bias. This objection can be met by postulating instead of vergence movements of the eye, a cortical shifting mechanism. That a passive cortical shifting mechanism exists has been indicated by the label-preserving mechanism during binocular retinal stabilization and slow pulling (Fender and Julesz 1967a). That an active cortical shifting mechanism operates is indicated by an observation of Derek Fender (personal communication). He claims that if a horizontal scale is presented to one eye and a pointer to the other eye, both under retinal stabilization, the subject can shift the pointer over the scale with increasing mental effort.

However, the serial model based on cortical shifting mechanism is isomorphic to the parallel model, which postulates the simultaneous existence of many neural depth planes, if we further assume some weighting of these depth planes. In addition to assuming that we attend to that depth plane which contains the largest density of points, we also assume a certain individual weighting function according to the observer's natural bias. These two models cannot be differentiated by psychological means, and the neurophysiological evidence of Barlow, et al. (1967) is at least one, if not two levels more peripheral.

Returning to the problem of perception time determination, in the experiments with ambiguous erasing stimuli the physical bias was "tailor made" such that for each subject the ambiguous two organizations were perceived with equal probabilities, provided the first stimulus was shorter than about 10 msec. We have seen that 50 msec was the minimum time to see a random-dot stereogram in its entirety. The difference between these two times—40 msec—is the "portrayal time." If the unambiguous stimulus lasted longer than the "attention time," the depth marker faded out and the test target (ambiguous stereogram) became independent from the adaptation target (unambiguous stereogram).

That the perception time for stereopsis was less than 50 msec was also evident from random-dot stereo movies in which each successive stereogram was uncorrelated. The resulting percept of dynamic noise segregated into two distinct depth planes could be easily obtained at 30 msec frame duration (33 frames/sec). The perceived apparent movement indicated that the successive frames partly erased the previous frames. In order to raise the movie speed and still remain in the best apparent movement region, the resolution of the stereograms had to be increased, too, which was beyond available technology. Otherwise the increased presentation rates produced simultaneity (the integration of several successive stereograms), which did not produce dynamic noise, and of course resulted in good stereopsis.

These very brief perception times are in contrast to the much longer perception times necessary for stereograms portraying complex surfaces, such as the two intersecting ellipsoids shown in figure 6.3-2*. It is interesting that an a priori knowledge of the expected shapes (even by verbal description) speeds up perception time considerably. It seems that this advanced knowledge helps shift attention to areas in various depth planes and by so doing, the search time for finding all the corresponding points is greatly diminished.

6.3-2* Random-dot stereogram portraying two intersecting transparent ellipsoid surfaces.

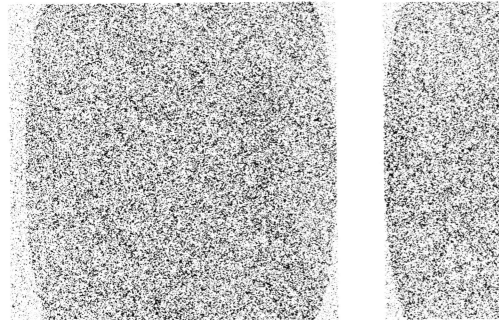
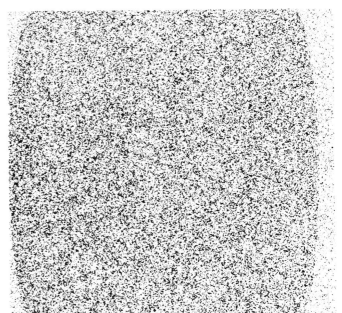

6.4 Effect of Increased Disparity

We briefly quoted an unpublished finding by Fender in which a retinally stabilized scale in one eye and a pointer in the other eye could be shifted from the fixation point with increased mental effort. Without contact lenses it is very difficult to reproduce this finding since with afterimages the binocular rivalry interferes with the observation. However, there are indirect methods to show how increased disparity requires increased effort. Random-dot stereograms permit an easy way to quantify this relationship.

Instead of the loose term "mental effort," we shall introduce the "minimum corresponding area size" that yields stereopsis for a given disparity value. In figure 6.4-1* a square of 3×3 picture elements is portrayed with 1, 2, 4, 6, and 8 disparity values, while in figure 6.4-2* the same disparity values are used, but the square size is 5×5

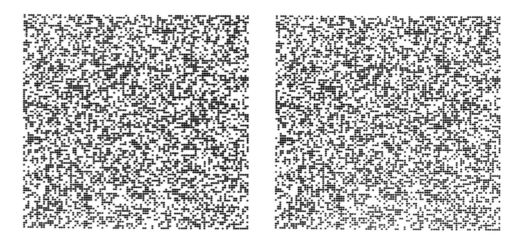

6.4-1* Random-dot stereogram containing 3×3 squares with 1, 2, 4, 6, and 8 picture element disparities.

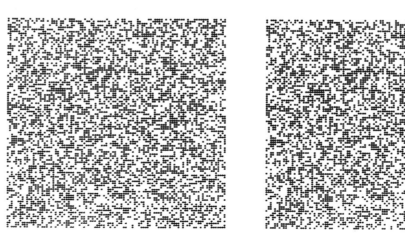

6.4-2* Random-dot stereogram containing 5×5 squares with 1, 2, 4, 6, and 8 picture element disparities.

picture elements. The reader can verify that for the smaller square only the smallest disparities will yield stereopsis, while for the larger square the disparity threshold for depth sensations is larger. There is an asymptotic disparity value regardless of area size above which no stereopsis can be obtained. If one would plot the minimum area size that yields stereopsis as a function of disparity, a bowl-shaped curve can be obtained. The minimum necessary area size is an indicator of the mental effort to be overcome. The exact quantification of this function is a very laborious task, but it might become necessary for testing accurate models of stereopsis. Work in this direction has been started in an M.S. thesis by Disbrow (1964). Sperling (1970) called this bowl-shaped curve "stereopsis energy." A similar curve with orders-of-magnitude larger disparity range (having a minimum at a fixation point determined by corresponding accommodation) can be postulated for vergence as proposed by Sperling (1970). In the absence of stimuli to help vergence motions, the necessary mental effort to converge or diverge from this neutral position Sperling calls "vergence energy." Vergence force is a derivative of this function. The similarity between the left and right retinal images is the image-disparity energy, derived by adding the difference between the two images over the whole retina with certain weightings. This image disparity energy (somewhat differently defined for stereopsis than for vergence) represents the external factors, while vergence and stereopsis energy is determined by internal factors. The image disparity energy adds perturbation to the bowl-shaped energy curves. The local minima of the sum of internal and external energy curves are the stable conditions of stereopsis and vergence in Sperling's model. A simple vertical line adds a different perturbation to the internal energy curves than a complex stereogram, which is in accordance with psychological evidence.

My basic objection to this model is the lack of an inherent structure. This is a question of taste and is discussed next.

6.5 A Model of Binocular Depth Perception

A model is not identical to the structure it wants to describe but is isomorphic to those features of the structure that are already known. The value of a model is the extent to which it is better matched to the heuristics of human thinking than is the original structure itself, or it is the extent to which it is more similar to structures that are familiar to us than the original evidence. Let me elucidate this statement using an ingenious example by Donald Michie as quoted by Shepard (1964): Two players alternately draw single chips from a set initially containing 9 chips (visibly numbered 1 through 9). The winner is the first player who possesses three chips that total 15. In this abstract number-theoretical form, people play the game rather badly, mastering it very slowly. Surprisingly, this game is formally isomorphic to tic-tac-toe (when we realize that a 3×3 magic square exists—[6,7,2], [1,5,9] and [8,3,4] are the rows—in which the rows, columns, and diagonals total 15), which people can master after a few trials. Thus, of two isomorphic structures only one is well matched to human capabilities. If the tic-tac-toe game is chosen as the model and the number-theoretical version as the original structure, one gets a much deeper insight into the game from the model than from the original structure.

Usually the correspondence between reality and model is not so complete as in the previous example, and of the many possible models we select that one which exhibits the greatest similarity in structure to the real phenomenon. However, if two models are isomorphic to each other it is wise to select the one which is most familiar as the canonic representation. Should this inherent familiarity permit valid predictions which are only implicitly present in the model, then the model goes beyond a convenient mnemotechnique; it becomes a useful scientific tool.

The simpler and more primitive a model is, the better it appeals to our intuition, and the deeper is the insight it provides. In the field of electricity or in acoustics it used to be popular to postulate mechanical or hydraulic models. Only as we became more familiar with electricity or acoustics did the use of these models recede. This newly acquired insight may reverse the roles, and an electronics expert often uses his familiarity with this field to portray mechanical or hydraulic phenomena by electrical circuits.

The model I propose for binocular depth perception uses the most familiar mechanical parts and phenomena such as springs, masses, frictions, rotations, displacements, and forces; for "remote interaction," magnets are used. The most important feature of the model is its ability to show the intricacies of local and global interactions on several levels. It exhibits many features of cooperative phenomena which the reader can follow without needing complicated mathematics. The basic structure of the model is as follows:

Imagine tiny magnetic dipoles (compass needles) suspended by balljoints at their centers so that the needles can rotate in any direction in or out from a plane. This plane contains a large array of these dipoles and will be called the cortical mapping of the retina, while the dipoles correspond to binocular units.[2] We will assume that stimulating the retina with black and white dots is equivalent to the magnetic dipoles (D_i) being oriented with their north (N) or south (S) poles turning out from the plane. After the magnetic dipoles assume the pattern of the stimulus, the tips of all adjacent dipoles (D_i and D_j) are coupled with two springs (k_{ij}). The dipoles are coupled with their left and right, as well as their top and bottom neighbors. The coupling (spring constant) between horizontal neighbors can be stronger than that between vertical neighbors. A similar ordering of the dipole array occurs as a result of stimulating the other retina.

Imagine the two ordered-magnetic-dipole arrays one behind the other in exact registration. If the left and right stimuli are identical, then the N-poles and S-poles of the corresponding dipoles interlock. This is illustrated by figure 6.5-1A in one-dimension showing a row of interlocked dipoles in the left and right arrays. Although the dipoles are able to turn in any spatial angle, only their horizontal rotation is sensed as depth. The maximum-sensed rotation angle corresponds to the limit of local patent stereopsis. For identical patterns when the arrays become aligned, all the correspond-

---

2. We will see how the dipoles in the left and right magnetic dipole arrays interact with each other. It is these interacting dipoles between the left and right arrays that correspond to cortical binocular units.

ing dipoles interlock and the integrated attraction force between the arrays is very large. The interlocking of two dipoles will be called local fusion, while the set of all interlocked dipoles will be called global fusion.

If the left and right pattern are uncorrelated, then the attraction between the dipole arrays is very weak, if not zero. Since 50% of the dipoles interlock owing to chance alone, they force their noninterlocked neighbors to face a dipole of the same polarity (see figure 6.5-1B). Because of this, 50% of the dipoles in the two arrays repulse each other, thus the net attraction between the arrays is zero. Whenever two dipoles of the same polarity are forced to face each other, we speak of local binocular rivalry. Thus uncorrelated complex patterns exert no attraction (global fusion) for each other. This also applies for two identical patterns that are far out of registration. They must be brought within the fusional limit (Panum's area) in order for the corresponding dipoles to be close enough to each other to turn and interlock.

Thus, identical dipole arrays must be brought within each others' near fields in

6.5-1 Spring-loaded dipole model under various conditions: (A) entire array interlocked (fused); (B) uncorrelated arrays 50% attract, 50% repulse; (C) interlocked arrays within the fusional limit; (D) surround interlocked (fused); (E) after interlocked surround the center area is aligned and becomes interlocked; (F) refined model in spring-suspended frame; both surround and square have interlocked dipoles.

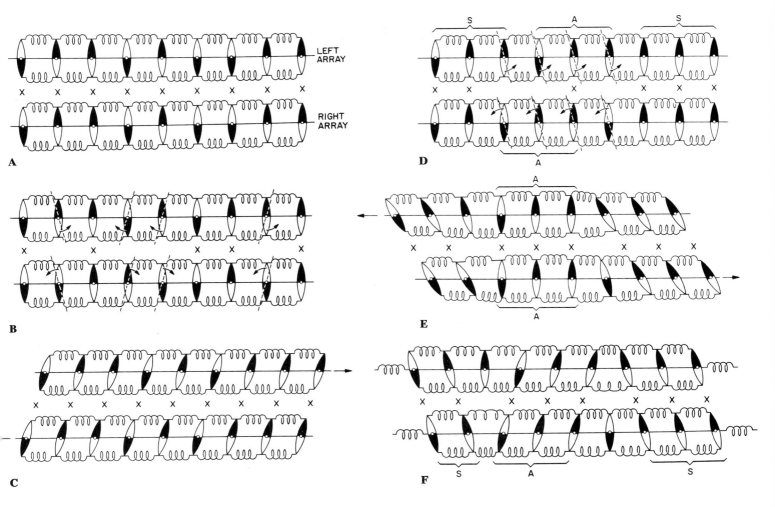

order to interlock, but after this they can be pulled apart by large distances without causing the dipoles to unlock, as shown in figure 6.5-1*C*. Because of the spring coupling, the interlocked dipoles act as a single hyper-dipole pair and turn together so that all the corresponding dipoles keep facing each other. Therefore, the left and right identical (or very similar) dipole arrays exert a large attraction force on each other even at distances that are orders of magnitude larger than the attraction between two dipoles, provided the number of dipoles is large. This is the essence of the model.

Let us follow the workings of the model: First, by some random shift we align certain similar dipole arrays which then interlock. But after these arrays are interlocked we do not have to concern ourselves further with them. We can search for other similar dipole arrays by performing horizontal shifts in the nasal or temporal direction. The already interlocked clusters of dipoles will remain interlocked. This procedure is shown for the case of a simple random-dot stereogram with a center square having different (nasal) disparity from its surround. If, owing to a chance shift, the surround is first aligned, the corresponding dipoles interlock, yet in the center area only 50% of the dipoles interlock while the other 50% repel each other, as shown in figure 6.5-1*D*. If the arrays are now shifted in the nasal direction in order to align the dipoles in the center area, the interlocked dipoles of the surround will turn and remain interlocked. However, the dipoles in the center area will not exert any global force during the pulling. Only after they are brought into semialignment (Panum's area) do they also interlock (see figure 6.5-1*E*).

Since stereopsis is the sensation of relative depth, we postulate that only differences between dipole rotations contribute to stereopsis. Thus, it is the same whether dipoles in the surround rotate while those in the center do not, or vice versa; that is, the percept does not depend on the sequence of alignments. This perceptual invariance can be incorporated easily in the model by some modification so that the dipoles occupy the same configuration independent of the past history of matching procedures. This modified model is shown in figure 6.5-1*F* and will be discussed later in detail.

As long as the disparity difference between adjacent areas is within the fusional limit, no vergence movement is necessary, and the different depth planes can be perceived in a brief flash (or under binocular retinal stabilization). However, if the disparity between square and surround is larger than the fusional limit, only one organization can be perceived in a tachistoscopic illumination, namely the one at which the observer converged. But what happens if one aligns the surround first and then changes his fixation and aligns the central square? When the surround is aligned, all its dipoles interlock and exhibit a very large attraction between the two arrays, while the dipoles of the center square partly attract and partly repulse each other with a net attraction force of zero. The arrays are then shifted with respect to each other (which corresponds to vergence motion of the eyes) such that the center square becomes aligned and its dipoles interlock. Because the surround has already been fused, the attraction between the corresponding dipoles in the two arrays and the spring coupling between neighboring dipoles causes the whole ensemble of dipoles in the surround to turn by the same amount, keeping all the dipoles interlocked. Thus when the center square aligns and its dipoles interlock, the dipoles of the surround remain interlocked, too. In turn it is now possible to have the surrounds aligned again without losing the interlocked state for the square. However, if the eyes are closed long enough, some interlocked dipoles may separate due to internal noise, and yet the surround may remain interlocked. The square may again interlock upon vergence movements. Of course, without the spring coupling between neighboring dipoles, no hysteresis phenomenon of this kind could have occurred. Without such a coupling, a set of dipoles would behave similarly to a single dipole. But if the dipoles are simultaneously coupled, then a previously obtained state becomes much more stable than it is for independent dipoles.

Although qualitatively the model could also explain the stability of fusion through $2°$ of slow pulling under binocular retinal stabilization, it might simplify matters to incorporate another degree of freedom in the model. Imagine that the two dipole arrays are suspended in a frame by horizontal springs (s_H) which permit a $\pm1°$ horizontal displacement (to compensate for the $2°$ hysteresis effect). This final model is illustrated by figure 6.5-2. As the two dipole arrays are brought into alignment and

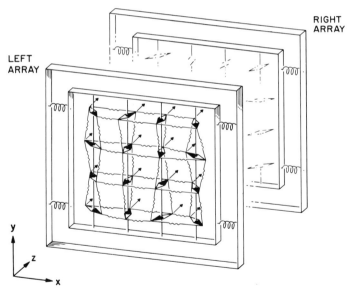

LEFT ARRAY

RIGHT ARRAY

6.5-2 Refined dipole model in which the dipole array is in a spring-suspended frame.

interlock, the attraction between them becomes so great that even after slow pulling, the two arrays not only remain interlocked but also aligned. This is achieved by displacing the arrays within their suspension frames. Since the arrays can move by only $1°$ each (totally by $2°$) before reaching the edge of the frames, the alignment of the arrays will cease at this point and with further pulling the dipoles suddenly unlock. This is the breakaway point. It is clear from the model that the two arrays must be brought in coarse alignment again for the dipoles to interlock. Without a great number of interlocked dipoles it is impossible to counteract the opposing forces of the springs (s_H).

If we further assume that in addition to the s_H resilience, the springs have an inertial mass (m_H) and a friction (r_H), it is easy to visualize what happens if the two arrays are suddenly pulled apart or a shield is briefly placed between them (corresponding to the occlusion of the eyes). Because of the m_H inertia a sudden pulling of the frames is not dampened by the springs, and the full force of the pulling acts on the arrays moving them apart beyond a critical distance. Once the dipole arrays are beyond this distance the attraction force between the interlocked dipoles is not adequate to keep the arrays aligned. If the separation is bigger than Panum's fusional limit then the dipoles will quickly unlock, owing to inherent noise (e.g., neural noise).

The finding, illustrated in figures 5.9-3 and 5.9-4, that binocular fusion is momentarily lost after fast pulling but returns after a few seconds, gives further insight into the workings of the spring-coupled dipoles and their parameters m_H, s_H, and r_H. The transient response of a simple oscillator depends on a single parameter $\alpha = ms/r^2$. If $\alpha > \frac{1}{4}$ (i.e., α is larger than the value for critically damped condition) then the time response will alternately overshoot and undershoot the final response state. If $\alpha < \frac{1}{4}$ no such oscillation takes place. For coupled oscillators the treatment is much more complex than for the single oscillator, yet the same oscillatory and nonoscillatory states are obtained. For the binocular fusion experiment under fast pulling it is necessary that the transient response of the needles be oscillatory in order to obtain refusion. Without such oscillation the end points of the corresponding needles would constantly drift away, and after a critical distance is reached the needles could not interlock again. However, if the needles rotate in an oscillatory manner, after unlocking, they might still return to their original positions or even beyond that so corresponding needles might point towards each other and interlock again (in spite of the fact that the left and right dipole arrays are at some distance from each other).

Occlusion has similar effects. If the two interlocked arrays are pulled apart by more than Panum's limit and a magnetic shield is placed between them, then the springs start to pull the arrays apart correspondingly faster the more they are loaded. The larger the original separation between the arrays is, the more the springs are loaded. Thus the heavily loaded springs can pull the arrays further apart during a given occlusion time than can the lesser-loaded springs. Therefore, the larger the original separation is, the shorter the occlusion time must be in order to preserve interlocking.

The spring suspension in a frame provides another advantage. In the simpler model we had to use an additional rule in order to make the surrounds in figure 6.5-1D and figure 6.5-1E identical. However, with the spring suspension, both cases become identical as depicted by figure 6.5-1F. Since the two arrays can move horizontally with respect to each other, they tend to occupy a relative position, which is the first moment of the interlocked-dipole orientations. If the square had the same number of dipoles as the surround, then the position between the centers of the two arrays would be at an angle, half way between the orientation of the two dipole organizations. Thus with effort (vergence) one could obtain the states of figure 6.5-1D or 6.5-1E, but the internal equilibrium would be the state of figure 6.5-1F.

This model can be further refined both in its structure and in its practical realization. The magnetic dipoles could be thin electromagnetics whose polarities might be determined by retinal stimulation. The retinal states are not restricted to black and white but can take intermediate values, corresponding to various magnetic dipole charges (dipole strength). The separation of the dipole array and the strength of the dipoles may increase at the periphery. The frames themselves could be suspended in a second common frame by springs introducing a force which must be overcome by vergence, and this could be regarded as the derivative of vergence energy. Accommodation might be introduced that would vary the density and size of the dipoles and might be coupled with the vergence mechanism.

A further refinement of the model may consist in choosing the dipoles not of a circular cross section but of elongated ellipses in all possible orientations and shapes corresponding to the various receptive fields of given orientations and spatial frequencies. Then only dipoles whose cross sections are similarly oriented will interlock, while those whose elongated cross sections are orthogonal will not.

Readers who have greater familiarity with electronics might prefer replacing magnetic dipoles with coupled oscillators having certain eigenfrequencies. The shifting of frequencies as a result of coupling could correspond to the rotation of the dipoles. Yet for the majority of readers, I believe the mechanical-magnetic model is preferable. Tangential to this is Sperling's model in which little balls roll down on bowl-shaped energy curves and may or may not fall into holes caused by various stimuli. (The ordinate is vergence or disparity energy; the abscissa is vergence or disparity in minutes of arc.) The model is a combination of mathematically defined curves and kinematic rules. The cooperative nature of the phenomena is hidden in the various correlative functions that define energy. The advantage of the magnetic dipole model is its portrayal of cooperative phenomena without mathematics. In a way the model is an analogue computer. To define the dipole model mathematically, one would certainly face grave problems.

The first model of cooperative phenomena was proposed by Ising (1925) in order to explain long term order, the sudden ordered states in ferromagnetic material at a certain energy level (Curie point). In this simple model only adjacent neighbors of dipoles were doubled and could have only two local states: parallel or antiparallel orientation. The coupling monotonically decreased with temperature but was such that it favored neighboring dipoles to line-up parallel. (Owing to the Pauli [exclusion] principle of quantum physics, the neighbors had a bias to assume antiparallel organization.) Ising conjectured that in his simple model there would be a critical temperature (coupling energy) under which the randomly oriented dipoles would suddenly form large domains of the same orientation. Theoretical physicists spent decades mathematically proving this simplest possible model. It turned out that for one-dimensional dipole arrays, no long-term order exists; only for two-dimensional or higher dimensional dipole arrays does it exist. Why the coupling with a left and right neighbor is not adequate for long-term order, but that an upper and lower neighbor must also exert influence, is beyond simple insight and requires great sophistication in topological thinking. Somewhat more realistic models of ferromagnetism are too complex for present mathematical techniques. The interested reader may consult the review articles by Newell and Montroll (1953). In my model the dipoles can have all orientations; in addition to their neighbors, they are also coupled with dipoles in another array. Nevertheless, my dipole model is much simpler than any Ising model of ferromagnetism. After all, the ordering of dipoles is accomplished by an external mechanism. It is simple to imagine the workings of this model, and it may not be very difficult to build such a model of a small dipole array in order to clarify points where intuition fails. For example, a few hundred tiny magnets could be suspended in gelatin, which might provide coupling between adjacent dipoles and also permit rotation. It remains for the reader to invent better models or realizations. Some attempts have been started in my laboratory to computer-simulate this model.

Let us test the model on ambiguous stereograms whose surfaces are far apart. In this case a large movement can cause the alignment of the dipole arrays such that it favors one or the other organization. The locking of all the dipoles according to one organization maintains this organization until a sudden vergence motion (which during global fusion requires great effort) breaks the organization; if vergence happens to bring the other organization into alignment the new organizations interlock. If the ambiguous stereograms contain both organizations within Panum's area, then no vergence motion is necessary, but a small cortical shift biases one of the organizations. This means that in addition to the vergence forces whose analogs act on the frames and pull them together or apart, a cortical shifting force is acting as well. Whether the analog of the cortical shifting force acts indirectly on the frames or directly on the arrays is an interesting question. It depends on the time constant of the cortical shifting force. If this force can act as suddenly as vergence, then it makes no difference whether it acts on the frame or on the array since the m_H inertia of the springs transmits the transient unattenuated. One can assume that a cortical discharge is at least as fast as vergence, which requires muscle contractions, and therefore we can assume that the analog of the cortical shifting force acts also on the frames.

Another phenomenon on which we will check the model is the difference between normal and binocularly stabilized vision in maintaining fusion during fast pulling. Can the model explain why during fast pulling for normal vision, fusion is maintained for large displacements, but it cannot be maintained for stabilized vision? First, we must realize that global stereopsis corresponds to the spring-suspended array together with the frame, while the retinae correspond to the dipole arrays themselves. During fast pulling under retinal stabilization the dipole patterns are displaced at such a rapid rate that the springs cannot compensate for this displacement. With slow target displacement under stabilized vision, as we already discussed, the springs permit the two arrays to remain interlocked. For fast target pulling under normal vision, however, the vergence mechanism can lock some contoured features of the moving target and can assure that with respect to the retinae the projections of the target are stationary. As long as the targets have monocularly perceivable contours, the vergence mechanism remains fixated on targets of considerable velocities. If one assumes that the vergence force is moving the frames in a position determined by the first moment (center of gravity) of the retinal patterns (dipole patterns), then a moving pattern remains positioned on the arrays.

For random-dot stereo movies in which dynamic noise portrays both the center square and its surround and the square moves in depth, no monocular contour drives the vergence mechanism. As long as the moving square stays within fusional limits one can follow the movement in depth. However, if the square with changing disparity values surpasses Panum's limit, one has difficulty fixating on the moving square, particularly if this movement follows an unpredictable trajectory. However, in most situations a monocular feature of movement can drive the vergence mechanism. This is even true for random-dot stereograms as long as their textures are static and not dynamic noise. I have not indicated in the model how the center of gravity of the dipole patterns can carry the frames with it. This is assumed to be part of the vergence forces.

For binocular depth perception (i.e., stereopsis, vergence, and accommodation) this two-stage model may suffice as a first approximation. It incorporates the difference field model as a special case, but the dipole model is more general than the difference field model and its essence is the coupling between adjacent dipoles. Without coupling, as exemplified by the difference field model, it was found that inhibition (subtraction or division) as a combining operation was more explanatory than facilitation (addition or multiplication). However, with coupling between adjacent elements a model based both on facilitation (attraction between corresponding magnets) and inhibition (repulsion between noncorresponding magnets) can explain most of the observed facts. This is more in agreement with neurophysiological findings. For instance, Pettigrew, Nikara, and Bishop (1968) studying binocular interaction on single units in cats' striate cortex found disparity sensitive units with a sharp peak at optimum value. For this optimal setting, most units showed facilitation or summation of the monocular responses with the minority showing occlusion or inhibition. When disparity changed from optimum value, binocular occlusion could be demonstrated in all units.

It should be noted that experiments, in which one image of a stereogram is expanded by 15% or rotated by 6° without loss of fusion, can be explained by the model. Since the dipoles are suspended in ball joints they can stay interlocked with another expanded or rotated array as long as the solid angle in which they must turn is within fusional limits. However, if the arrays are initially interlocked, much larger expansions-contractions or rotations can be tolerated (see, for instance, stereo movie 6 in § 5.10.) Indeed, if the dipoles are already interlocked, then a slow expansion or rotation will only cause the interlocked dipoles to turn, but they will still face each other. The global interlocking collapses only if the expansion or rotation increases the distance between interlocked dipoles so much that the attraction between them is not adequate to counteract the pulling of the coupling springs. As is expected, dipoles of the periphery first will unlock, but through the coupling springs they will quickly pull on the dipoles in the center areas and unlock them.

As noted, magnetic dipoles and cooperative phenomena have been linked for decades since Ising's proposal. An interesting explanation of cooperative phenomena, long-term order, and Ising models was given by Cowan (1965). It should also be noted that Cragg and Temperley (1954) already constructed a model for neuronal assemblies based on an analogy between the cerebral and the ferromagnetic structures. The novelty of the magnetic dipole model of stereopsis is in its particular construction and not that magnetic dipoles are used.

A few phenomena, however, cannot be explained by the model and require further elaboration. For instance, during voluntary eye-movements the patterns on the retinae change rapidly, yet the patterns do not appear to change position or depth. We have already discussed the afference-copy model of Holst and Mittelstaedt (1950) and noted that it is basically Helmholtz's "effort of the will." Teuber's "corollary discharge" is a neurological hypothesis for the same translational-constancy process. In my model it must be assumed that the difference signal between reafference and efference copy of disjunctive voluntary eye-movements acts as a force on the frames of the dipole arrays. This force then compensates for any pattern shift on the arrays, thus the interlocked dipoles remain invariant. That translational constancy is not inherent in the model is acceptable, since the model is about stereopsis, while translational invariance is a general phenomenon in visual perception and concerns retinal-cortical representations.

On the other hand, there is a phenomenon in stereopsis that is not inherently evident from the model. Vertically expanding one image of a random-dot stereogram pair (while the other image is kept the same size) produces the percept of tilted depth planes after fusion. But the tilt is opposite to the horizontally expanded case. This is shown in figures 6.5-3* and 6.5-4*. In figure 6.5-3* the left image of the basic random-dot stereogram of figure 2.4-1* is horizontally expanded by 15%, while in figure 6.5-4* the left image of 2.4-1* is vertically expanded by 15%. An obvious "explanation" might be to assume that a two-dimensional zooming process tries to make the vertical extent of the stereo pairs identical and in so doing contracts the horizontal dimension of the left image by 15%. This contraction in the horizontal dimension indeed produces the observed tilt.

Unfortunately the magnetic dipole model cannot explain the perceptual findings during expansion-contraction. Even the uniform two-dimensional expansion is not adequately explained by the model. As shown in figure 2.8-8* the left image of figure 2.4-1* is expanded in both dimensions by 15% with the centers aligned. Interestingly, the fused percept is the same as if no expansion had been produced; the surround and the center square lie in two parallel depth planes. From the dipole model one would expect that owing to expansion the magnets would have to turn by increasing amounts when going from the center to the periphery. This would result in the percept of bowl-shaped surfaces instead of the planes that are actually perceived. It is hard to judge whether the size constancy mechanism should be incorporated in an external model or should be an inherent property of a good model of stereopsis. An external zooming model that tries to keep the vertical dimensions matched to the left and right retinal images is one solution. However, an internal zooming model might also be proposed. Suppose that the dipoles are not suspended in a rigid array, but in a rubber sheet. In the case of uniform expansion in two dimensions of the left array, the interlocked dipoles at the periphery—owing to their spring loading—try to become parallel to their neighbors closer to the center. This force will expand the right rubber-dipole-array by the same amount. Thus in the case of figure 2.4-1*, both the center and the surround will have parallel interlocked-dipoles. It is less clear why a vertical expansion will cause horizontal contraction, although it is possible to visualize how the expansion of a rubber sheet in one dimension may result in shrinkage in the perpendicular direction.

In spite of the versatility of the dipole model, there are a few phenomena for which the difference-field model (AUTOMAP-1) is more successful. The main limitation of the dipole model is that search for fusion is a serial process, whereas we have seen in § 6.3 that stereoscopic localization is probably based on a parallel search process. In ambiguous stereograms areas that contain the most dots of the same disparity are

6.5-3*   Random-dot stereogram in which one image is expanded in the horizontal direction by 15%. The planes of the surround and center square appear tilted.

6.5-4*   Random-dot stereogram in which one image is expanded in the vertical direction by 15%. Tilt of planes is opposite of figure 6.5-3*.

usually fused first, while areas with the least points are fused last. This is easily accounted for by a parallel model—for instance, the difference-field model—but is very difficult to incorporate in a serial-search type model.

This finding is so important for the model of stereopsis that the argument for a parallel search process will be verified by two experiments. In figure 6.5-5* an ambiguous random-dot stereogram is shown. When such a stereogram is presented in a brief flash (under 100 msec duration) a center square is seen in depth either in front or behind but only in *one* way for a given subject. This natural bias can be overcome by a slight physical bias of 3–10% depending on the subject. For instance, figure 6.5-6* is similar to figure 6.5-5*, except that one organization contains only 90% of fusible dots whereas the other has 100%. This 10% bias is probably enough for most readers to overcome their natural bias during a tachistoscopic flash (Julesz

6.5-5* Ambiguous stereogram that contains a center square either in front of or behind the surround.

6.5-6* Ambiguous stereogram, similar to figure 6.5-5* except that one of the depth organizations is biased by 10%.

1964b). Since 90% fusion is adequate to stop further search and yet the visual system seeks the 100% solution, one has to conclude that a parallel process evaluates the cross-correlation between various binocular disparities and selects the disparity value which results in a maximum fit.

One way to improve the dipole model is to add a preprocessing stage that affects the direction of the horizontal shifts. For instance, the cross-correlation between the left and right retinal projections is computed. If the maximum value of this cross-correlation can be obtained by a nasal shift or a temporal shift this will initiate a corresponding cortical shift. If the maximum cross-correlation occurs at zero shift, no shift takes places. Such a bias caused by a cortical shifting mechanism can pre-select the largest areas to be fused first, and the remaining areas may be fused by a nasal-or-temporalward shift that aligns them owing to chance. This global cross-correlation can be a separate mechanism from the dipole arrays. However, one could incorporate it in the dipole-array model by postulating an initial quick shift that sweeps through the entire disparity range. If this sweep is faster than the time constants of the dipoles, the dipoles will not interlock, yet the average of their slight clockwise or counterclockwise turn will measure the cross-correlation between the left and right arrays.

6.6 Extension of the Model to Perceptual Learning

For more complex phenomena, including cognitive levels, one could try multiple-stage models. The first stage of local percepts could be represented by an array of coupled dipoles. This array could be suspended in a frame by springs representing the second stage. However, several such frames could form an array of frames which would be suspended in a hyperframe by springs. Each hierarchy of frames would correspond to the next global level and each corresponding spring might represent a certain invariance (concept), similar to our second stage, whose spring represented the preservation of fusion. Instead of a left and right interlocking dipole array, the two hyperarrays might correspond to the incoming stimulus and its stored representation. For complex levels the stored representation might lack the local elements and may only contain the hyperframes. Hyperarrays suspended in hyperframes might be themselves coupled with their neighbors. Of course the coupling and geometry might differ from stage to stage. Without this coupling between elements at each stage, the model would be similar to several others, for instance the pandemonium model by Selfridge (1958). However, this coupling between local and global levels results in a very rich structure. A model like this could account for the long time necessary for a global percept to form and for its stability. In order to destroy a global organization after it is obtained, one must try to unlock the bonds at a local level. Only then will the local elements form new bonds which in turn will result in new hyperarrays.

This dependency on a global organization for local reorganization can be observed in many perceptual and cognitive tasks. For instance, in order to reverse a Necker-cube one has to inspect (on the local level) a Y-shaped vertex. If this is denied under retinal stabilization, one cannot reverse the Necker-cube; it remains in the old organization (Pritchard 1961). Similarly, on a much higher level, in order to switch to the other organization in Boring's young-wife-mother-in-law figure one must shift his attention to some critical local features. Without trying to emphasize these local features one stays locked in the old organization.

The importance of changing local levels in order to switch global states might explain why adaptation, the weakening of organization with prolonged use, occurs at the earliest stages in the CNS. Without adapting to the earliest feature extractors, one could never notice ambiguities. Without adaptation there would be no figure and ground reversals and one would never be able to enjoy a good pun or joke. According to the writer Koestler (1967) the basis of a good joke is the sudden realization that a situation in a certain frame of reference has an unexpected meaning in another frame of reference. Perhaps the basis of creativity is to be able to quickly adapt to certain obvious organizations and try to form some unexpected new ones from the same local data.

Going back to the model of binocular depth perception, one can now understand why it takes so long to perceive a random-dot stereogram portraying a very complex surface, but once fused, to be able to repeat the same task in a very short time. In order to obtain fusion, one has to shift areas in registration—to interlock and proceed with other areas of succession. If one proceeds in the wrong sequence, it will take a very long time until by chance alone all the corresponding dipoles become interlocked. However, if one learns the proper sequence of vergence movements or cortical shifts, the step-by-step interlocking of areas follows rapidly. For instance, in order to fuse the stereogram of a deeply recessed half ellipsoid as shown in figure 6.6-1*, one tries to

6.6-1* Complex random-dot stereogram portraying a recessed ellipsoid.

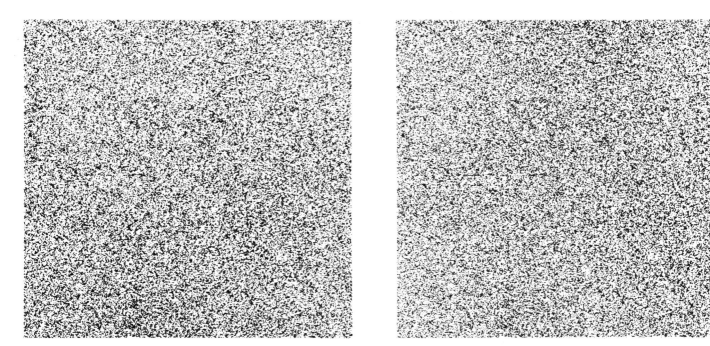

fuse the surround first. Soon he will have the surround interlocked and notice that a center disk-shaped area is still unlocked. If the observer knows that a divergence motion (or temporalward shift) is needed, he can save considerable time by skipping convergence motions (or nasal shifts). With more complex surfaces there occur several such "strategic decisions"; at each stage a wrong shift can prolong perception. The learning task is similar to exploring a maze. At each strategic junction one must learn whether a nasal or temporalward shift should be made. The correct sequence of these forks and the direction is apparently learned unconsciously and rapidly. Also, it is retained for a considerable time. If someone has fused several dozen complex random-dot stereograms of different surfaces, he will be able to fuse each of them a second time quite fast, particularly if he knows a priori which of them was presented.

The perceptual learning might be even simpler than outlined in the previous paragraph. It is conceivable that one learns only the range of binocular disparity values and the number of quantization levels in a given stereogram. For instance, if one tries to fuse figure 6.6-1* (which has a 1,000 $\times$ 1,000 dot resolution with a binocular disparity range of 0–100 dots in 1 dot quantization steps), fusion might take a minute at the first trial. However, at the second trial, after the range of disparity values has been learned, fusion will occur in a few seconds. If one tries to fuse another stereogram that portrays a different surface from the first but has the same disparity range and number of quantization levels, it is possible to fuse this second stereogram considerably faster than without previous learning. The exact study of this learning is a difficult undertaking since each subject can be used only once. However, such a program is now being pursued in my laboratory.

There are other improvements of a simpler nature that occur after thousands of trials. Several experiments I have carried out over the years and have repeated with a predoctoral student, Cary Lu, in 1967 indicate that after a large number of tachistoscopic presentations of random-dot stereograms *beyond* the subject's initial fusional limit, the stereogram will be gradually fused. It is as if the subject were taught to increase his cortical shift. Usually when vergence motions are used there is no need for such ability. However, in fusing the afterimages, subjects are deprived of their use of vergence, and these effective stimuli near the disparity limit of fusion exert adequate attraction force to pull the two arrays together. Thus, gradually the observer's "mental muscles" become strengthened by constant exercise. This suggestion, that forced use may result in neurophysiological changes, is mere speculation, although some evidence has been quoted earlier when reviewing a finding by Chow (1969).

I think that the dipole model for stereopsis is a model matched to our present psychophysical and neurophysiological knowledge. It is much more sophisticated than most models of stereopsis are, yet it would be relatively easy to propose somewhat more complex models. I only fear that such an extension would be premature at present. I refrained from building a neurophysiological model. The interested reader will find such a physical-neurophysiological method in a work by Sperling (1970). However, the next section deals with some implications of my model to neurophysiology.

This ends my attempts at model building. Yet, I have to clarify a general problem of how to find a model that is matched to the complexity of a given field of knowledge. It is well-known that a model which is too general, and is unable to cope with the detailed structures of experimental facts, is useless. However, the opposite, that a prematurely detailed model can be of less use than a simpler, less structured one, is less known. Let me illustrate this by an example.

A few years after holograms were invented, we (Julesz and Pennington 1965) noted certain similarities between holograms and human memory; both use the same principle of "equidistributing information." We were very impressed by the following fact. If a few years ago someone had presented us with a black box having a few, seemingly ad hoc properties (such as the ability to store several images without interfering with each other; the closer an input image resembled its stored version the sharper the recall; the fact that an input in the form of a segment of a stored image may retrieve the entire memory; the invariance against horizontal and vertical shifts of the input; the insensitivity of this storage against large destruction of the black box, etc.) we would have answered without hesitation that the box contained a piece of cortex of some higher animal. Yet, now a few years later we would not be so certain and would give a somewhat different answer. We would say that the box contained either a piece of cortex *or* a hologram (of the diffusion, Fourier-transform type). We also noted that the essence, the equidistributed information mapping, could be achieved in many ways. One could map the same information repeatedly, many times by actual wiring or by computer copying. The information could be duplicated by some biological cell division scheme such as having the "blue print" of ourselves in the form of chromosomes in each cell of our bodies. A fly's-eye lens array would also equidistribute an input picture in thousands of copies. Obviously, such densely repeated structures would exhibit many of the above-mentioned properties from the insensitivity of storage against extirpation of these structures to the invariance of positional translations. We can see that equidistributing the input information by using diffraction patterns of coherent light rays is just one of the possible techniques. Yet, there are several authors who took the hologram analogy literally and actually were trying to search for phase-coherent neural events that would produce holograms in the central nervous system. Such a detailed structural analogy may or may not be discovered. On the other hand, to note some fundamental, general similarities between holograms and memories has a strong heuristic appeal. To the skeptics (who do not understand the usefulness of such analog models) let me note that holistic models strongly influenced neurophysiology as indicated by Lashley's search for an "engram" or by Pollen et al.'s recent search for spatio-temporal-coherence in cortical cells (see p. 13).

6.7 Implications of the Model for the Neurophysiology of Binocular Fusion and Rivalry

After reviewing the psychological evidence of stereopsis and discussing a model based on these findings we are now in a position to relate these results to neurophysiology. The basic impact of the findings using random-dot stereograms was to convince neurophysiologists that binocular phenomena are simpler than monocular ones and that it is more profitable to study binocular vision first and postpone the study of the problem of monocular form perception.

However, the original demonstrations of random-dot stereograms not only showed that binocular depth perception in man is simpler than monocular form-perception but also indicated that it must be a very early process in the central nervous system using only point-by-point or feature-by-feature comparisons between the left and right fields. This conjecture of early processing has been somewhat set back by Hubel and Wiesel's findings that the input to the fourth layer of Area 17 has preponderantly monocular units in the monkey and has many monocular units in the cat. When using inverse cyclopean techniques the question came up, why these monocular units do not process monocular shape information after binocular fusion scrambles it.

In addition to these problems, the question of the relation between binocular fusion and rivalry was an unsettled matter. In the magnetic dipole model the attraction and repulsion of magnets with opposite or similar polarities assumed that rivalry and fusion are two sides of the same process. In spite of renewed interest by psychologists in the question of binocular rivalry, no neurophysiological evidence existed which would have clarified the role of rivalry in fusion.

The long-awaited new findings came just recently from Bishop's laboratory. It turned out that the majority of simple units (stellate cells in Layer IV in Area 17 of the cat) which were believed monocular are in fact exquisitely binocular as shown by Henry, Bishop, and Coombs (1969). While these units may be discharged exclusively by the dominant eye, it is still the case that a receptive field may be plotted which is activated by the nondominant eye, the field being largely inhibitory with a very small region of subliminal excitation. This observation was missed by Pettigrew, Nikara, and Bishop (1968) as well since most simple cells have little or no spontaneous activity. Furthermore, the sharp zone of excitation by the other eye is very hard to find. Whether the monocular simple units in the monkey cortex will turn out to be binocularly driven too, remains to be seen. Yet this finding, if true for the monkey as well, may explain many psychological observations. First of all it clarifies why the closing of one eye has no effect on vision other than loss of stereopsis. As long as the two eyes receive similar information, or the nondominant eye is closed, these simple units behave the same way. Only when the nondominant eye receives an image dissimilar to that in the dominant eye does the inhibitory influence show, and binocular rivalry in the form of suppression becomes evident. Only while the dominant eye is stimulated can the nondominant eye influence the simple unit by a narrow peak of subliminal excitation (0.3° across) and a broad range of powerful inhibition (2° across). The subliminal excitation requires the same precise stimulus shape as the dominant eye does, but the requirement for inhibition is much more general. Much the same properties as those just described are displayed by simple binocular units that are discharged by the separate stimulation of each eye as has been described by Pettigrew, et al. (1968) and by Joshua and Bishop (1969). Thus, when the nondominant eye elicited only weak responses it had nevertheless very strongly inhibited the dominant eye. The zone of binocular inhibition had approximately the same dimensions as the range of receptive-field disparities. It should be noted that the range of receptive-field positions within a cortical column (Hubel and Wiesel 1968) is of the same order of magnitude as the range of receptive field disparities (Joshua and Bishop 1969). Bishop (1969a) claims that receptive field disparity, zone of binocular inhibition and random receptive field position, are of approximately the same magnitude and might have a common basis.

The conjecture that suppression in binocular rivalry may be part of the binocular fusion mechanism is not new in psychology. Kaufman (1963) measured the amount of suppression in binocular rivalry. One eye was presented with a horizontal line the other eye with two vertical lines. The horizontal line segment enclosed between the vertical lines in the binocular view was rivalrous with maximum suppression times up to 14 min of arc separation between the vertical lines, quickly tapering off and reaching the minimum at 2° of arc. The parallel with the binocular inhibitory effects on simple units is apparent.

These findings are important for the dipole model. The missing evidence of combining attraction (facilitation) and repulsion (inhibition) in a basic element is now provided. In return, the psychologically derived model can tell the neurophysiologist what to look for. Obviously, the simple disparity units cannot solve the ambiguity problem of stereopsis and cannot explain the many hysteresis phenomena. These have to be clarified by the complex and hypercomplex disparity units which are less well-known. Pettigrew, et al. (1968) described a complex cell which had 6° of arc wide receptive fields but less than a degree shift reduced markedly the binocular response. If a complex cell of Hubel and Wiesel (1962) is considered as having as input a set of simple cells of the same orientation, one must also assume that these simple cells must have similar binocular disparities. One could imagine complex units that would integrate simple units of all orientations but of the same disparities. Such units we have already conjectured in § 3.9 to explain point-by-point comparison between the left and right fields of figure 3.9-1*. Some complex units have been found by Pettigrew, et al. (1968) that have a multimodal binocular response curve with a wide range of binocular facilitation. Thus for these cells binocular disparity is generalized. The possible complex properties of higher-order units might defy imagination, and therefore the dipole-model may suggest some needed restrictions.

7.1 Cyclopean Phenomena Revisited

In this chapter I return to the main topic of the book—the tracing of the visual system by cyclopean stimulation. Some of the findings introduced in chapter 2 will be further elaborated on, and new cyclopean phenomena will be reviewed. Furthermore, most results obtained by using random-dot stereograms will be checked using random-dot cinematograms. Since I have shown that local movement perception occurs after stereopsis and that global movement perception occurs before stereopsis, it is not obvious that stimulation of these different cyclopean retinae will result in similar results. But these cyclopean retinae are probably very close, and it is unlikely that highly central phenomena might differ when evoked by stereograms or cinematograms.

My practical reason for checking the stereoscopic results with cinematograms is because everyone, except those who are almost blind, can perceive random-dot cinematograms, while at least 2% of the population cannot perceive random-dot stereograms at all, and more than 15% have difficulty with complex stereograms. This is my personal observation with several thousand observers who were in my audiences at scientific meetings, lectures, and popular demonstrations. With random-dot cinematograms I have never met anyone who could not see a small area of dynamic noise in a surround of static noise. It is easier to demonstrate the cyclopean phenomena in motion than in depth for large audiences at a meeting, since 16-mm movie projectors and standard screens are readily available, while stereo-projectors or vectograph slides, aluminum screens, and individual polarizing glasses are necessary for a stereoscopic mass demonstration. For printed material the opposite is true. A stereogram on a printed page can be readily fused by people who have learned to dissociate their accommodation from their convergence motions. Furthermore, a prism (wedge) in front of one eye can greatly aid fusion. It is even possible to print a stereogram as a red-green anaglyph and provide red-green goggles for a few cents —the technique applied in this book. On the other hand, to present the simplest stroboscopic movement by alternating two images in rapid succession is very clumsy, if not impossible in a book. It might be feasible to print on the margin of several consecutive pages a succession of movie frames and have the reader quickly turn the pages to produce apparent movement, but this is a primitive solution. In our age of xeroxing, one could copy two images of a stereogram or a cinematogram and glue them together in exact alignment with a rubber-band between them. By twisting the rubber-band and pulling on its end, one can rapidly rotate the two sides and produce stroboscopic movement.

In this chapter some of the random-dot stereograms will be repeated as random-dot cinematograms. The only difference is that the portrayed information will be uncorrelated (dynamic) noise surrounded by correlated (static) noise. Also, the random-dot stereograms can be presented monocularly in temporal succession, producing a figure of static noise moving horizontally between two positions in a static surround. For all these examples the moving static noise or the dynamic noise yield the same cyclopean phenomena. The dynamic-noise technique produced by only two alternating frames is not as vivid as a succession of several frames of uncorrelated areas, but it gives satisfactory results.

All the cyclopean movement phenomena reported in this book were produced by movie strips in an infinite loop, containing at least twenty partly uncorrelated frames in succession. However, prior to this, a quick pilot study was conducted using only two frames in rapid alternation. Instead of turning the two sides of the aligned pair by a rubber-band, I used a special electronic timer triggering low inertia light bulbs that illuminated the two slides in succession. One of the slides was mounted in a vernier-driven frame enabling three-degrees-of-freedom alignment of horizontal and vertical translation and rotation. The exactly aligned sides were viewed through a half-silvered mirror that superimposed them. The driving circuit flashed each slide three times before switching to the other slide, in order to eliminate flicker. The setting of the frame-repetition rate was critical in order to obtain optimal stroboscopic movement for various disparity values. Both the opticomechanical parts and the electronics were designed by R. A. Payne; the circuitry of the timer-driver units is shown in figure 7.1-1. Those readers who have a stereo-projector can build a simpler system by rotating a disk in front of the two lenses having openings on the disk such that it alternately occludes one of the lenses. (Of course, no polarizers need to be used.) It is also possible to present a random-dot stereogram alternately to the left or right eyes, respectively (Dodwell and Engel 1963). Above a certain alternation rate corresponding to the monocular simultaneity region, of course, the stereogram is perceived in

7.1-1 Circuitry of electronic tachistoscope designed by R. A. Payne.

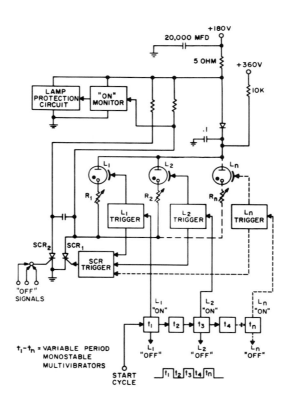

depth. At lower rates, it is possible to produce apparent movement between corresponding shifted areas in the left and right eye's views. I observed that there is an intermediate frame repetition rate at which it is possible to perceive in vivid depth either a center square moving or at a standstill. These two states alternate at will but can be perceived only one at a time. These alternations between the two percepts of stroboscopic movement and stereopsis indicate a close similarity between the two processes. Some relations between stroboscopic movement perception and stereopsis were discussed in § 3.7.

Throughout this treatise we have assumed that all cyclopean experiments that were produced by random-dot stereograms could be repeated monocularly by random-dot cinematograms and would produce similar results (except for phenomena that may reside between the cyclopean retinae of stereopsis and movement perception).

Indeed all the random-dot stereograms in this book, when presented monocularly in temporal alteration, can produce apparent movement and can portray cyclopeanly the phenomenon in question. Anstis (1970) tried a series of random-dot stereograms with decreased correlations (Julesz 1960a,b) to be shown alternately and found that movement ceased exactly at the same level of correlation where stereopsis ceased.

However, although the perception of random-dot cinematograms is similar to stereograms, there are important differences. The most important difference is that stereopsis requires a dense match of adjacent elements of the same disparity, while movement perception does not. While a left and right uncorrelated random-dot array (of 50% identical elements owing to chance) cannot be fused, the same arrays in temporal succession are perceived as dynamic noise. Each element appears to move into its nearest neighbor in a successive frame. The fact that stereopsis can only tolerate small horizontal displacements, while movement perception can work on larger displacements in any direction is such a well-known difference that it does not necessitate further elaboration.

On the other hand, Anstis (1970) made an interesting observation which is different for the two perceptual processes. While Julesz (1960a) showed that a positive and negative image of a random-dot stereogram does not yield fusion, Anstis demonstrated that alternate positive and negative frames of a movie scene will result in an inverse phi-phenomenon that moves in the opposite direction from the original sequence. This explanation is based on the priority of brightness cues over form cues. In this respect movement perception is again similar to stereopsis. First, a point-by-point comparison takes place between the brightness values of successive frames, regardless of any form cues. Anstis even took the difference field model of stereopsis (Julesz 1960a) and applied it to movement perception (by removing the emphasis on clusters of adjacent points with the same displacements).

It is hard to assess at this time to what extent the matching processes of movement perception and stereopsis are similar. Since stereopsis does not require memory while movement perception does, the two processes are not easily comparable!

In order to measure visual memory span, in a pilot study with Chiarucci, we tried the visual counterpart of the auditory repeated-noise experiment by Guttman and Julesz (1965). We repeated a sequence of random-dot arrays $A_1, A_2, \ldots A_N; A_1,$

$A_2, \ldots A_N; A_1, A_2, \ldots A_N; \ldots$ and asked for what N value (length of cycle) was the observer unaware that the dynamic noise was periodic. It appears that $N = 8 - 12$ (0.3–0.5 sec) is long enough to prevent the observer from noticing the recurrence of the same local movements.

Computer-generated movies in real-time recently became feasible. As a result, I am sure that in the 1970s an expansion of knowledge will occur in the study of apparent movement perception similar to that which took place in the 1960s in binocular depth perception.

7.2 Localization of Optical Illusions

Interest in optical illusions has not diminished with time. Within fifty years following the first report in a scientific publication on a geometrical optical illusion (Oppel 1854) about two-hundred papers by well-known physicists and physiologists were devoted to illusory figures. The interested reader will find an excellent summary of nineteenth-century contributions in *Experimental Psychology* by Titchener (1902). It remains a favorite preoccupation of workers in perception. There is nothing really special about optical illusions, since each percept is an "illusion." When light energy of a certain wavelength impinges on our retinal receptors, it does not inherently contain any indication that it will be perceived by some organisms in color of a given brightness, hue, and saturation. Similarly, two flat projections of an object on our retinae have no compelling reason to be perceived in vivid depth. Indeed, the fact that certain human beings with two functioning eyes do not possess stereopsis indicates the illusory nature of depth perception. Why is it then that the departure of geometrical facts from their perceptions is regarded as something special? The only explanation I can offer is that geometrical illusions are believed to serve as a good illustration for perceptual constancy phenomena. Most of these constancies are taken for granted by the layman, who often does not understand that there is any problem in explaining how he perceives his hand the same size while he moves it back and forth. After all, it is the same hand, and it does not occur to him that its retinal projection might have changed by an order of magnitude. However, when he notices that a vertical line segment which he judged to be longer than a horizontal one is physically of the same size, he suddenly becomes aware that reality and its perception are not identical. Perhaps sophisticated workers in perception are interested in optical illusions because of similar experiences they encountered in their childhood.

In spite of the vast, constantly growing literature of optical illusions, no satisfactory theory explains them. Probably several of the illusions are based on different perceptual processes. One recurrent explanation is that these illusions are the result of eye-movements. For instance, in the case of the Müller-Lyer illusion, our eyes make an unbroken sweep across the length of one figure but are prevented from doing the same for the other. Unfortunately for this explanation, tachistoscopic exposures too brief to permit eye-movements result in the same illusion.

According to the perspective theory of optical illusions, which is another recurrent theory from Thiery (1896) through Tausch (1954) and Gregory (1963), a hidden monocular depth cue exists owing to linear perspective. By an unconscious inference of depth, it is proposed that we perform a size scaling according to the "size-distance invariance" hypothesis. We have already discussed how vergence motions change the apparent size of afterimages. Its more general form, known as Emmert's law, does not link vergence with depth, but states for monocularly apparent depth the following: for a given proximal stimulus, as apparent distance increases, apparent size should increase and vice versa.

Such reasoning has been proposed for the famous moon illusion as well, by Kaufman and Rock (1962). In addition to the Müller-Lyer figures, the Ponzo illusion and Hering illusion can be easily interpreted as perspective drawings in three dimensions. According to Gregory (1963) the depth cues in the illusory figures are not apparent because of the texture cues always present on a printed page, but can be made visible in dark rooms by using luminous wires to portray the figures. There are opponents to this theory. It has been pointed out by Rudel and Teuber (1963) and by Over (1968) that a tactile counterpart of the optical illusions exists, whereas no perspective cues (due to projections) do exist in tactile space. There are also several geometrical illusions, such as the famous Poggendorff illusion (a break in a line, partly hidden), which are difficult to imagine as a perspective drawing. Humphrey and Morgan (1965) even constructed some counterexamples. Their modified version of the Ponzo illusion is shown in figure 7.2-1, while figure 7.2-2 illustrates the original Ponzo

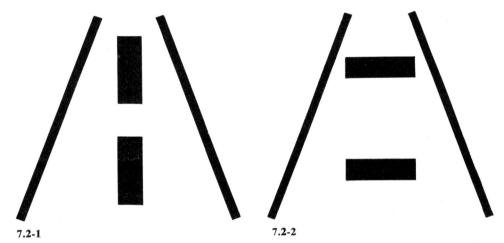

7.2-1

7.2-2

7.2-1 A variation of the Ponzo illusion after Humphrey and Morgan. No illusion is experienced.

7.2-2 The classical Ponzo illusion.

illusion. Gregory explains the original Ponzo illusion—the fact that the upper line appears longer than the lower one—by saying that the upper line lies in a part of the figure which the converging lines indicate to be more distant so that perceptual scaling magnifies its apparent size. As it stands, this explanation predicts the expansion of any line drawn in the upper (more "distant") part of the figure, regardless of its orientation. Yet for two vertical line segments, as shown in figure 7.2-1, the illusion does not occur.

I am not trying to propose a new theory of optical illusions. However, the cyclopean counterpart of the best known illusions gives important insights into the underlying processes. First of all, most of the illusions are perceived under cyclopean conditions, thus their central origin (after stereopsis or movement perception) can be established. Usually the illusions are strongest (when portrayed by random-dot stereograms) if both the test figures and inducing figures are of the same depth.[1] For example, in the cyclopean Ponzo illusion demonstrated in figure 7.2-3*, the two horizontal rectangles (test figures) have the same disparity as the converging bars (inducing figures). Under these conditions the perspective cues are so strong that the two horizontal rectangles appear at different depth planes; the lower bar seems to be behind the upper bar and the converging bars. The cyclopean Ponzo illusion appears to be as strong as the classical one of figure 7.2-2. However, when the test bars have less disparity than the converging bars, the perspective depth sensation and the perceived illusion are reduced as shown in figure 7.2-4*. Whether this finding is a further vindication of the perspective theory is not certain. After all, some lateral interaction in the cortex could produce some of these illusions, and there might be reduced inhibition between cells that extract different disparity values.

In all these manipulations of disparity one should be cautious about keeping the disparities of the test figures the same. If the test figures were of different disparities, then Emmert's law would operate, reducing the size of the nearer figure and increasing

7.2-3*   The cyclopean Ponzo illusion. Effect is similar to the classical case.

1. That disparity changes reduce illusions when portrayed stereoscopically was already noted by Thiery (1896). A detailed account of optical illusions and depth perception is given in a book by Klix (1962).

that of the more distant one. The effects caused by Emmert's law can be tested, however, for illusory figures which cannot be separated into test and inducing figures. Such is the case in the well-known vertical-horizontal illusion which is more appropriately also called the bisection illusion. (The classical bisection illusion is shown in figure 7.2-5, and turning of the page by 90° does not make the new vertical bar appear longer than the new horizontal bar as the "vertical-horizontal" name of the illusion might suggest.) In figures 7.2-6*, 7.2-7*, and 7.2-8*, the cyclopean bisection illusion is reproduced under three conditions. That the cyclopean bisection illusion is as strong as the classical one was shown by Julesz (1968d). The modified cyclopean bisection illusions of figures 7.2-7* and 7.2-8* are, however, first demonstrated here, and the fact that the vertical bar is in front or behind the horizontal bar has slight if any noticeable effect on the illusion of the vertical bar being much longer. That depth changes do not affect this illusion indicates that neither Emmert's law is operating nor perspective interpretation is possible. Thus the bisection illusion is probably very different from the Ponzo illusion and some other famous illusions to be discussed next for which disparity changes have a great effect.

In figures 7.2-9, 7.2-10*, and 7.2-11* the classical, cyclopean, and modified versions of the cyclopean Ebbinghaus illusion are shown. When the cyclopean Ebbinghaus illusion portrays all figural areas of the same depth, as in figure 7.2-10* the illusion is experienced. The illusory effect is greatly reduced when the test figures are behind the inducing figures.

7.2-4* Modified cyclopean Ponzo illusion. The test and inducing parts of the figure are at different depths. Illusory effect greatly diminished.

7.2-5 The classical vertical-horizontal illusion.

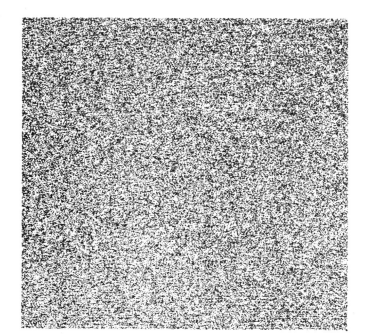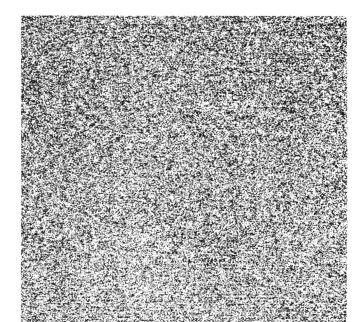

7.2-6*

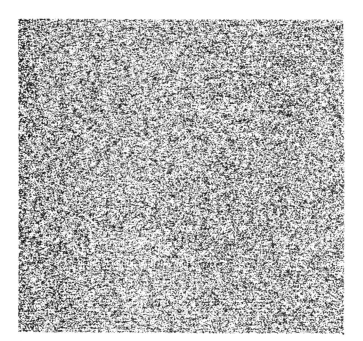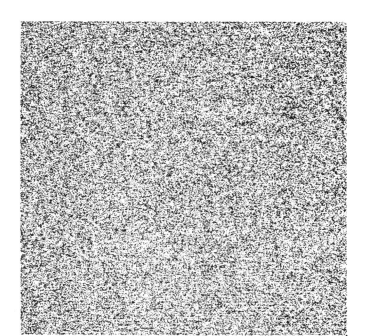

7.2-7*

7.2-8*

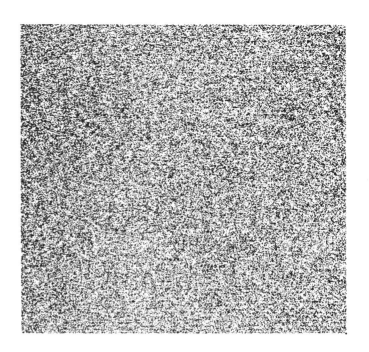

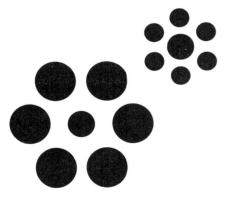

7.2-9

7.2-6* The cyclopean vertical-horizontal illusion. Illusory effect similar to the classical case.

7.2-7* Modified cyclopean vertical-horizontal illusion. Vertical bar is closer than the horizontal one.

7.2-8* Modified cyclopean vertical-horizontal illusion. Vertical bar is farther away than the horizontal one.

7.2-9 The classical Ebbinghaus illusion.

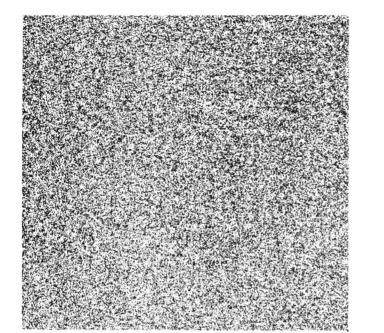

7.2-10*

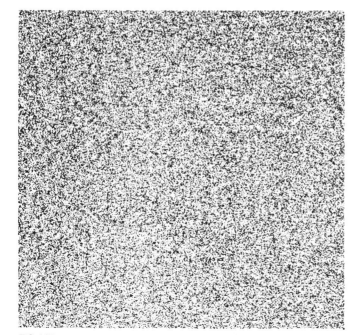

7.2-11*

In figures 7.2-12, 7.2-13* and 7.2-14* the Poggendorff illusion is shown in the classical, cyclopean, and modified cyclopean versions. Again the cyclopean illusion of seeing the line broken is present when each figural area is at the same depth, yet is greatly reduced or even disappears (sometimes even reverses) when the test figures are behind the inducing figure. This is the more interesting since the Poggendorff illusion does not lend itself to perspective explanations.

The Müller-Lyer illusion has been shown already in figure 2.6-2* as a random-dot stereogram; it is now presented as a cinematogram. In figure 7.2-15* the same illusion can be experienced when the figures are presented as dynamic noise in a static surround. The interested reader might view the anaglyph of figure 7.2-15* through the glasses while illuminating it in rapid succession by a red and a green projector. The modified Müller-Lyer illusion is shown in figure 7.2-16*. Here the test bars are *above* the inducing arrowheads causing a great reduction in the illusion.

Besides the Müller-Lyer illusion the other illusions tried could be experienced when portrayed by random-dot cinematograms. However, in order to test the modified cyclopean illusions by cinematograms, one has to portray the inducing figures and test figures by different movement gradients. If the test figure changes its uncorrelated texture at each frame, while the inducing figure has identical successive frames, then the dynamic noise for the latter changes at half the rate of the former. This yields discrimination between the test and inducing figures. Such movies were tried recently by Chiarucci and me, and we found that a movement gradient will have the same result in reducing illusory effects as a change in depth between the inducing and test figures will have.

These experiments do not suggest a new theory to explain geometrical illusions, but they do clarify certain points. It is interesting to know that, for instance, the Müller-Lyer illusion is not the result of some lateral inhibition in retinal ganglion cells, as several theories proclaimed. That some of the illusions are central was first demonstrated by Papert (1961), using random-dot stereograms. Hochberg (1963) further elaborated on this idea. Schiller and Wiener (1962) combined the illusory figures from two different left and right outlined drawings by binocular viewing, using short and long exposures. They concluded that the binocular combination of the test elements and inducing elements produced all the main optical illusions indicating also their central origin. However, their technique could not overcome binocular rivalry. Notably in the Zöllner figure (which has almost a complete overlap of test and inducing figures), the binocular rivalry reduced the illusion, particularly for long exposures.

The cyclopean technique has the advantage over the Schiller and Wiener technique in that it produces fusion without rivalry. As a last cyclopean experiment, the Zöllner illusion (fig. 7.2-17) is demonstrated. In figure 7.2-18* the test figures (parallel lines) are at the same depth as the inducing figures (stripes). The illusion is greatly reduced in figure 7.2-18*.

One limitation of the cyclopean method becomes apparent from these experiments. One can only portray figures that have adequate areas, and random-dot correlograms are not suitable for creating very thin outlined drawings. On the other hand, some of

7.2-10* The cyclopean Ebbinghaus illusion. Illusory effect similar to classical one.

7.2-11* Modified cyclopean Ebbinghaus illusion. Test and inducing parts of figure at different depths. Illusory effect greatly reduced.

7.2-12 The classical Poggendorff illusion.

Pages 232–33

7.2-13* The cyclopean Poggendorff illusion. Illusory effect similar to classical case.

7.2-14* Modified cyclopean Poggendorff illusion. Test figures are behind inducing figure. Illusory effect disappears.

7.2-15* Müller-Lyer illusion portrayed by a random-dot cinematogram. For viewing see text.

7.2-16* Modified cyclopean Müller-Lyer illusion. Test figure closer than the inducing arrowheads. Illusory effect greatly reduced.

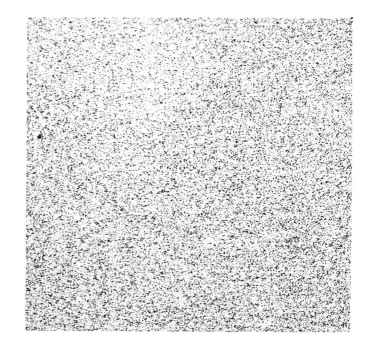

7.2-13*

7.2-14*

7.2-15*

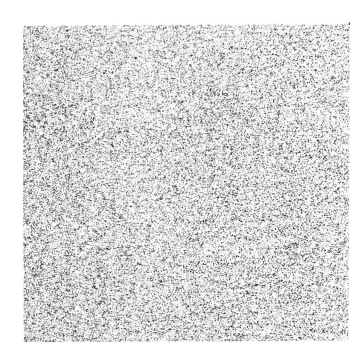

7.2-16*

7.2-17   The classical Zöllner illusion.

the classical illusions are the strongest for very thin outlines. This is particularly true for the classical Ponzo (fig. 7.2-2) and classical Zöllner (fig. 7.2-17) illusory figures as presented in this book. For both cases the illusion is less than it used to be in textbooks that use thin lines. This fact might partly account for the weakness of the cyclopean Zöllner illusion (fig. 7.2-18*). Classical illusory figures, however, that are based on area rich figures, such as the Ebbinghaus illusion, produce strong illusions when cyclopeanly portrayed.

What conclusions can one draw from this section? Are the proponents of perspective theory, such as Tausch (1954), Holst (1957), or Gregory (1963), right? The only statement we can make at present about our findings (that with disparity changes the perceived illusions diminish), is that they are in agreement with the perspective theory. Indeed, the larger the disparity between the test and inducing figures, the less these cyclopean figures can be seen in perspective. However, an even simpler explanation can be offered in order to explain reduced optical illusions with increased disparity differences. One might conjecture that the optical illusions are the result of lateral interaction of the inducing figures on the test figure. The closer the inducing figure is to the test figure the greater this influence. However, for random-dot stereograms this influence is exerted in three-dimensions, and with differences in disparity the interaction distance between the inducing and test figures becomes too large and the optical illusion disappears. I will return to this question briefly in § 9.3. But already it is apparent that optical illusions might be just a name for several different perceptual processes. For instance, it is hard to give a perspective interpretation for the vertical-horizontal illusion, and yet it is one of the strongest illusions. Nevertheless, that all the illusions tried were central and depended on depth (disparity) cues is important.

It would require a vast volume to review the literature on illusions alone, which might be a premature undertaking anyway, since no explanations exist as of yet that were not discredited by some counterexample. Nevertheless, the sophisticated reader will find in the previous demonstrations several new results that might lend support to or disprove existing theories on illusions. My reason for explicitly mentioning the perspective and inappropriate size-constancy theories was not because of their validity, but because of their claimed generality. There are many counterexperiments that cast serious doubt on these theories (see, for example, Fisher 1968), and yet no other explanation has been given that would account for several illusions that appear in two-dimensional figures. Although it might be possible that the illusions have no common mechanism, the cyclopean phenomena reported in this section reveal two general tendencies: First, all the illusions tried were also experienced under cyclopean conditions and in most cases (except for the Zöllner illusion that was very weak) without reduction of the illusory percept. Second, the illusion disappeared or became reduced in all cases (except the horizontal-vertical illusion) when the inducing and test figures had different binocular disparities. Perhaps it is not without significance that the two anomalous illusions, the horizontal-vertical and the Zöllner illusions, have no perspective interpretations. On the other hand, the cyclopean Poggendorf illusion behaves like the other illusions, although no convincing perspective explanation might be suggested.

Unfortunately, time did not permit quantification of the dependence of the illusions on the difference between disparities of the test and inducing figures when cyclopeanly portrayed. One would expect a certain critical distance in the depth dimension beyond which the illusory effect rapidly diminishes. If such psychophysical functions were

7.2-18* The cyclopean Zöllner illusion.

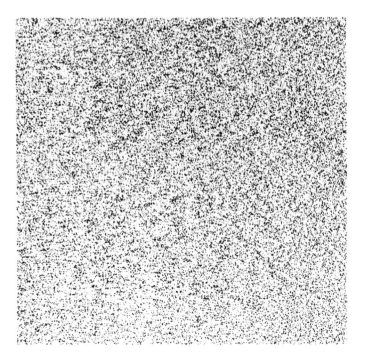
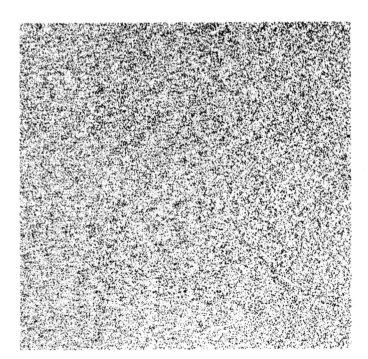

determined for the various illusions, several new theories could be tested. One obvious assumption might be that the influence of the inducing figures on the test figures occurs in three-dimensions, and the unit of distance is much smaller in the third dimension, similar to the finer stereoscopic acuity threshold which is an order of magnitude better than visual acuity in the x-y plane.

If one abandons attempts for generality, there are several specific explanations for a given illusion. For instance, the Müller-Lyer illusion has sometimes been explained by "confusion" theories (Woodworth 1938, p. 645). Several variants of the confusion theory exist (for an early review see Titchener 1902); however, all postulate that observer's judgment is not based on the length of the test figure alone, but also on the length of the entire figure, including the arrowheads. A recent test of the confusion theory was carried out by Erlebacher and Sekuler (1969) who first varied the angles of the arrowheads. In a control experiment the length of the arrowheads was concomitantly varied with the angles so as to keep the distance between the ends of the arrowheads constant. Their results were in agreement with the confusion theory. If the inducing figure is at a different depth from the test figures (such as in fig. 7.2-16*), then the distances between the arrowheads are less confused with the length of the test bars, than if both were at the same depth level. I included this example to show that most "explanations" are of this nonrigorous kind. They are usually so unspecified that it is impossible to disprove them (or, as a matter of fact, to prove them)! We are still very far from a rigorous hypothesis of a given optical illusion that is concrete enough to be verified or disproved by experimentation.

As we have seen, all the optical illusions we have tried are the result of processes that occur after global stereopsis. It is interesting to note that the opposite assumption was made by Lau half a century ago. Influenced by Gestalt psychology, Lau (1922) used optical illusions to show their influence on stereopsis. In a typical experiment of his, one image of a stereogram consists of equally spaced vertical lines, while the other image contains an identical vertical grating to which have been added the slanting cross lines that make this image a Zöllner illusory figure. Lau claimed that, when stereoscopically viewed, the monocularly perceived illusion (that causes the successive parallel lines to appear alternately inclined) yields a corresponding depth percept—that is, the vertical lines appear to tip alternately top toward and top away from the observer. However, as Ogle (1962) pointed out, neither he, nor other workers in the field were able to perceive this alleged depth. My own experiences agree with Ogle's. The fused vertical lines clearly appear to lie in the plane of the printed page. My purpose of citing Lau is twofold. First, I want to warn the reader that the "Lau-effect" probably does not exist. But more importantly, I want to emphasize that Lau's method of using a phenomenon (i.e., a given optical illusion) in order to evoke another phenomenon (stereopsis) can be regarded as a cyclopean idea. It is really too bad that the particular combination did not yield convincing results. Had Lau succeeded in his effort, I think cyclopean psychology would have been developed right then.

That the Zöllner illusion is greatly reduced under cyclopean stimulation might also be explained by the conjecture of Blakemore, Carpenter, and Georgeson (1970).

They showed in a perceptual study that two lines of different orientation interact with each other so that they seem to be displaced from one another in orientation. This perceived shift in orientation depends on the angle between the two lines and can amount to $\pm 2°$, which is adequate to explain the Zöllner (and Hering) type of illusions. They assumed that this perceptual effect can be explained in terms of mutual inhibition between neighboring columns (edge detectors) in the visual cortex. Were the majority of these edge detectors in Area 17 before global stereopsis, one could understand why the cyclopean Zöllner illusion is much weaker than its classical counterpart. In this case, however, one is at a loss to understand why Lau's experiment does not work, the more so since the stereopsis units in the cat and monkey are orientation-sensitive. Of course, the failure of the Lau experiment might be due to binocular rivalry, and the lateral inhibition theory of edge detectors might be the correct explanation of the Zöllner illusion.

7.3 "Cyclopean Depth" Sensation

What would happen if perspective drawings were produced by random-dot stereograms or cinematograms; would they be perceived in depth? This question is particularly interesting for random-dot stereograms since each of the vertices of a cyclopean Necker-cube are at the same depth level. If we could reverse perceptually a cyclopean Necker-cube, then some of its vertices would be perceived closer to us than others. If stereoscopic depth perception were coupled with this perspective-depth sensation, then those vertices which seem closer to us than those determined by disparity would disappear. After all, "closer" means larger nasal disparity, but at any other disparity value than that used for the cyclopean presentation, no fusion—and thus no cyclopean figure—can be experienced.

This ingenious experiment was carried out by Hochberg (1963, 1966). He portrayed a Necker-cube as a random-dot stereogram and demonstrated that it can be seen in depth and can also be reversed. In order to produce cyclopean-reversible Necker-cubes, the width of the random-dot bars used for portrayal must be thin, otherwise the crossing points between bars become stably locked in depth. However, if the bars are too thin, they are difficult to fuse, particularly the vertical ones. (That horizontal thin bars are easier to fuse than vertical thin bars is obvious from the model. For horizontal bars, many picture elements contribute to fusion which is facilitated by coupling between adjacent elements. For vertical bars, only a few picture elements along a horizontal line contribute to fusion and because of coupling only a few adjacent rows having a few picture elements will aid fusion.) A compromise solution is shown in figure 7.3-1. The resolution is high, in order to have many picture elements along the width of the narrow bars. Moreover, the intersections are avoided, which aids reversals. With some practice, each side of the cube can be perceived in depth, and it is also possible to perceive the figure as a perspective cube and reverse it.

This demonstration shows that perspective depth is independent of stereopsis and is a central depth sensation. Thus perspective depth is a cyclopean depth. That perspective depth is not the same as stereopsis is not surprising. After all, human beings

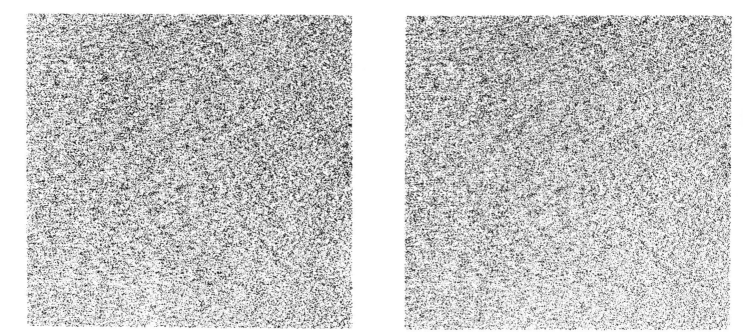

7.3-1 Cyclopean Necker-cube with no intersections in order to facilitate reversals. (An anaglyph of a similar cyclopean Necker-cube but with intersections is given in fig. 5.6-5*.)

with no stereopsis have spatial perception for perspective, two-dimensional drawings, and photographs. Furthermore, monocular motion parallax is a powerful depth cue still available for them, which might help to calibrate the cyclopean depth mechanism. On the other hand, the finding that this central depth sensation does not feed back to the stereopsis mechanism is somewhat unexpected.

The same cyclopean depth sensation is evoked when the Necker-cube is portrayed by a random-dot cinematogram using moving static texture. Here again, that the sides appearing closer to the observer move at the same speed as the ones further behind is against monocular motion parallax cues, yet perspective depth is perceived. Thus this cyclopean-depth sensation is independent of stereopsis or motion parallax and is the result of perspective cues.

The problem of cyclopean depth is strongly related to mental imagery and to the problem of being able to spatially transform three-dimensional perspective images in one's mind. Introspection does suggest that, in order to determine whether two differently oriented solid objects are identical in shape, we may imagine one of the objects rotated into the same orientation as the other. Evidence that such a rotation can be carried out in a purely mental way comes from a communication by Shepard and Metzler (1971). They report that the time to recognize that two perspective line drawings are drawings of the same three-dimensional object is a linearly increasing function of the angular difference in the portrayed orientations of the two objects and is largely independent whether the rotation is performed in the plane of the picture or in depth.

When I began to write this book I never guessed that the anecdotal eidetic memory would reportedly be evoked before stereopsis and, therefore, that it would belong to my subject matter. It was known—as a matter of fact, served as its definition—that eidetic images are built up as a result of scanning eye-movements in contrast to the usual afterimages and can be projected on a wall and inspected by eye-movements without moving the eidetic image. This, of course, makes it certain that eidetic storage cannot be retinal. Yet the entire field of eidetic phenomena was regarded as enigmatic, though the work of Haber and Haber (1964) demonstrated that a fair number of children and a very few adults are eidetikers and can recall images for several minutes.

In my seminar at the M.I.T. psychology department two researchers, C. F. Stromeyer and J. Psotka from Harvard surprised us all by reporting some experiments they undertook. During that spring of 1969 they experimented with a twenty-three-year-old, very bright and educated lady who had an unusual eidetic ability. The fortunate coincidence was the finding of an adult eidetiker of unusually vivid and long-lasting imagery together with sophisticated investigators fortified by an incredulous group of famous experts on visual perception of the Cambridge scientific community who suggested control experiments of all sorts.

In this section I shall deviate somewhat from a fundamental principle, namely that only such findings will be reviewed that I have tested myself and which can be experienced by almost everyone. Inspired by Helmholtz's monumental *Physiological Optics,* I tried mainly to report (and often demonstrate) phenomena I have checked myself; only in special cases, such as anatomy or neurophysiology, would I consent to the passive role of a reviewer. However, the importance of the findings which I shall review and the method to test the eidetiker by random-dot stereograms caused me to include this section.

It would be improper to go into all the details; now that some first results were reported by Stromeyer and Psotka (1970) and Stromeyer (1970). I shall review only two of their experiments because of their cyclopean nature. The other experiments are also most revealing yet introspective in nature and therefore less acceptable for the sceptics.

The first class of experiments was based on an eidetic version of the Land demonstration (Stromeyer 1970). In the classical Land demonstration two matrices of, say, 3×3 squares of randomly selected and uncorrelated gray scales are superimposed on each other, one projected in red light, the other in green light. The combined image results in a "gamut" of colors (in addition to red and green). The interested reader is referred to the publications by Land (1959, 1964).

The eidetic version is as follows. The red matrix was scanned by the eidetiker for two minutes one day, a green record was scanned for two minutes the following day. Then the subject was asked to superimpose the two eidetic images that had 24-hours delay between their formation. The resulting gamuts of colors were exceedingly accurate and identical to the classical, simultaneously superimposed matrices. This finding is the more interesting since the Land colors change very much with the brightness of the red or green projected colors. This indicates that eidetic images do not lose their quality in time, except that their recall becomes more difficult.

While two (or more) monocularly presented images to the same eye produced long-lasting, stable eidetic images, in contrast, dichoptic presentation did interfere with the building up of eidetic images. When one matrix was presented to one eye and the other uncorrelated matrix to the other eye, the produced eidetic images caused binocular rivalry and lasted only 10–15 sec.

The second class of experiments used random-dot stereograms of a 100×100 matrix of black and white elements. One matrix of the stereogram was presented to one eye of the eidetiker for 2 min. The subject rested 10 min and then superimposed the eidetic image of the first image on the other image of the stereogram viewed with the other eye. The subject correctly reported an inverted T-shaped area receding in depth from the textured surround. The subject had no way of knowing the cyclopean shape except by superimposing the eidetic image on the corresponding real image. The subject remarked that during the rest period the eidetic image appeared as sharp as the real image and did not blur with time. After this observation, the separation between stereograms was extended to 24 hours. The subject scanned with the right eye a 100×100 picture element stereogram for four 3-min periods separated by 1-min rests. Twenty-four hours later the other image of the stereogram was viewed with the left eye, and the eidetic image superimposed. The subject correctly reported a textured square raised out from the textured background. What is more, the subject was able to accurately adjust pointers to indicate the correct positions of the borders and the depth of the center square. This is the more remarkable, since blurring one of the images under normal viewing conditions causes the corners of the center square to be seen as rounded off.

This experiment indicates that eidetic images are represented in the nervous system before stereopsis, since the stereo mechanism can produce depth from eidetic images. This explains why the dichoptic Land experiment evoked binocular rivalry. Besides localizing the site of eidetic-image representation, these cyclopean experiments show several remarkable properties of the eidetiker. First, that eidetic images can remain eye-specific and thus produce stereopsis. It also shows that the eidetic image can take the place of the real image. After all, during the stereoscopic experiment the eidetic eye was looking at a blank field, and instead of the expected binocular rivalry, fusion took place. But most importantly, the difference between a 3×3 matrix and a 100×100 matrix is more than quantitative. That 10,000 picture elements of a random-dot texture could be precisely stored for 24 hours without loss of detail argues that at least a select few of Homo sapiens have a detailed texture memory.

During the period this book was in the process of publication, Stromeyer (1970) reported a new finding of considerable importance with the same eidetiker. He reported that turning of the head by 90° down to the side (while the left eidetic image was being built up) preserved the eidetic image in the vertical position as shown by the fact that this image could be fused stereoscopically with a right image during normal head position.[2] Thus, some vestibular compensation of the eidetic image occurred

---

2. That head rotation is compensated before stereopsis is in agreement with recent neurophysiological evidence. Horn and Hill (1969) found simple receptive fields in Area 17 of the cat whose

before stereopsis. Stromeyer and Psotka (1970) performed several control experiments to show that the eidetic image was eye specific. Whenever the eidetic image of the left stimulus array was compared with the other array but presented to the *same* (left) eye, the uncorrelated areas became apparent but never yielded a depth sensation. Another important experiment was reported by Stromeyer who showed his eidetiker the test figures of figure 8.1-1. In this random-dot stereogram, one eye (say, the right) is presented with the same matrix, while the other eye is presented with different images yielding different cyclopean figures. The eidetiker first built up an eidetic image of the left matrix L_A, which we will call L_A*. Later, she simultaneously viewed L_B with her left eye and R (the common right image) with her right eye. She saw figure B in depth (as we would see it in ordinary stereoscopic viewing). While being presented with L_B and R, she could call up the eidetic image L_A*, which selectively suppressed L_B, which she still viewed with her left eye. With much surprise she witnessed figure B in depth blending into figure A (which she had never seen before).

The question of how many random-dot arrays she can retain simultaneously was also investigated. Here the major difficulty is to mark the various random-dot eidetic images, since monocularly they appear all alike. It turns out that by viewing the various random-dot images through different color filters, the eidetic images can be labeled by their color. Thus, it is sufficient to show the common contralateral image in different colors in order to selectively choose the corresponding eidetic image from the stored set having the same color. Stromeyer (1970) reported that she could simultaneously store four eidetic images of 100×100 dot arrays.

It should be mentioned that Stromeyer's eidetiker could recall these complex eidetic images at will, but often once they appeared they could be seen for only a few seconds. In the classical tests on eidetikers the task consisted of describing many features of the eidetic image which often required minutes. One advantage of random-dot stereograms is that the time required to perceive depth can be made to vary (depending on surface complexity) from fractions of a second to a minute. Thus the perception time for stereopsis can be chosen shorter than the time the eidetic image is retained. This permits the discovery of excellent eidetikers, who would be otherwise missed by the long scrutinizing procedure.

These experiments appear incredible. Yet, plasticity of depth perception in eidetikers has been reported by Jaensch (1930) and Kroh (1922). There are several anecdotal stories about eidetikers who could imagine a beard on a face with the chin actually disappearing, a feat which Stromeyer's eidetiker could easily accomplish. Stromeyer and Psotka (1970) quote an even more astounding feat reported by Stratton (1917). The "Shass Pollaks" allegedly can report accurately from memory what word appears in what position of every page of each of the twelve volumes of the Babylonian Talmud. The cyclopean experiments reviewed in this section merely show the objectivity of a new technique that helps to quantify the fidelity and dura-

---

receptive field axis (with respect to the head axis) changed after tilting the cat. The compensatory torsion varied from cell to cell and sometimes gave a near complete compensation for the bodily rotation.

tion of eidetic imagery which existed previously only in the province of anecdotal accounts.

The site of eidetic storage is not known. That eidetic images require a relatively long scanning time to be built up and that the eidetic image remains during this process argues for a nonretinal process. That eidetic imaging occurs before stereopsis is consistent with this argument, since between the retina and the stage where global stereopsis may occur there are probably several synaptic junctions. The site of memory may be as early as the LGN, but one could imagine that the site of eidetic storage is highly central. The eidetiker may have a functional efferent pathway to the periphery (i.e., LGN or Area 17) which is inoperative for the rest of us. An even more speculative hypothesis might suggest an eidetic pathway through the pulvinar to Area 18, whose anatomical existence was only recently established; its function, however, is not yet understood. Whatever bizarre hypotheses one suggests, each is more plausible than the suggestion that the site of eidetic storage is retinal.

I did not venture into this field to proffer unwarranted hypotheses, but rather to show the reader the usefulness of some unfakable tests in special cases where no other foolproof evidence can be obtained. It is also a good example for the tracing of various processes in the CNS. Since eidetic images exist prior to stereopsis and can produce Land colors and since Land colors cannot be usually obtained binocularly (or if they can, the colors are very unsaturated), the mechanism of Land-colors appears to reside before stereopsis; and eidetic images are evoked before Land-color processing.

Let me conclude this strange section with a final remark. One might ask about the relevance of findings which exist for only a few children and a very select group of individuals among billions of people. The answer is simple. One should not forget that these eidetikers are otherwise normal human beings with a nervous system similar to ours. How do they differ from us? Is it a recessive gene, producing an atavistic visual system which might have been common in a previous generation? Is it a mutation, which is not reinforced by natural selection, since the survival value of eidetic imagery is questionable? Is it some chemical or neural change that occurs during childhood under rare conditions? Is it a latent skill which potentially all of us possessed but failed to learn during a critical period of maturation? Besides these questions, an even more provocative one can be raised. Could it be that all of us possess eidetic memory, but our recall mechanism is defective? Could it be that the detailed images experienced in hypnogenic states or drug-induced hallucinations, or direct electrical stimulation during brain surgery (Penfield (1952)), are a sudden access to eidetic recall?[3] I can only hope that some of these problems will be solved in the not too distant future.

I am aware that this section will be read by many colleagues with disbelief. If they

3. Some of the strange persisting or recurring visual imagery of brain-injured patients, called "palinopsia," may also indicate the existence of detailed texture memory (see Bender, et al. 1968). The palinoptic images are reportedly of high detail and great vividness and reoccur minutes, hours or sometimes even years after the stimulus is removed. Palinopsia is similar to visual hallucinations described by patients with hysteria or schizophrenia but is the result of lesions usually situated in the parieto-occipital lobe. Palinoptic images differ in some properties from

7.4-1 Left half of a stimulus pair for testing eidetic imagery; this half should be shown to one eye in order to be memorized, and the right half (fig. 7.4-2) should be shown to the same eye later.

7.4-2 Right half of stimulus pair for testing eidetic imagery (fig. 7.4-1 is the left half).

remain skeptical, there is no way to convince them but to encourage them to find eidetikers of their own. At least, this section tells how to do it. Since the possibility exists that the eidetiker may have poor stereopsis and might be missed when tested by a random-dot stereogram, a monocular test pair is provided. Figures 7.4-1 and 7.4-2 consist of two random-dot arrays (of 100×100 dots). The two arrays are

both ordinary afterimages and eidetic images. For instance, palinoptic images are always positive (unlike afterimages that would be negative under similar stimulation), but move with the eye-movements (unlike eidetic images that remain stationary). The cited paper by Bender, et al. gives several neurological references on palinopsia. Although eidetic imagery is clearly different from palinoptic imagery the reported details and long persistence of both kinds of imagery are very similar. Perhaps cyclopean techniques may be used in order to quantify the amount of information of palinoptic imagery as well.

identical, except for a cyclopean form. The cyclopean form is *not* shifted, but the dots that portray it are each other's complements in the two arrays. If any of the arrays is scanned for a few minutes with one eye (or both eyes) and then the other array is presented to the same eye (or both eyes), an eidetiker should be able to register the eidetic image with the second image and see the cyclopean form as a gray shape in a black-and-white random surround. In order to keep this test double blind, figures 7.4-1 and 7.4-2 are printed one above the other (thus they cannot accidentally be fused by the two eyes). Double-blind experiments are proposed to rule out the unlikely event of ESP.

Although figures 7.4-1 and 7.4-2 could be eidetically combined—at least in theory—this appears to be a more difficult feat than the stereoscopic fusion of two arrays of the same complexity (provided the eidetiker has good stereopsis). After all, for stereoscopic fusion the two arrays do not have to be in exact registration, but suffice to fall within Panum's fusional area. On the other hand, eidetic and physical images of the correlograms of figures 7.4-1 and 7.4-2, when monocularly presented, have to be in exact alignment within a fraction of a picture element to get the cyclopean message.

In order to help the interested reader in finding eidetikers, a much simpler random-dot correlogram is presented in figures 7.4-3 and 7.4-4. They were designed by J. O. Merritt (Harvard) and used by him for screening prospective eidetikers. The test is similar to the previous one, except we can assume that the eidetic storage of a few dozen large random dots and their proper alignment is much easier than that of ten thousand tiny dots. Only if an eidetiker is found who can perform this test should one proceed with the more complex correlograms and test whether the subject is a "super-

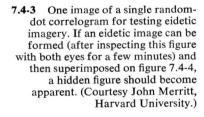

7.4-3 One image of a single random-dot correlogram for testing eidetic imagery. If an eidetic image can be formed (after inspecting this figure with both eyes for a few minutes) and then superimposed on figure 7.4-4, a hidden figure should become apparent. (Courtesy John Merritt, Harvard University.)

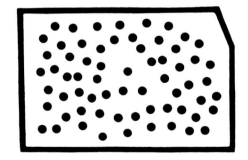

7.4-4 Other image of random-dot correlogram for testing eidetic imagery. See instructions with figure 7.4-3. (Courtesy John Merritt, Harvard University.)

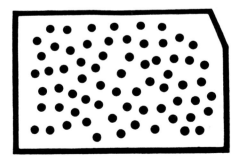

eidetiker." It appears that eidetic talents cover a wide spectrum from Haber's best subjects to Stromeyer's eidetiker. It is an intriguing possibility to rank-order them by the complexity of the random-dot correlograms they could store and match.

The existence of a detailed texture memory is particularly interesting for the study of cyclopean perception since it provides a new strongly cyclopean retina. It is possible to paint a cyclopean form by combining an eidetic image with an actual image (using one eye) such that these images by themselves do not contain the cyclopean form. Haber (1969) reported a similar experiment with his eidetikers where a drawing of a ship was stored as an eidetic image and later superimposed with a physical drawing containing some random lines (or another figure), which the eidetikers recognized as a human face. However, he did not exploit this demonstration for process localization.

With the eidetic-cyclopean techniques one can repeat all the experiments reported in this book and many others. Unfortunately, the site of the eidetic-cyclopean retina is *before* that of stereopsis, thus all the phenomena we have localized as being central (by using random-dot stereograms) are certainly central with respect to the eidetic-cyclopean retina. However, since no efferent pathway is known to exist to the retina, it is almost certain that the eidetic-cyclopean retina must lie after the anatomical retina. Therefore, one can tell apart retinal from nonretinal phenomena.

Stromeyer communicated to me a few eidetic-cyclopean experiments in November 1969. He tested several perceptual phenomena by eidetic-cyclopean stimulation. He portrayed the Benussi-Koffka ring, Hermann-Hering grid, the spiral aftereffect, Mach-band-like phenomena and the O'Brien (Cornsweet) disk phenomenon by eidetic-cyclopean stimulation.[4]

According to Stromeyer all the above listed phenomena could be obtained under eidetic-cyclopean stimulation. This is not surprising since I showed with random-dot stereograms that the Benussi (Koffka) ring can be cut by a cyclopean edge or a central component of the movement aftereffect exists. The implication of these findings that suggest that the Hermann-Hering grid phenomenon is not retinal or that some Mach-band-like phenomena are not retinal is also not unexpected. Inhibitory interactions in the LGN or in Area 17 might account for these phenomena. Nevertheless, what is surprising is the implication that eidetic memory is not only before global stereopsis, but perhaps much earlier.

I included this reference to an eidetic-cyclopean technique only for didactic reasons. Until several eidetikers are found and give similar results, these introspective reports are anecdotal. Thus the question whether lateral inhibition—responsible for

---

4. The Cornsweet disk phenomenon is a special case of earlier findings by O'Brien (1958) and is reviewed by Ratliff (1965, p. 75).

The Hermann-Hering grid, shown in figure 2.7-1, was produced in two steps. First only every second white square of figure 2.7-1 was presented in a black surround (corresponding to the white squares of a checkerboard) and stored as an eidetic image. Later the remaining white squares were physically presented (corresponding to the black squares of a checkerboard), and the subject was instructed to align her eidetic and physical images such as to obtain the grid of figure 2.7-1. The other stimuli were portrayed by similar methods and the design for these experiments is left for the reader as an exercise.

the Hermann-Hering grid and Mach-band-like phenomena—occurs in retinal ganglion cells, LGN, Area 17 (or some other sites before stereopsis) or in several of these stages simultaneously, remains to be answered.

Let me conclude this section with a final remark. One might accept these startling findings without reservations but doubt their relevance to physiological psychology, by noting that they were observed only on a few very special individuals. A possible answer to this argument is simply to define psychology broad enough to encompass the widest spectrum of human abilities. Were we to limit ourselves to the study of typical human abilities and exclude the unique feats of an eidetiker, or a simultan-blind chess master, or a J. S. Bach, our grasp of the mind would never reach over mediocrity.

The detailed coverage of these reports might convey the false impression that I have more reason to believe in these incredible feats than the reader has. In fact, I have never verified these findings myself! I believe that eidetic abilities must cover a continuum, and even simple random-dot correlograms might be too difficult stimuli for most eidetikers. So I must caution the reader to regard these reports with extreme care until other eidetikers of similar talents are found.

For cyclopean perception, per se, the existence of a cyclopean retina for eidetic imagery is of secondary importance if it occurs before stereopsis, as reported. There are other cyclopean retinae that can be stimulated before stereopsis and are accessible in any normal individual. However, for vision in general, there could hardly be a more important problem to be tackled than the existence of a detailed texture memory, and eidetic imagery might be one promising way to go about it. The main thing is that the right methodology exists now for establishing and quantifying eidetic memory. If some of my readers decide to follow up this work, particularly by testing young people, artists, and other special groups with high yield of prospective eidetikers, this section will not have been included in vain!

7.5 Cyclopean Movement Detectors

We have frequently discussed the conjecture that movement perception might be processed at several stages. In Hubel and Wiesel's findings in the monkey, the movement detectors in Area 17 are monocular, while in Area 18 and higher the movement detectors become increasingly binocular. Thus for classical stimuli containing compatible motion information on both monocular and binocular levels it is impossible to separate the action of monocular and binocular movement detectors. However, for random-dot stereogram movies it is possible to stimulate only the binocular movement detectors and determine whether the known rules of movement perception apply.

Julesz and Payne (1968) showed that stroboscopic movement perception of random-dot stereograms differs from the perception of classical stimuli both quantitatively and qualitatively, provided that the former stimuli are presented such that in addition to the elimination of all monocular form cues, all monocular movement cues are removed as well. Thus, while the frames of the movies contain correlated left and right stereoscopic images, successive frames are uncorrelated. When random-dot stereograms are presented this way and a binocularly perceivable grid is portrayed in translational or rotational movement, a new perceptual response can be experienced between the classical optimal movement and simultaneity regions. In this new region a single grid is seen at a standstill. This binocular standstill should be contrasted with simultaneity in which the two alternate stimuli are seen as super-imposed.

With this summary let us review in greater detail the differences between monocular and binocular stroboscopic movement perception.

Stroboscopic motion (also known as apparent or beta motion) can be experienced when a pair of slightly displaced stimuli are presented in rapid, temporal succession. As a result of the pioneering work of Exner (1875), Wertheimer (1912), Korte (1915), and Neuhaus (1930), the best conditions for stroboscopic movement perception are well-established. When the temporal interval between the onset of the two stimuli is longer than that required for optimal movement perception, the stimuli are perceived in succession. On the other hand, when the two stimuli are presented at briefer intervals than those which give stroboscopic motion, the experience is that of seeing the two stimuli simultaneously. Quantitative values for succession, optimal movement, and simultaneity thresholds depend on the spatial disparity between the two stimuli, the luminance of the earlier and later stimulus, the exposure duration of the stimuli, and the temporal pause between the stimuli according to Korte's laws (Korte 1915). Quantitative values also depend upon whether the two stimuli short sequence DS_1DS_2D or as a long periodic series $DS_1DS_2DS_1DS_2. \ldots$ In the experiments to be reported only the latter presentation is used.

In addition to these three perceptual responses, some researchers have introduced some further subdivisions of part movements, between optimal movement and simultaneity, in which S_1 and S_2 would be seen either as moving as a single entity or as two entities standing still. In all the reported cases (using ordinary line drawings) no mention has been made of subjects experiencing a single stimulus at a standstill between the simultaneity and optimal movement regions.

In our experiments we used random-dot stereograms in figure 2.8-5*. The binocularly correlated textures were selected to portray a vertical grid in the format of figure 2.8-4. When figure 2.8-5* is stereoscopically fused, vertical stripes are seen above a background.

In order to study stroboscopic movement perception with random-dot stereograms, two stereograms such as figure 2.8-5* were alternately presented, such that they portrayed binocular stripes in different locations (translated or rotated). For details I refer the reader to the experimental procedure, but I want to point out an important aspect of our experiments. It is *not* sufficient to change only the binocular shapes (global organization) while using the same random dots (local organization) in the two stereograms. In this case the correlation in time between texture points in successive left images and right images produces movement cues both monocularly and binocularly. The monocular movement cues also yield movement parallax, thus the basic rule of random-dot stereoscopic technique—the absence of all monocular cues—is violated.

In order to remove all monocular cues for stroboscopic movement studies, a special technique (Julesz 1966), was employed. Here the random-dot stereograms S_1 and S_2 each have a left and right image which of course are properly correlated (and portray some binocular shape), but the random-dot textures of S_1 and S_2 are uncorrelated. When such stereograms are presented in temporal succession and are viewed monocularly (no shape cues are present), the texture changes constantly and the patterns appear similar to the visual noise seen on a television receiver. When viewed stereoscopically, however, the binocularly correlated areas are seen at a different depth from that of the surround. The illusion of depth can be easily obtained in excess of 40 frames/sec. If the global organization (e.g., the stripe in fig. 2.8-5*) is the same in S_1 and S_2, the stripe is seen in vivid depth at a standstill, although the covering texture is dynamic noise. If the global organization is the same in S_1 and S_2 but with different disparities, it is easy to produce an optimal movement in depth in spite of the dynamic noise. If the global organization is the same in S_1 and S_2, it will have the same shape and disparity. Under these conditions, if the binocular shape is shifted or rotated in its position in S_1 relative to S_2, it will be possible to experience optimal movement. Furthermore, no monocular movement parallax can be experienced, and since the binocular shapes (e.g., grid) are at the same depth level, no change in depth can be experienced either. In our studies we used this stimulus condition exclusively, except for a control experiment.

In summary, we produced an artificial situation which was devoid of all monocular form and depth cues, including movement parallax. In addition, the contour of the binocularly produced shapes moved independently of the covering texture. For this class of stimuli, as our findings indicate, stroboscopic-movement perception differs from the classical type, not only quantitatively but qualitatively as well. Between the optimal movement and simultaneity region a new perceptual response is obtained.

We compared our results with those obtained by two-dimensional classical line targets (of similar format) only for calibration. One might argue that perhaps a three-dimensional classical line stimulus might have been a better choice for comparison with the random-dot stereograms. This argument misses an important point. It is not depth which produces the new phenomena to be reported here, but the independence between local movement (texture) and global movement (binocular shapes). When each point in the random texture follows the movement of the binocular shapes as if it were a solid sheet (i.e., S_1 and S_2 have correlated textures) the psychological response resembles the classical results (as shown later in the binocular control-experiment).

The experiments were divided into two groups: (a) classical (monocularly perceivable) gratings and (b) random-dot stereograms portraying gratings only when binocularly fused. Within both groups two stimulus configurations were used: a horizontal or a vertical grating, and two classes of movement (translation or rotation). Thus altogether eight cases were studied: monocular vertical—monocularly perceivable vertical grid with horizontal translation (M.V.); monocular horizontal (M.H.); binocular horizontal (B.H.); binocular vertical (B.V.); monocular vertical rotating (M.V.R.); binocular vertical rotating (B.V.R.); monocular horizontal rotating (M.H.R.); and binocular horizontal rotating (B.H.R.).

The classical grids have the same grid constants as the stereograms. The classical stimuli consist of eight black and white alternate bars as shown in figure 2.8-6. The random-dot stereograms as shown in figure 2.8-5* consist of 100×100 black and white picture elements selected at random. The left and right images are identical point-by-point except for vertical bars, 6 picture elements wide, which are shifted by 2 picture elements in the horizontal direction in one of the images. Only the even bars are shifted; as a result of this binocular disparity, when figure 2.8-5* is stereoscopically fused the binocular percept is of alternate bars in front of a surround. When monocularly viewed, of course, figure 2.8-5* gives the impression of an entirely random texture. Stimuli such as figures 2.8-4 and 2.8-5* were generated to portray horizontal grids as well.

For both the classical stimuli and the stereograms, the displacement between S_1 and S_2 was 4 picture elements (corresponding to 21 min arc) (thus 2 picture elements of the 6 picture-element-wide bars overlapped). For the vertical grids, the displacement was in the horizontal direction, and for the horizontal grid in the vertical direction.

For the rotational movement, S_1 containing a vertical (or horizontal) grid was tilted by $+6°$ from the vertical while S_2 was tilted by $—6°$, thus the grids of the two stimuli intersected at a $12°$ angle. This arrangement is illustrated by figure 2.8-6 which also shows the temporal course of the stimulation.

In figure 2.8-6, $S_1 = A$ and $S_2 = B$ has a duration $0.1T$, and the dark field D is $0.4T$, where T is the duration of a full cycle (that is, grid A rotating into position B and rotating back into the original position A). The stimuli were projected by two projectors by means of electronic strobe lights and triggering devices. In order to minimize the trapezoidal distortions resulting from off-axis projection, a stereoscopic projector with two optical systems in close proximity was converted to be used as a double projector with two independent strobe lights.

The random-dot stereograms S_1 and S_2 were made in vectograph format, so an ordinary projector could present them on an aluminum screen for direct stereoscopic viewing by means of polaroid glasses. The double-projector system projects S_1 and S_2 as two stereoscopic image pairs in exact alignment except for the 4 picture elements (21 min arc) translation or $12°$ rotation. The only difference between the use of classical stimuli and random-dot stereograms is the following: For classical stimuli, S_1 and S_2 can be exact copies of each other, and then one of them is displaced by 4 picture elements or rotated by $12°$. For random-dot stereograms, on the other hand, it is desirable to have S_1 and S_2 generated from uncorrelated random dots. This prevents monocular motion cues between local picture elements. An obvious way to assure uncorrelated stereograms is to use for S_1 a stereogram such as figure 2.8-5* and use for S_2 the same stereogram turned upside down. (The reader can obtain the same binocular percept when stereoscopically viewing figure 2.8-5* by turning the page by $180°$. The only difference will be the position of the black and white picture elements.) When such uncorrelated stereograms are presented in appropriate temporal succession, the monocular percept is one of dynamic noise—that is, the local black and white picture elements are perceived as performing random stroboscopic motion with small random displacements. On the other hand, in the stereoscopic view the global binocular organization can be perceived.

In the experiments four subjects with good binocular fusion were used. The luminance for both the classical and random-dot targets was kept identical at 0.32 ftL. The size of the stimulus was 17.9 cm and it was viewed from 118 cm distance, thus subtending 8.9° of arc. Observations on the classical targets verified the familiar findings and are given only for control purposes. Out of the three possible perceptual responses, only the optimal-movement threshold and the simultaneity threshold are plotted in figure 7.5-1. Thresholds were randomly approached from either above or below. The small standard deviation in the perceptual responses for any given subject is remarkable, since they strongly depend on various individual criteria. Deviation among the subjects was considerable and is probably due to the different perceptual criteria selected, but this is of no great consequence.

7.5-1 Stroboscopic movement perception thresholds averaged over four subjects (*A–D*) for eight stimulus conditions. (From Julesz and Payne 1968.)

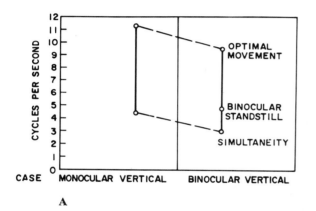

A

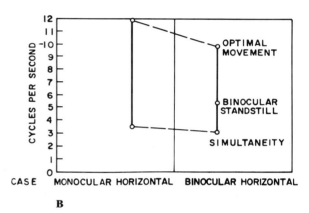

B

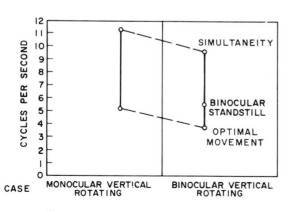

C

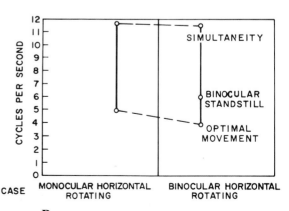

D

For the monocularly perceivable stimulus conditions the perceptual response for optimal movement and simultaneity are well-defined; the former clusters around 4 Hz, while the latter clusters above 10 Hz. In all these cases either the grid was seen as a single grid moving laterally (or rotating) back and forth, or the percept was of two superimposed grids at standstill. It should be noted that the classical grid was viewed binocularly in all these experiments.

The results for the eight stimulus conditions and averaged over the four subjects are shown in figure 7.5-1. For instance, note for random-dot stereograms the condition for optimal movement. The alternate bars in front of the surround are moving laterally or are rotating completely detached from the texture which covers them and is seen as dynamic noise. In the simultaneity region the bars in the two stereograms superimpose; and, since they are in the same depth plane, the result is the formation of a depth plane in front of the surround. Optimal movement and simultaneity thresholds for the random-dot stereograms are different from the classical cases, which is not surprising, since these thresholds depend on the stimulus contrast and contrast has a different connotation for random-dot stereograms.

An unexpected finding was obtained for random-dot stereograms between the optimal movement and simultaneity regions. A new percept could be experienced. The laterally moving (or rotating) vertical (or horizontal) grid in depth seemed to stop, and a single grid was experienced at standstill. This absence of movement was the more pronounced since the frame of the stereograms (which could be monocularly seen) was perceived as moving optimally, and the random-dots were perceived as dynamic noise. For instance, as shown in figure 7.5-1C for the B.V.R. case, the subjects experienced optimal movement of the vertical grid as a pendulum-like back and forth rotation by $\pm 6°$ around 4 Hz. When the repetition rate of S_1 and S_2 was increased, the extent of the swinging grid seemed to be reduced and at about 6 Hz the grid was experienced to be at a standstill in the vertical position. From figure 7.5-1C it is apparent that for the equivalent monocular case (M.V.R.) 6 Hz corresponds to optimal movement, and indeed the frame of the stereograms was vividly perceived as oscillating between $+6°$ and $-6°$; while the grid in front of the surround was seen at rest.

This finding extends the perceptual responses for stroboscopic movement into a new region which we named "binocular standstill." This region always lies between the optimal movement and simultaneity regions; and it exists for all four stimulus conditions we have tried (B.V., B.H., B.V.R., and B.H.R.) as given in figure 7.5-1. The problem of how binocular disparity affects binocular standstill and other thresholds in random-dot stereograms has not been studied.

It should be mentioned that for a strong binocular standstill to be experienced, the random-dot textures in S_1 and S_2 have to be uncorrelated. If the textures are correlated, then a strong monocular cue is produced. The textures (both of the grid in depth and of the background) are perceived as oscillating laterally in exact synchrony with the stripes themselves and can be seen monocularly as well. This "classical" movement condition interferes with the binocular standstill-phenomenon. Thus we can conclude that the binocular standstill phenomenon is not the result of binocular depth perception alone but requires the separation of object movement from texture movement. It should also be noted that the optimal movement thresholds are much higher for the binocular control experiment than for the cases of B.V. and M.V.

In classical situations, monocular and binocular movement cues are inseparable. The random-dot stereoscopic technique permits us to produce movement cues which exist only binocularly. It is possible that the simple movement detectors found in the visual cortex of the cat and monkey by Hubel and Wiesel (1962, 1968) are different from the higher-level moment analyzers which have to occur after the pattern-matching operation (on which the stereopsis of random-dot stereograms is based).

Our findings corroborate this hypothesis. It appears that movement perception for random-dot stereograms is qualitatively different from the classical results. In the absence of all monocular cues, the binocular movement detectors (higher than the pattern-matching process) seem to operate differently from the simple movement detectors. It seems as if this higher process would interpolate (with some inertia) a certain position between the positions of the two given stimulus sequences. As long as the process can follow the time course of the stimulation, the interpolated solution will be seen as moving. As the stimulation time gets shorter, the amplitude of the interpolated moving percept is reduced. When the threshold is reached, the process takes the mean value between the two extreme positions of S_1 and S_2, and the binocular standstill situation is experienced.

Some recent experiments already indicate that the local monocular and binocular processes which take place prior to the global pattern-matching process (which is necessary to perceive random-dot stereograms) are quite different from those processes which operate afterward (Fender and Julesz 1967a). This study provides another example which indicates that these higher-order processes can detect stroboscopic movement, but this is based on mechanisms which are different from the more peripheral monocular and binocular movement detectors. When cues exist for which both mechanisms can operate, the peripheral mechanism usually dominates, and the operation of the higher-order mechanism is concealed. Only with the random-dot stereoscopic technique is it possible to eliminate those cues which can be utilized by the peripheral processes and thus to study the higher-order processes.

It is most revealing that for the binocular control experiment (when the stereograms S_1 and S_2 are correlated in time having similar random-dot textures), the binocular standstill-phenomenon cannot be experienced. Thus the binocular standstill-phenomenon is not the result of binocular depth perception alone but requires the separation of perceived object movement (global organization) from texture movement (local organization). Since in real life, objects and their textures usually move together, it is not surprising that the reported phenomenon has not been noticed until now.

7.6 Pulfrich Phenomenon without Monocular Cues

When a pendulum bob is swung across the field of view and observed with an attenuator over one eye, the path of the pendulum appears to be elliptical in depth. As the density of the attenuator is increased, the apparent depth of the pendulum path is also increased. If the attenuator is changed from one eye to the other, the forward part of the ellipse will be seen on the opposite swing of the bob. This is the Pulfrich phenomenon, discovered by Pulfrich (1922), who assumed that attenuating one eye in effect puts a delay line in the system for that eye. This explanation fits the data collected by Lit (1949), but the first direct test of the delay line hypothesis was provided by Julesz and White (1969) using random-dot stereogram movies.

In the Julesz and White experiments a computer-generated film loop was prepared in which each frame was a random-dot stereogram. Previous research has shown (see § 5.10 and Julesz and Payne 1968) that depth is easily seen with rapid cinematic presentation of a succession of such statistically independent random-dot stereograms. Although these movies appear as visual noise when monocularly viewed, when they are viewed binocularly, this dynamic noise separates into a square seen in front of (or behind) the surround. The film loop used in the demonstration was identical with the previous case except that the stereoscopic pairs were displaced by one frame. Because of this displacement, no simultaneous pair in the loop of statistically independent matrices could form a stereoscopic pair, but every pair formed by the left image in frame n, and the right image in frame $n + 1$ could form a stereoscopic pair as shown in figure 7.6-1. Indeed, the left matrix on frame n if paired with the right

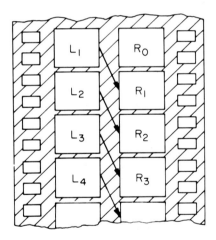

7.6-1 Schematic illustration of the way the Pulfrich phenomenon was tested with random-dot stereogram movies. (From Julesz and White 1969.)

matrix on frame $n + 1$ would portray a square in front of the surrounding matrix. The left and right matrices of each frame were presented separately to the left and right eyes by polarizing filters; however, the eye receiving the "leading" matrix was looking through an attenuator.

In accordance with the delay-line hypothesis, observers reported stereopsis when the eye receiving the leading matrix was attenuated by the proper amount. The experiments were conducted at several film speeds and with various illuminations, the dimmest luminance being 0.01 ftL, and the brightest 0.1 ftL. The horizontal extent of the viewed image was 30° of arc. Observers were asked to watch the film, while slowly adjusting the attenuation of a continuously variable neutral density wedge until the display appeared to change. At slower speeds (12 frames/sec) they had little difficulty in reporting stereopsis once the attenuator was adjusted to sufficient density. They first reported a scintillating rectangular field of dots; and then as the attenuator was adjusted to greater density, they would remark that there was "something different about an area in the middle," and with further attenuation they would report, often with surprise, the presence of a center square in front of the surround. Optimal attenuation could be easily set, and at 12 frames/sec was about 12 dB (for 0.01 ftL) and 15 dB (for 0.1 ftL).

At faster film speeds the adjustment of the attenuator became less critical and had to be decreased to achieve stereopsis. At 19 frames/sec an attenuator of 5 dB gave the best stereopsis, but even without attenuation the center square could be perceived in depth. At 24 frames/sec strong stereopsis could be obtained without attenuation, and only 2 dB attenuation was required for optimum stereopsis, although the improvement was slight. The perception of depth at high film speeds without attenuation of the leading frame, was possible with either eye receiving the leading frames. Similar results occurred at higher illumination (0.1 ftL) except that at 19 frames/sec the optimal attenuation was 12 dB and without attenuation stereopsis was difficult to obtain. Yet at 24 frames/sec no attenuation was required for stereopsis.

While the basic finding seems to confirm the delay-line hypothesis, the result that at faster film speeds attenuation is not critical or is not even necessary was explained by a simple model of Julesz and White (1969). We assumed that the stereoscopic depth mechanism has an inertia with a time constant T. As a result of this temporal inertia, two or more successive frames will be retained and averaged, thus T will be also referred to as integration time. The higher the film speeds the more frames of the left and right images will be averaged and the more correlated frames will be contained between them. For example, if at a certain film speed four frames are averaged with assumedly equal weights, then at a given point in time the left eye integral $L = L_1 + L_2 + L_3 + L_4$ and the right eye integral $R = R_0 + R_1 + R_2 + R_3$ will contain between them three binocularly correlated matrices (L_1R_1), (L_2R_2), and (L_3R_3) out of the four. Adding a one-frame delay by the attenuator merely increases the correlations from three frames to four. But at slow speeds, where the number of averaged frames is small, the averages of the left and right images will have inadequate correlation, and one-frame delay (caused by attenuation) will result in 100% correlation so that a dramatic increase in stereopsis will result. From the findings one might also assume that T decreases with increased illumination.

Despite the simplicity of this summation model it is hard to explain the fact that stereopsis without attenuation is obtained for film speeds as slow as 16–24 frames/sec which corresponds to the optimum apparent (stroboscopic) movement region. At higher speeds, where simultaneity is experienced, the summation model seems very plausible because many successive frames are actually being perceived as superimposed. In the optimum region, however, the texture elements in successive frames are seen not as superimposed but as moving, so that the individual frames must be preserved. A further complication is that the individual dots in the matrices are seen moving in the various depth planes, which means that after binocular processing the successive frames are not summed. One possible assumption might be that stereopsis and the binocular combination of two eyes' views into a single percept are two different mechanisms. The time constant for stereopsis might be shorter than for binocular combination. Even if the correct model remains elusive, it seems that the explanation of the Pulfrich phenomenon requires a short-term visual memory having two time constants, one being the differential delay time between signals from the two eyes (depending on the brightness of the stimuli), the other being the monocular integration time.[5]

This technique can serve to measure propagation times in the optic tract for various color receptor outputs. In this case the left and right eyes are presented with random-dot stereograms of different hues. We have not done this experiment so far. However, instead of brightness changes between the two eyes, we used random-dot stereograms of the same brightness but the leading frame was composed of dots having diameters smaller than those of the other image. This diameter change had the same effect as brightness change.

7.7 Cyclopean Contours and Closure Phenomena

One of the original motivations for developing the technique of random-dot stereograms was my interest in the problem of defining contours. In 1959, I published a paper on a method of television signal encoding (Julesz 1959) that transmitted only the local extremal values of the signals and at the receiver a linear brightness function was interpolated between two adjacent transmitted samples. Since television signals are linearly scanned, this scheme corresponded to the detection of one-dimensional edges. The purpose of this scheme was television bandwidth reduction, and the obtained savings were rather modest (about a factor of two). It seemed more promising to define a contour in two-dimensions and to try to interpolate some simple two-dimensional surfaces between adjacent contours. A simple definition of a contour as the local maxima of the gradients goes back to Mach (1897), but for complex scenes with textures it yields very poor results. Without semantics no simple mathematical definition can describe the various boundaries between different areas (i.e., the hairline of a face, or the contour of a furry animal against a complex background).

In order to define a contour one must be able to separate the overlapping or contiguous (as bricks in a brick wall) objects from each other. The Gestaltists strongly pursued this problem of "unum-duo" organization which, for their outline drawings, was particularly difficult to study. Objects occurring in real life can be separated from each other owing to differences in their contours or in their spatial coordinates. Two touching objects with the same texture can be told apart only by some familiarity cues. If, for instance, a wooden block is placed on another piece of wood having the same texture, it is difficult to determine whether one or two objects are being shown. However, if a wooden statuette is placed on a wooden desk top, one usually assumes

---

5. This monocular integration time is in the same range as the visual short-term memory of Sperling (1960). Whether these memory processes are actually identical, remains to be seen.

that they are two different objects. On the other hand, if the textures of the objects are different or they are geometrically separated, one does not need familiarity cues to tell them apart.

This observation led me to the problem of two-dimensional texture discrimination and into three-dimensional vision. With random-dot stereograms I wanted to prove that one can separate objects and produce contours without using form recognition. My motivation has been explained in detail elsewhere (Julesz 1965a). While I still have no idea how contours of complex monocular images might be extracted, it is relatively easy to develop an automatic procedure that can separate the objects and thus define their contours from a corresponding stereoscopic image pair. (The problem of automatic depth localization will be discussed in § 8.4.)

The importance of random-dot stereograms for contour perception was immediately recognized by psychologists. White (1962) emphasized that here was a rare instance of visual contour without the usual abrupt gradient in brightness. Lawson and Gulick (1967) devoted considerable effort to anomalous contour (without brightness gradient) and to the problem of whether stereopsis is necessary to induce anomalous contours. They stressed the following historical facts:

Recently, Ogle (1959) presented a theory of stereoscopic vision based upon experiments of the late nineteenth century as well as recent investigations (Ogle, 1950, 1952, 1953). It is representative of the role of retinal disparity in depth perception. According to Ogle, ". . . in every case stereoscopic depth depends on the disparity between images of identifiable contours" (1959, p. 380). Clearly, this is not so in the Julesz situation. *Instead of contours giving rise to depth, rather, it is depth that gives rise to contours.*

The importance of this fact is that contour formation can occur in the absence of retinal processes. Lawson and Gulick (1967) also pointed out that the notion of anomalous contour had some previous history. Schumann (1904) reported findings of what he termed "subjective contours" with figures in which an abrupt brightness gradient defining an edge was interrupted over angular gaps as large as 5°. Some of his figures gave the impression of a continuous contour which extended across the physical interruption. The appearance of this subjective contour depended on contrast and form, and on this basis Schumann concluded that contours can arise from both retinal and central processes. Schumann was more concerned with closure phenomena than with contour formation, and there are some clear differences between subjective contours of Schumann and the anomalous contours produced with random-dot stereograms. Schumann's contours appear only when the point of fixation does not lie upon the contours, while Julesz's contours remain regardless of the point of fixation and far exceed Schumann's contours in clarity and sharpness.

In an interesting experiment Shipley (1965) combined the two types of nonclassical contours into a single figure. A random-dot stereogram similar to Shipley's is shown in figure 7.7-1*. Here the basic random-dot stereogram, with a center square in depth, is interrupted by a horizontal uniformly white stripe that cuts both the center square and its surround in half. In spite of the fact that neither binocular disparity nor brightness gradient is provided in this uniform area of interruption, a crisp

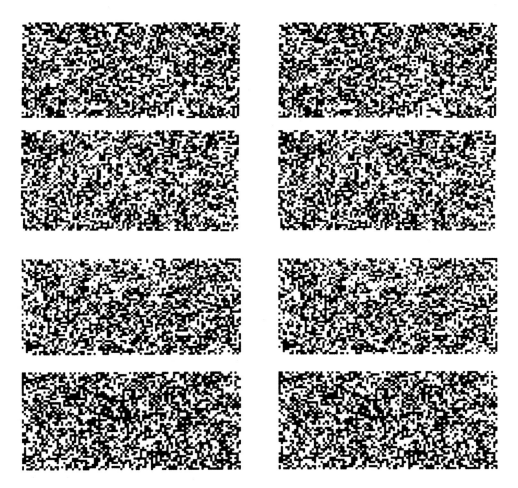

7.7-1* Anomalous contour after Shipley. Closure of cyclopean square across white gap can be experienced.

7.7-2* Similar to figure 7.7-1*, however, the diamond-shaped cyclopean figure does not give rise to a corner-shaped anomalous contour.

contour is apparent that seems to be a continuation of the cyclopean contours. Furthermore, the blank area between these contours of Shipley is perceived in depth as a part of the upper and lower parts of the center square, respectively. This contour phenomenon appears to be the cyclopean counterpart of Schumann's contour. It is the strongest when the eyes are not fixated at the locale of the contour. However, since the cyclopean figures in random-dot stereograms have no contrast changes, it is clear that the subjective contour and closure effect of Schumann is a central process, occurring after stereopsis is obtained. Whether this central process is the result of some Gestalt factors, as Shipley contends, is by no means clear. For instance, it is much more difficult to obtain closure of a cyclopean diamond shape and impossible to obtain a subjective contour in a corner shape, as demonstrated by figure 7.7-2*. Nevertheless, this subjective contour phenomenon is of a much more central level than the contours produced by random-dot stereograms and it would not be surprising if some gestalt factors entered, whatever they may be. On the other hand, for anomalous contours produced by random-dot stereograms no gestalt factors operate (as pointed out by Julesz 1960a). Indeed, close inspection of the cyclopean vertical edges of the center square in figure 2.4-1* reveals that these edges are not straight but are jittered by one or two picture elements. The reason for this is simply that each dot at the neighborhood of the vertical edges has a 50% probability (by chance alone) that it can belong either to the square or to the surround. If gestalt factors, particularly good gestalt were operative, one should have perceived straight edges and not jagged edges. Thus, stereopsis-produced anomalous contours are processed at cortical levels, but still not so centrally that gestalt considerations might have any role.

Of course, all these nonclassical contours are meant in a global sense. This obvious fact is often forgotten, which leads to some confusion. For instance, Kaufman (1965) argues that in random-dot stereograms the contours that surround dot-clusters in the corresponding two half-fields are similar. Although figure 3.9-1* which was generated a decade ago (Julesz 1960) argues against this view (since the contours that surround the dot-clusters are very dissimilar in the two views and yet global stereopsis can be obtained), the main issue is quite different. Even if some local contours are correlated in the two images (and the binocular edge detectors in the cortex exploit this fact), the cyclopean forms and their contours are not determined by these local contours. These cyclopean contours are only the results of finding corresponding areas of similar textures in the two views. Whether or not local brightness contours line up along a cyclopean contour is quite irrelevant.

In my terminology, anomalous and subjective contours are called cyclopean contours. They can be produced by binocular disparity as in figure 2.4-1*, by areas of binocular fusion or rivalry when trying to stereoscopically fuse figure 2.3-5. They can also be produced by movement gradients, when 2.3-5 is monocularly viewed in rapid temporal alternations or by interpolation between existing contours (classical or cyclopean) such as in the closure phenomena of Schumann or Shipley. Kaufman (1965) produced several patterns similar to figure 2.3-5 in which a correlated area was inserted in uncorrelated surrounds which were often horizontally and vertically shifted in the two eyes' views. The correlated areas appear in front of the uncorrelated areas that yield binocular rivalry; because of this Kaufman concluded that binocular disparity is not necessary to produce anomalous contours. This is certainly true; however, the correlated areas perceived above the uncorrelated ones are not at a stable depth. This depth without disparity will be discussed in the next section.

At this point it should be noted that cyclopean contours produced by random-dot stereograms behave, in many respects, similarly to classical contours. From cutting the Benussi (Koffka) ring perceptually in half to the portrayal of illusory figures and perspective outlined drawings, cyclopean contours are indistinguishable from classical contours. The only difference is a technical one. Classical contours based on brightness gradients can be drawn very thin while cyclopean contours must be a few picture elements wide to produce stable depth or movement gradients. Of course, with fine picture resolution, the cyclopean lines can be made quite thin but still require at least two or three dots in width. For an observer with about 1 min of arc visual acuity, the minimum thickness of the cyclopean lines is about 3 min of arc.

There are some phenomena, however, for which cyclopean edges behave differently from the classical ones. I discussed such a case in § 7.5, where the perception of cyclopean edges in motion differs qualitatively and quantitatively from the strobo-scopic motion perception of classical contours. Furthermore, experiments with cyclo-pean optical illusory figures at different depths indicate that the influence between cyclopean contours may take place in a three-dimensional space.

7.8 Perception of Undertermined Areas

The problem of how undetermined areas (i.e., areas without disparity information) are localized in depth started with the pioneering work of Panum (1858) known as Panum's limiting case. As shown in figure 7.8-1* this stimulus is comprised of two

vertical lines presented to one eye, and one vertical line presented to the other eye. For figure 7.8-1* the percept after stereoscopic fusion is that of the right-hand line being forward and the left-hand line back. Thus, Panum's limiting case provides the minimal stimulus conditions for the occurrence of "stereopsis." Optimal stereopsis occurs when the space between the lines is within Panum's fusional area. Whether this "stereopsis" is identical to stereopsis in the usual sense will now be discussed.

Since only one of the lines has a corresponding representation in the other eye's view it is somewhat of a puzzle why the line without disparity is seen in depth. Of course, one can regard the single line in one view as being two interposed lines, one hiding the other, in which case the riddle is resolved. There are a large number of variants of Panum's limiting case in the literature, customarily in the form of single-line drawings. Such line drawings are degenerate forms of real-life images and as such are particularly unsuitable for getting better insight into this phenomenon. Indeed, in random-dot stereograms such as figure 2.4-1* areas on the left and right side of the center square (denoted by X and Y in figure 2.4-2) are presented to only one eye. However, as stereoscopic viewing of figure 2.4-1* shows, these undetermined areas are seen as being at the depth level of the farthest areas that are determined in depth. Thus, when the center square is in front of its surround, the "no-man's-land" is perceived as being the continuation of the surround; when the square is seen behind the surround (by changing the red-green viewers) the center square appears as a rectangle since the undetermined areas are seen as being continuations of the center square.

This tendency to perceive binocularly correlated areas in front of uncorrelated areas is even more clearly apparent when figure 2.3-5 is stereoscopically viewed. Here, only the center T-shaped area is correlated in the two eyes' views while the surround merges with the printed page and forms a depth plane behind. This tendency to perceive correlated areas as being in front of uncorrelated areas is probably the result of the simple geometrical fact that usually objects closest to us (thus unobscured by other objects) are seen by both eyes, while uncorrelated views are usually the result of interposing objects in one eye's view. This way of perceiving correlated areas is a highly central phenomenon, since it occurs after stereopsis. With ambiguous stereograms, such as figure 6.2-8*, it is possible to study this phenomenon. What is remarkable about figure 6.2-8* is that one of the ambiguous surfaces is a plane close to the observer in which each texture element casts a projection on both retinae. Nevertheless, the other organization, a wedge behind the page of the printed page, can be also obtained with ease, and for some observers this wedge is the preferred organization. Thus, stereopsis overrides the tendency to perceive the correlated areas in front in the case where stereopsis cues of a competing organization are provided.

This tendency of perceiving undetermined areas at the same depth as the depth of adjacent determined areas has been generalized by Gogel (1965) in his "adjacency principle" for other perceptual cues besides depth. This principle is similar to the Gestaltist's view which regarded adjacency (proximity) as one basic cue for perceptual grouping. Since the demonstrations by the Gestaltists were usually two-dimensional drawings, Gogel's idea is, therefore, a useful reminder that proximity must be regarded as a three-dimensional entity. However, we have seen how such general principles have to be modified for each specific case, which raises the question whether it is worthwhile at all to state them.

A very nice demonstration of how undetermined areas can be perceived in depth was given by Lawson and Gulick (1967). They stripped the random-dot stereogram to its minimum. The essence of their demonstration is shown by figures 7.8-2* and

7.8-2* Classical case of stereopsis after Lawson and Gulick for which each dot has a well-defined disparity. The white surfaces within the inner and outer square are perceived at the depth of the printed page. (From Lawson and Gulick 1967.)

7.8-3* Anomalous contour by Lawson and Gulick. Fusion gives rise to the inner square surface in front of the printed page and anomalous contours (without disparity). (From Lawson and Gulick 1967.)

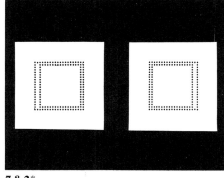

7.8-2*

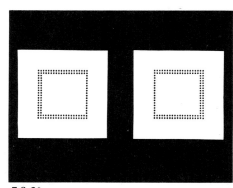

7.8-3*

well-defined disparity. Fusion yields depth; the dots of the inner square appear in front of the dots of the outer square. However, the white surfaces within the inner and outer squares are perceived at the depth of the printed page. On the other hand, in figure 7.8-3* the left and right inner vertical dotted array has no corresponding disparity. Fusion gives rise to the inner square surface in part of the printed page and anomalous contours. Although the authors have other explanations, I regard this as an extension of Panum's limiting case. The two vertical dotted arrays on the temporal sides give rise to depth as in Panum's limiting case. But this depth is not identical to normal stereopsis, since the uniform undetermined areas in-between are carried with the undetermined vertical contours.

This and the previously cited examples clearly show that perceiving undetermined areas in depth is not identical to stereopsis proper. The explanation to regard the Lawson and Gulick figures as Panum's limiting case is further emphasized by the study of Lawson and Mount (1967) in which figure 7.8-3* was modified by omitting the horizontal dotted arrays. Even under these conditions depth and anomalous contours were seen, demonstrating that lateral disparity alone is adequate.

In all these examples areas without disparity were surrounded with areas that possessed disparities. Whether depth can be perceived in the total lack of disparity is a question that is often posed but never is substantiated. Attempts to produce stereopsis by one eye have always failed, and usually the investigators were misled by some disguised cues that gave rise to movement parallax. I have discussed experiments in which stereopsis could be obtained from alternate presentation of disparate images to the left and right eyes, respectively (Efron 1957, Dodwell and Engel 1963). However, reports that stereopsis can be obtained by alternately presenting disparate images monocularly (e.g., Stasiak 1965) were repudiated by Engel (1966) using random-dot stereograms. He concluded that without monocular depth cues no stereopsis could arise from monocular presentation.

7.9 Cyclopean Aftereffects

Visual aftereffects like optical illusions are popular research topics for workers in perception. Some of these aftereffects, such as the "waterfall illusion" belong to a class of phenomena more technically and more generally known as the visual aftereffects of movement (VAM), and had been reported by Addams (1834) two years after Necker (1832) described one of the first optical illusions. These aftereffects were regarded for many years as a special case of optical illusions, and the interested reader is referred to Thompson (1880) who in his article "Optical Illusions of Motion" reviews many such illusions observed by Brewster, Wheatstone, Faraday, Plateau, and others. The many other aftereffects from the "tilt aftereffect" of Gibson (1937), through the "figural aftereffects" of Köhler and Wallach (1944) to the "color-edge aftereffects" of McCollough (1965) and "color-movement aftereffects" of Hepler (1968), are important additions to the contributions of the last century.

Although the discovery of cortical-edge and moving-edge detectors and their presumed adaptation provides a reasonable hypothesis for several of the aftereffects, it is not at all certain whether all the aftereffects are central. Furthermore, with increased neurophysiological sophistication an Area 17 process is less central than an Area 18 or Area 19 process, and it would be instructive to localize some of the aftereffects with this precision. Some of the aftereffects last for hours or days (which might imply central phenomena at work), yet are retinal-position specific (which might suggest peripheral mechanisms). For instance, Masland (1969) observed a long-term spiral-aftereffect of motion. He showed that if an observer fixates a rotating spiral figure for 15 min a negative aftereffect of motion is experienced when the observer inspects a stationary spiral 20 hours later. When the adaptation spiral is rotating clockwise at 80 rev/min so that it appears to contract, the stationary test spiral appears to expand and only a counterclockwise rotation with 1.3 rev/min will abolish this expansion. This speed of abolishing the aftereffect remains almost constant whether the test run occurs 15 min or 24 hours after adaptation. The illusory motion is seen only when the test spiral falls upon the portion of the retina which had been stimulated by the rotating spiral. The stationary test stimulus has to fall within about 1.5° of arc of the location of the adapting spiral in order to obtain an aftereffect. Masland suggests that the physiological basis of this aftereffect might be related to the findings by Morrell (1967) and Chow, et al. (1968), which show evidence of plasticity in firing patterns of single units of cat visual cortex and LGN.

The spiral-aftereffect of Masland is much weaker than the classical spiral-aftereffect that fades away within a minute. This classical spiral-aftereffect is very strong. After adapting for a few minutes to a rotating spiral that appears contracting (expanding) any textured surface appears expanding (contracting). (The reader can easily verify this by drawing a 3–5 revolution spiral and placing it on a turntable (with the usual 33⅓–78 rev/min) and fixating at the center for ten minutes). However, this aftereffect transfers very weakly to the other eye. By the way, Exner in the last century observed that he could adapt one eye to contraction and the other eye to expansion and observe the opposite aftereffects for each eye at the same time. In the light of physiological evidence of binocularly driven moving-edge detectors in the monkey cortex it is very mysterious why the spiral-aftereffect does not yield a strong interocular transfer. Although when I tried to observe interocular transfer of the spiral aftereffect, it was very weak, there are many distinguished researchers who report strong interocular transfer. One of the recent such reports is by Freud (1964). Furthermore, Barlow and Brindley (1963) and Scott and Wood (1966) reported that interocular transfer persists even when the previously adapted eye is pressure blinded, and concluded that cortical processes are at work. Richards and Smith (1969) observed that the diameter of the adapted area did not change size with convergence movements of the eyes. From this lack of zooming they assumed that the spiral aftereffect is not processed in the geniculostriate system but in the optic tectum (superior colliculus). Although this conjecture is based on another conjecture (that the zooming process occurs in the LGN) it shows that there is no consensus of where the side of the spiral aftereffect may be, and even the problem of whether it is entirely central is not settled. Therefore cyclopean techniques can be advantageously used in order to tell peripheral and central processes apart.

Besides cylopean techniques, binocular-transfer experiments can sometimes establish whether peripheral or central processes are at work. Even simpler arguments are often used to localize aftereffects. Kohler (1964) regards peripheral aftereffects as being superficial and central aftereffects as being far-reaching. He claims that as long as the stimuli are of short duration, their effects can be only peripheral, restricted to a limited area. Only prolonged stimulation can bring about stable, long-lasting aftereffects that affect central visual areas. His definition, based on short and long stimulations, might be useful in prism adaptation experiments for which lasting perceptual changes require several days of adaptation, but we shall not use such methods. We simply produce an aftereffect by a random-dot stereogram, and if we are able to experience it, then the process occurs after stereopsis. Such a cyclopean localization by stereopsis is particularly informative in the case of aftereffects, since most aftereffects are probably not retinal; and to show their central origin is not very informative. However, to show that they occur after stereopsis—that is, most likely after Area 17— is more than trivial.

It is interesting that the first cyclopean aftereffect tried was the waterfall illusion by Papert (1964). History repeats itself. By using random-dot stereogram movies of temporally uncorrelated frames, he produced moving bars in depth that could not be perceived by one eye alone. He claimed an aftereffect of opposite motion when the moving cyclopean contours were substituted by a uniform field. My own observations with similar techniques confirm the existence of the cyclopean waterfall illusion, yet I found the effect much weaker than obtainable by classical stimulation. It appears that movement extractors prior to and after stereopsis have their share in this aftereffect of movement.

It is possible to repeat all the aftereffects under cyclopean stimulation, as we did for optical illusions, and in the successful cases separate the central and peripheral components. Such a program would be quite laborious and difficult. An attempt to localize the McCollough effect by cyclopean techniques by Stromeyer and myself failed. The only reason why I review a negative experiment is that it shows some of the difficulties of cyclopean techniques. We presented a random-line stereogram of 70×70 horizontal and vertical line segments. The center square can be seen after fusion even when one of the images is rotated. According to the perceptual criterion used (rounding off the corners of the center square or loss of depth) one can tolerate approximately $\pm 6°$ of rotation. One would expect that coloring the corresponding line segments with the same color would improve fusion (i.e., increase the tolerated rotation) while opposite coloring would reduce fusion. We used the McCollough effect for coloring the line segments, by adapting subjects to vertical red bars and horizontal green bars in both eyes (or oppositely in the contralateral eye). The obtained color aftereffect had such a low color saturation that its effect on fusion could not be measured. If greater saturation had been produced one would have been able to determine whether the color aftereffect occurred before stereopsis or after. The failure of this experiment illustrates that stereopsis often requires stimulus parameters that have to be stronger than those which can be produced by aftereffects.

A refinement of the previous experiment is obtained by the use of the stereogram shown in figure 7.9-1. Here a 50×50 cell array is identical in the left and right field with a center square of 20×20 cells, having a one cell disparity. The only departure from the usual stimuli is that each cell instead of being randomly black or white consists of randomly chosen vertical or horizontal grids. In figure 7.9-2 the grid orientations in corresponding cells of the two images are orthogonal to each other. Because of this, figure 7.9-2 cannot be fused and strong binocular rivalry is experienced. Would it be possible to color the corresponding cells with the same hues and overcome the rivalry owing to the orthogonal gratings? If the answer is affirmative could the McCollough phenomenon be used to color the corresponding cells? The first question can be answered immediately in the affirmative. We demonstrated several experiments (e.g., figure 3.9-5*) in which similarity of brightness could overcome binocular rivalry arising from orthogonal line segments. (In § 3.5, one of my experiments was reviewed in which the rivalry between a negative and positive stereoscopic image pair could be overcome by using correlated hues.) The possibility of answering the second question depends on three requirements. First, the colors of the aftereffect must be adequately saturated to counteract rivalry. Second, the McCollough effect has to occur before stereopsis. Third, it should be possible to adapt one eye to horizontal-green and vertical-red gratings and the other eye to vertical-green and horizontal-red gratings. This third requirement assumes that the aftereffect occurs prior to Area 18 where the majority of receptive fields are binocularly driven by edges that have the same orientation on both retinae. Unfortunately, the low saturation of the colors interferes with the experiment. This experiment is included here only because subjects differ in the intensity of this aftereffect. (Perhaps the reader might try to adapt his left and right eye with perpendicular gratings of alternate red and green colors, respectively, and obtain stereopsis afterwards when fusing figure 7.9-2.)

Another unsuccessful attempt is shown in figure 7.9-3. Here the array consists of only 33×33 cells. The cells of the left image of figure 7.9-3 consist of horizontal and vertical gratings, while the cells in the right image are uniformly colored by unsaturated red and green. It is easy to produce the McCollough color aftereffect for the left image, and it appears similarly colored to the physically colored right image. However, the saturation is not adequate to overcome the rivalry between the striped and the uniform cells, though I observed some fleeting moments of fusion. The random mesh at the boundary of the cells is identical in the left and right images and is intended to prevent fusion by an artifact. Without such a mesh the horizontal and vertical gratings in adjacent cells are perceived as clusters, respectively, with clear boundaries between these clusters. These boundaries between horizontally and vertically striped areas are identical to the boundaries between the red and green areas in the right field. Thus the high frequency spectrum is similar in the two arrays and fusion is easily obtained. However, the random mesh effectively breaks up these clusters, and no fusion takes place prior to adaptation.

Our failure to produce global stereopsis by utilizing the McCollough effect is not conclusive. I am confident that by further tricks the desired result will be obtained. My belief is based on the fact that each eye respectively can be adapted by orthogonal gratings to an opposite aftereffect. Thus, the McCollough effect occurs before

7.9-1   Random-dot stereogram for testing the McCollough aftereffect. The 50×50 array of cells is composed of horizontal and vertical gratings.

7.9-2   Same as figure 7.9-1 except the corresponding cells in the left and right images contain orthogonal grids.

7.9-1

7.9-2

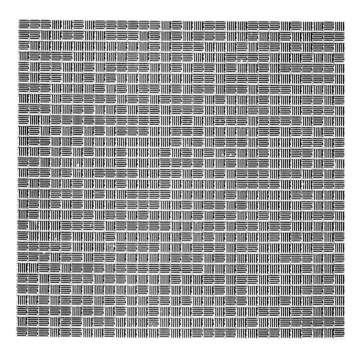

7.9-3 Stereogram for testing the McCollough aftereffect. The left image should be projected and viewed as it is. The right image should be built-up from two projected images in exact registration; one of the images should be projected through a red filter, the other through a green filter. The resulting red and green right image, and the black and white left image should be viewed stereoscopically with the help of prisms.

stereoscopic combination. That the aftereffect is not strong enough to overcome binocular rivalry is probably the result of the procedures we used.

In this context, I have to mention a recent experiment by Stromeyer (1971). He was able to obtain Land colors with the McCollough effect. He used a stimulus similar to one field of figure 7.9-1 except the array had much fewer cells and the grids in each cell had varying contrast. This contrast change was produced by making gratings of different grays. Adaptation due to the McCollough effect produced pink and green aftercolors of various saturations which were superimposed on a matrix of square cells having the randomly chosen gray values. Thus a red or green aftereffect was superimposed on the random gray cells. Subjects reported a gamut of colors. It is known that a gray-red and gray-green system of random arrays produces Land colors (Land 1959, 1964), but the discovery that McCollough colors can give rise to Land colors is remarkable. In our terminology this means that the McCollough aftereffect occurs prior to the Land effect. On the other hand, the Land effect probably occurs before stereopsis, since it is very difficult if not impossible to induce Land colors by binocular mixing. The question of binocular Land colors is as controversial as binocular color mixing used to be in Helmholtz's time. Under low luminance levels some researchers reported Land colors by binocular mixing of the low and high spectrum channels, but the obtained colors are very unsaturated. The difference between binocular Land colors and monocular Land colors is so striking that one may question whether they are the result of the same phenomenon. Yet, until this question is settled we cannot say for a certainty that the Land effect occurs before stereopsis.

There is the possibility of finding the site of the Land effect by producing random-color-dot stereograms with the help of Land colors. If one could produce Land colors by monocular mixing of a red and a green random array and produce the *same* color array in the other eye by monocular mixing of white and red random arrays, and finally could stereoscopically fuse the two identically colored arrays, the crucial demonstration would have been performed. Nevertheless, it would be difficult to produce identical Land color arrays and simultaneously keep their brightness distribution constant or uncorrelated in the two fields.

In § 3.4 the Blakemore and Sutton (1969) aftereffect was reviewed. They observed a shift in perceived spatial frequency of a grating after prolonged adaptation and this aftereffect is an important addition to the list of aftereffects. Nevertheless, all these aftereffects (from Gibson to McCollough, Hapler, Blakemore and Sutton) are produced by *edges* of a given orientation, color, spatial frequency and movement. The question arises as to whether one could produce a genuine cyclopean aftereffect without edges.

At first thought, one might believe that many of these aftereffects are central since Köhler and Emery (1947) demonstrated that Gibson's tilted line aftereffect and the figural aftereffects of Köhler and Wallach (1944) can be perceived in depth, after adaptation to stereograms. They adapted in quick alternation the observer's left and right eye. The alternation was not rapid enough for stereopsis to occur, yet it was adequately fast to expect monocular adaptation. Under these conditions no three-dimensional aftereffect was observed. From this result they concluded that the tested aftereffects were genuinely three-dimensional.

However, Köhler and Wallach did not check whether this alternate monocular adaptation would abolish the monocular (two-dimensional) aftereffects. Without being able to show that after prolonged alternate adaptation each eye separately builds up a different aftereffect, the question of whether a real three-dimensional aftereffect exists is not answered.

A truly three-dimensional and cyclopean aftereffect was produced by Blakemore and Julesz (1971) just before closing this book. If the reader fuses the random-dot stereogram of figure 7.9-4* for a minute by fixating at the center fixation mark, he will see a square above this mark as being behind it. The rest of figure 7.9-4* belongs to the background. It is interesting to note that after a few seconds of steady fixation the depth effect fades out. In order to prevent this depth fade-out, the reader is asked to scan back and forth along the horizontal fixation mark. (It is interesting to note that during this depth-fade-out the random patterns often appear to have some regular organization.)

After adaptation to figure 7.9-4* the reader should quickly fuse the test stereogram of figure 7.9-5*. In spite of the fact that here the two squares above and below the fixation mark are at the same depth as the mark, they are perceived at different depths. The square that took the place of the adapting square closer to the observer than the fixation mark, appears to be behind the mark and the other square appears to be in front of the mark. The adaptation and test stereograms have different textures. This aftereffect can be obtained after only 5 sec of adaptation, but prolonged adaptation in-

7.9-4* Adaptation stereogram that demonstrates the Blakemore and Julesz (1971) three-dimensional aftereffect. After fixating at the center mark during fusion for about 1 min, the test stereogram of figure 7.9-5* should be viewed.

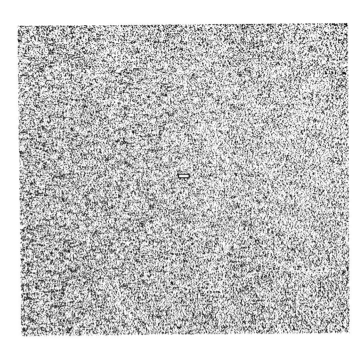

creases the duration of the aftereffect. For instance, 5 min adaptation time produces a 0.5 min long depth-aftereffect.

For details, the reader may consult Blakemore and Julesz (1971). It should suffice here that we found a genuine cyclopean aftereffect that does not require edges in order to be evoked. This three-dimensional aftereffect seems to be quite general. For instance, if the reader views the anaglyph of figure 6.6-1* (which portrays a recessed ellipsoid) with the red and green glasses and then suddenly removes the glasses, the flat background appears to bulge out somewhat (coming closer to the observer).

7.9-5* Test stereogram that demonstrates the Blakemore and Julesz (1971) three-dimensional aftereffect. After adaptation to figure 7.9-4*, when fixating at the center mark in depth, the upper and lower squares appear briefly at different depths. (From Julesz and Blakemore 1971.)

8 Miscellaneous Uses of Cyclopean Methods

8.1 Objective Tests for Stereopsis

While there is a large literature on color deficiency and it is common practice to screen the population for color defects both by military and industrial organizations, there are very few similar attempts yet in assessing deficiencies of stereopsis. This striking difference might have been brought about by two factors. First, color is widely used in our civilization for coding, from warning lights to electrical wiring layouts, whereas depth cues are not used for coding signals. Second, the Ishihara color test charts permitted an unfakable test of color deficiencies, while prior to the development of random-dot stereograms the many monocular depth cues present in ordinary stereo photographs interfered with the testing of stereopsis abilities.

Since random-dot stereograms recently have become more widely available, some work has been started in screening certain groups of the population for stereoscopic depth perception. At Temple University, Stauffer's group training medical students in stereo X-ray fluoroscopy used random-dot stereograms in prescreening applicants. In a large number of talks before professional and lay audiences where I demonstrated random-dot stereograms, I took rough statistics of the percent of participants who could not obtain stereopsis. If the stereograms are correctly aligned by the help of stereoscopes or polarizing techniques, then the population can be divided in three categories. The majority of the population has good stereopsis, and all the stereograms in this book will yield stereopsis for them. About 2% of the population is stereopsis-blind, and even the basic stereogram of figure 2.4-1* will not yield a sensation of depth for them. This group, of course, includes persons with strabismus or some other known functional inabilities of one eye or of eye-coordination. However, there are always a few instances when no obvious deficiency can be detected and yet no stereopsis occurs. The quoted 2% is only a first guess and will require careful mass screening by optometrists, ophthalmologists, and experimental psychologists.

The third class, roughly about 15% of the population has stereopsis-deficiency. They often see the basic random-dot stereogram in depth or at least notice that a center square with rounded corners is different from its surround; however, they have difficulty with the more complex random-dot stereograms.[1] After further tests, many of these subjects showed strong astigmatism which had not been corrected. When their astigmatism was corrected, they improved dramatically and could obtain stereopsis for all test plates. Yet, there is a subclass within the stereopsis-deficients who have no astigmatism and yet cannot fuse complex stereograms. Whether they could improve with training is an open question which should be tested under carefully controlled conditions. There is a remote possibility that their inadequate stereoscopic performance is the result of poor viewing habits. Perhaps they learned to rely too much on monocular depth cues and possess a very dominant eye.

In this context some recent work by Chow (1969) is of importance. He sutured an eye of new-born kittens and (as expected from Hubel and Wiesel's work) after a critical period of a few months, layers of neural units in the LGN that corresponded

---

1. In a recent study, Richards (1970) screened 150 students at M.I.T. and found that 4% had no stereopsis and 10% had great difficulty in seeing depth in a random-dot stereogram.

to that eye showed a permanent, abnormal histology. Hubel and Wiesel also showed that the binocularly driven units disappeared in the cortex; all cortical units became monocularly dominant. Behaviorally, these cats used only their formerly unsutured eye. However, Chow modified this procedure by taking adult cats who had an eye sutured from birth, opening the sutured eye, and suturing the previously unsutured eye, then imposed pattern recognition training on the functionally "blind" eye. After long, forced use of this eye, cats were able to learn pattern recognition tasks, and what is more, the previously distinguishable layers of the LGN that corresponded to the ipsilateral and contralateral eyes became indistinguishable in histological preparations. Although there is no evidence that any recovery took place in the cortical neurons too, nevertheless, it is important that even some drastic interventions may not be irreversible but might improve with forced use. I am not suggesting that congenitally strabistic patients will regain stereopsis after corrective surgery in their adult years. On the other hand, it is not inconceivable that humans who belong to the stereopsis-defective class might profit from intensive training with complex stereograms.

This section will try to launch an intensive program in assessing stereoscopic deficiencies in the general population. Such knowledge may be crucial for those deficient individuals who are planning to enter a profession where deficiency or complete lack of stereopsis would be a great handicap. It is not yet clear whether a truck driver, jet pilot, surgeon, mechanic, can do an outstanding performance in his field without stereopsis, but it is clear that the fewer monocular familiarity cues available the greater the handicap becomes.

In order to quantify and normalize stereopsis deficiencies, I provide here a few test plates which I hope will be used as standards by my colleagues. The first series is composed of a common left field, and four corresponding right fields, portraying in depth (in front of the surround) a square, diamond, cross, and disk, respectively (as is shown in figure 8.1). Only the diamond is demonstrated as an anaglyph for the reader.[2] Figure 8.1-1 can be either Xeroxed or a Polaroid slide can be made of all the five matrices which can be viewed in a stereoscope or projected by prisms and polarizing techniques. The center matrix of figure 8.1-1 is the common left image while the south, east, west, and north matrices, when stereoscopically fused with the center matrix, yield the cyclopean square, diamond, cross, and disk as one rotates the book accordingly. As discussed later these slides can be also used for testing stereopsis in monkeys and other animals. The importance of these tests is that it is not necessary to report that an object is perceived in depth; it is sufficient if the shape of the cyclopean figure evokes an appropriate response.

Nevertheless, people show some bias as to the direction they can more easily perceive cyclopean figures; that is, in front of or behind the surround. Therefore, it would be advisable to test the subjects with the pseudoscopic view of figure 8.1-1 by presenting the common matrix to the right eye.

Test slides, such as figure 8.1-1, will reveal stereopsis blindness. However, to quantify stereopsis deficiency, another series of tests are presented in figure 8.1-2*. Here

2. See the anaglyph of figure A, which faces the first page of the Preface.

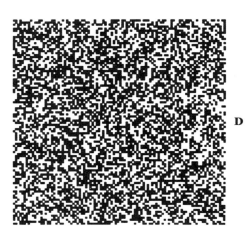

8.1-1   Test figures for stereopsis. Center matrix is common left image for all cases. By rotating the book clockwise, the horizontal stereogram pair, with the center matrix viewed by the left eye, will portray a cyclopean (*A*) square, (*B*) cross, (*C*) diamond, (*D*) disk, respectively.

8.1-2* Test figures for determining stereopsis deficiency. The stereograms have diminishing binocular correlation: *A*, 100%; *B*, 90%; *C*, 80%; *D*, 70%; *E*, 60%; *F*, 50%; and *G*, 40%.

A

B

C

**Miscellaneous Uses of
Cyclopean Methods**

8.1-2* cont'd.

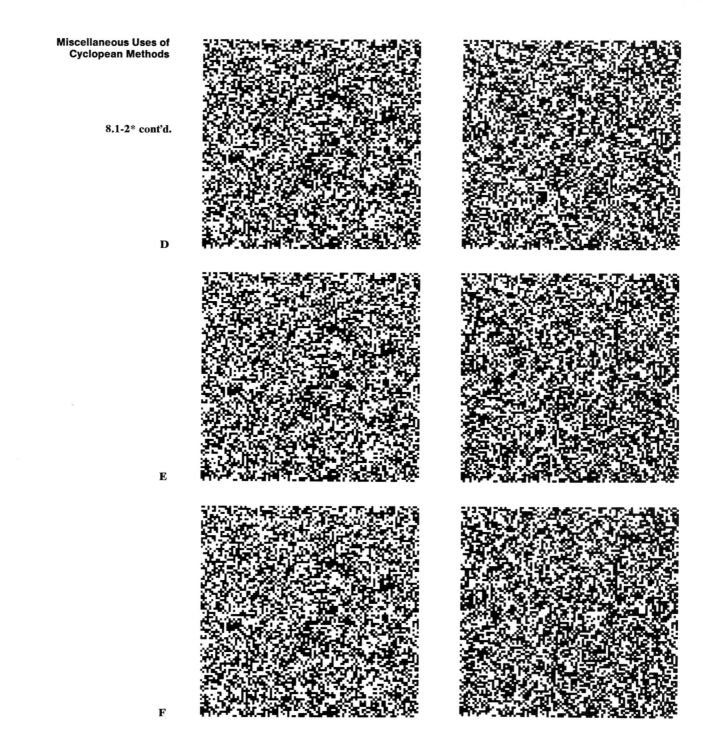

D

E

F

G

a center square of 50 × 50 cells is shown in an array of 100 × 100 cells. The surround has a 2-cell nasal disparity (in order to prevent seeing the cyclopean square for exactly registered stereograms when using poor quality polarizing filters or when rotating the head) and the center square has 4-cell nasal disparity. In figure 8.1-2*, *A* is 100% correlated, but the rest have increasing amounts of complemented cells that cannot be fused: *B,* 10%; *C,* 20%; *D,* 30%; *E,* 40%; *F,* 50%; and *G,* 60%. A 10-cell wide margin is kept unperturbed in order to help registration. Performance of subjects strongly depends on the perceptual criterion used. First, the corners of the cyclopean square disappear, but a rounded off area in the center is still perceived in depth. Loss of stereopsis gradually increases with increasing noise. More and more dots appear at other depth planes than that of the square or its surround. Finally, it is impossible to detect an area in the center as being different from the surround. One of the best ways to proceed is to determine that noise level where the cyclopean square is still recognizable as a square.

Another procedure is to show the stereograms of figure 8.1-2* as a random series of nasal or temporalward disparities. The criterion of stereopsis can be chosen to be 98% correct identification of the center square seen in front of or behind its surround. Presentation time should be chosen as 100 msec, and brightness level can be 0.1 ftL. Under such stimulus conditions, and with the stimuli presented at a 10° of arc viewing angle, it would be possible for researchers and clinicians to normalize stereopsis deficiency. While a normal observer with good stereopsis could achieve 98% correct depth responses for, say figure 8.1-2*E*, a poor subject would be able to perform similarly only for, say figure 8.1-2*B*. Very poor subjects could not perform at criterion even for figure 8.1-2*A*, but, say, only at a 70% correct response level. Since 70% is still better than guessing by chance alone (which corresponds to 50% correct response level) such observers would not be regarded as entirely stereopsis blind. I hope the reader, particularly if he is a clinician, becomes interested in cooperating in such a mass program of screening his patients with this method and might publish his findings by adhering to the stimulus conditions specified herewith.

8.2   Clinical Uses of Random-Dot Stereograms

Clinical methods seek to diagnose or alleviate certain disorders. In § 8.1 we discussed how random-dot stereograms yield an objective, unfakable test for stereopsis and help to detect stereopsis-deficiencies. As such, random-dot stereograms in the hands of optometrists and opthalmologists can serve as a quick and reliable tool for diagnosing amblyopia or other impairment in binocular vision.[3] However, this section tries to extend the applicability of random-dot stereograms to less obvious clinical tests.

A novel application of random-dot stereograms was recently reported by Carmon and Bechtoldt (1969) who found marked perceptual deficits in patients with unilateral right hemispheric lesions. When random-dot stereograms were presented to patients with unilateral left hemispheric lesions, performance was almost identical to normal controls with no brain damage. Performance was measured either in response time or number of errors in identifying the position of the cyclopean depth figure. Optimum performance was obtained for random-dot stereograms with equal black-white dot density and deteriorated at 10% (90%) black-white density by a factor of two; but patients with right hemisphere lesions had a four times poorer performance for all density values than did normals or patients with left lesions. The importance of random-dot stereograms in this application can be appreciated after Kimura (1963) found only hemispheric difference when the stimuli presented to patients for recognition were unfamiliar figures. The finding of Carmon and Bechtoldt is in accordance with mounting evidence that the left or (in Broca's usage) "dominant" hemisphere is involved with speech and symbolic processing while the right hemisphere processes spatial perceptual data of several modalities. A report by Teuber (1962) of a preponderance of visual abnormalities and hallucinations as well as deficits in sound localization in patients with right hemispheric lesions is consistent with these findings. The work of Milner (1964) and Dorff, et al. (1965) also showed perceptual dominance for tachistoscopically presented visual stimuli. The perceptual superiority of the right hemisphere for spatial tasks was strikingly demonstrated by Gazzaniga, et al. (1965) on corpus-callosum-severed patients. Several defects of stereopsis have been reported by Bender and Teuber (1947, 1948) in men with penetrating missile wounds which affected the right hemisphere, but stereopsis was spared in cases with left hemispheric lesions (i.e., Case A29 in Teuber, et al. 1960).

Ordinary stereograms are aided by so many monocular familiarity cues that subtle changes between left and right hemispheric lesions might not show because of the fast perception time. However, for random-dot stereograms the lack of all monocular depth cues considerably increases the perception time needed for stereopsis thus providing a more sensitive indicator for impairments in stereopsis.

Another clinical application of random-dot stereograms might be in stereo X-ray

---

3. Amblyopia (more properly, *amblyopia ex anopsia*) which means deterioration of vision through lack of use is a deficiency of visual acuity in one eye—usually in a strabismic eye. For an excellent review see McLaughlin (1964).

fluoroscopy. Julesz (1966) proposed that instead of the usual contrast material that uniformly covers the internal surfaces of cavities to be X-rayed, it might be perhaps more advantageous to randomly speckle the contrast material. This would provide a textured lining of the cavity walls and thus may yield binocular disparity for the entire area. I do not know whether this idea has ever been implemented but it might prove useful.

A pilot study of mine in which random-dot stereograms were presented to a hypnotized male subject might be of some clinical interest. A subject under deep hypnotic trance was shown a random-dot stereogram (like figure 2.4-1*) in vectograph format while viewing via polarizing glasses. The subject, sitting in a chair, was instructed to raise his right arm when the center square appeared in front of the surround, lower his arm when the square appeared behind the surround, and rest his arm on his knees when the square appeared flush with the surround. Effectiveness of the command was verified by several trials. When the vectograph was shown in normal position (nasal disparity) the subject would raise his arm without hesitation or delay; turning the vectograph upside down (temporalward disparity) caused the subject to lower his arm; turning the vectograph by 90 degrees (vertical disparity) elicited the response of resting his arm on his knees. My impression was as if a conditioned reflex had been established between disparity and arm movement as a result of hypnotic suggestion. If an ambiguous random-dot stereogram was presented (with both nasal and temporal disparities), the subject would usually lower his arm according to his natural bias of perceiving the center square behind. After these preliminaries, the subject was verbally instructed by the hypnotist to report verbally what he had perceived, but prior to the presentation of the vectograph the hypnotist would suggest to the subject that he would see a center square in vivid depth, say, behind the surround. However, the vectograph shown would portray depth oppositely to the verbal suggestion. Strangely enough, the subject would always accept the suggestion and report accordingly; for instance, he would tell the hypnotist that he sees a square deep behind, yet his arm would be raised in agreement with the disparity of the stimulus. Even for ambiguous stereograms, for which one might have expected that a hypnotic command would have counteracted natural bias (by starting with convergence or divergence of the eyes), the command had no effect on the arm position. While the subject always reported depth according to previous suggestion, nevertheless, raising or lowering his arm was independent of suggestions and probably depended only on how he actually perceived the ambiguous stereogram. The verification of this experiment by many subjects and several hypnotists would be most desirable since this observation has several implications about the hypnotic state itself.

We noted elsewhere the finding of Bower (1968) using random-dot stereograms that newborn babies have demonstrable stereopsis. We also know from Hubel and Wiesel (1965b) that kittens with artificially induced strabismus at a critical period after birth will have their binocular cortical neurons irreversibly converted into monocular ones. This neurophysiological finding together with the common belief of clinicians that congenitally strabistic patients whose strabismus is corrected at a late age by surgical intervention will not regain stereopsis but retain a dominant eye constitutes a pessimistic prognosis. (However, I tested a thirty-year-old woman who correctly reported the disguised figures in random-dot stereograms after a successful operation correcting her severe strabismus. Although she claimed that she was strabismic from birth, the possibility exists that she was involved in some accident after a few months of coordinated binocular vision.) Loss of binocularity is less severe and not irreversible if strabismus occurs in later life. This might suggest a clinical program of early diagnosis of stereopsis blindness together with trying to stimulate the strabismic eyes by means of prism stereoscopes using random-dot stereograms. With such a device one can stimulate corresponding points on the two retinae. Perhaps a brief training program of a few minutes each day might preserve stereopsis for a later less critical period when surgical intervention may be easier. Some of these ideas are not new; however, random-dot stereograms might prove to be a more powerful stimulus class than ordinary stereoscopic slides.

Random-dot stereograms and cyclopean techniques in general might become a research tool in determining the active sites of psychotomimetic drugs. The finding that macaque monkeys can perceive random-dot stereograms might permit the development of screening tests for the evaluation of drug potency. The interested reader is referred to the discussion following a talk of mine (Julesz 1970a). At this workshop meeting on psychotomimetic drugs, I learned that under the influence of LSD a visual image is often not erased by a successive image after a new fixation. The past and new images appear as being superimposed. This phenomenon could be quantified with random-dot correlograms (similarly to the testing of eidetic memory). Two test images, such as figure 5.5-2* can be presented to the same eye in temporal succession with variable delay. The random-dot surround in one image is the complement (negative) of the other one. If the subject retains the first image at the time the second one is presented the surround will average to a uniform gray field, while the center area appears textured. Increased temporal separation can determine the limit of this image persistence and its dependence on drug potency.

8.3 Neurophysiological Probing during Cyclopean Stimulation

Random-dot movies portraying moving edges by either depth or movement gradients might be useful stimuli for neurophysiological studies. The possibility of skipping several synaptic levels and portraying moving edges of desired extent and velocity on a certain cortical reference plane might be of interest. The most flexible experimental setup requires computation in real-time and portrayal on a video monitor permitting instantaneous changes in the cyclopean stimulus parameters. Nevertheless, a continuous loop of 800–1,000 frames of a computer-generated movie can be easily projected by standard 16-mm movie projectors. Such movies portraying a static (dynamic) vertical bar moving horizontally in a dynamic (static) textured surround are easy to generate, and copies are available from me on request. For this two-dimensional case a Dove prism over the projection lens can change the orientation of the bar. Bar size and aspect ratio are variables, similar to the classical case, and so is velocity of movement. On the other hand, texture density and speed of dynamic-noise changes by having new, uncorrelated areas at each frame (or identically re-

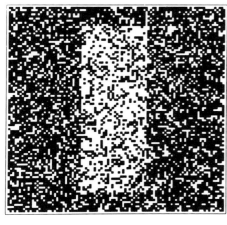

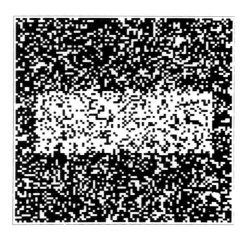

8.3-1   Random-dot pattern portraying
vertical bar for monocular conditioning of pattern discrimination.

8.3-2   Random-dot pattern portraying
horizontal bar for monocular conditioning of pattern discrimination.

8.3-1

8.3-2

peating every second or third frame) are new parameters with possible neurophysiological significance.

Random-dot stereogram movies might be useful stimuli for studying stereopsis.
Since Bough (1970) established stereopsis in the macaque monkey using random-
dot stereograms in a psychological experiment, it is fairly certain that random-dot
stereogram movies will elicit firing global disparity units in the monkey cortex. Such
movies portraying a vertical bar oscillating from left to right and back to the left can
be projected by any ordinary 16-mm movie projector using a commercially available
stereoscopic prism adapter with polarizers. The use of special projectors with widely
variable speeds is preferable. These and similar techniques are now in the process of
being used by several teams of neurophysiologists.

One bottleneck in this program is our ignorance of stereopsis capabilities in animals other than the monkey. Does, for instance, the cat perceive random-dot stereograms in depth? Besides the cat there are many animal species that have good stereopsis together with good visual acuity, as for instance the owl; yet we cannot even guess
whether they could perceive depth in random-dot stereograms.

In order to determine stereopsis of random-dot stereograms in animals, the following four test plates are provided: The first two test patterns shown in figure 8.3-1 and
8.3-2, are monocular and are intended for the operant conditioning of animals' horizontal-vertical discrimination. The reason why the horizontal and vertical bars are
portrayed by texture-density gradients and not just by brightness gradients is to adapt
the animals to textured stimuli. After the desired discrimination performance is obtained, the monocular stimuli are replaced by the two random-dot stereograms (figs.
8.3-3* and 8.3-4*) that portray the same horizontal and vertical bars in depth when
stereoscopically fused. The red-green anaglyphs shown are better than polarized
slides, since the separation between the left and right images is not critical for head tilt.
The animal tested must be provided with red and green viewers, and some care has to
be exerted that no unwanted cues interfere. For instance, it is advisable to check

8.3-3*   Random-dot stereogram
portraying a cyclopean vertical bar
in depth.

8.3-4*   Random-dot stereogram
portraying a cyclopean horizontal bar
in depth.

whether some minute scratch or dust particle does not serve as a cue, and it is advisable to interchange and rotate the left and right images by 180°. One can hope for a generalization of the learned horizontal-vertical discrimination to cyclopean stimuli, without really testing for depth.

8.4   Toward the Automation of Binocular Depth Perception

The demonstration that stereoscopic depth can be perceived without prior monocular form recognition is particularly important to engineers who work on the problem of automatic localization of depth. The realization that stereopsis is a simpler process than previously believed, (and thus the enigmatic problems of form recognition can be avoided) must have had some heuristic value. The appearance of devices in the last decade that automate contour-map plotting were the result of advances in computer technology and electro-optical methods. However, the belief that the solution to this problem is within our reach (in contrast to some premature attempts at automatic language translation or automatic recognition of complex forms) is a direct consequence of psychological and neurophysiological evidence collected in the last few years.

Besides this general contribution that binocular localization in depth is not based on semantics, the psychological findings obtained by random-dot stereograms have some concrete and detailed consequences for the automation of depth perception. The interested reader is referred to a paper of mine (Julesz 1962b) in which a computer simulation program of contour-map plotting, called AUTOMAP-1 is described. This program was written in cooperation with Miss Joan E. Miller at Bell Laboratories and is based on the difference field model of stereopsis, described in § 4.5. The details of this algorithm surpass the scope of this book; however, the basic idea of this routine serves to illustrate basic departures from the usual methods used for automated plotters.

Unlike the many devices which have been invented for automatic aerial map compilation, AUTOMAP-1 is devoid of the inherent limitations of conventional cross-correlation techniques. In most commercially available devices, two identical moving zones of constant size are compared in the left and right fields in a search for a best fit according to some similarity criteria. However, such methods are very restrictive. With small zone size, slight differences in the two fields due to reflections and perspective eliminate similarities; with increased zone size the resolution becomes inadequate at boundaries of objects at different depths. Only by varying the size and shape of the zone to match the objects could satisfactory results be obtained. But the monocular recognition of objects brings us back to the unsolved problems of semantics.

AUTOMAP-1 avoids the problem inherent in zone matching by simultaneously processing the entire presentation. The left and right visual fields are combined by using some simple point-by-point operation and all further processings are performed on this combined field. This processing of the combined field is based solely on simple cluster seeking techniques (of adjacent points having similar disparity values). These clusters immediately give the cross-sections (contour lines) of objects in a natural way. AUTOMAP-1 used difference fields between the left and right images with increasing amounts of horizontal shifts prior to subtraction. After finding clusters of minimum values in the difference fields the corresponding points in the original left and right fields were removed and the same difference field operations were iterated but the cluster seeking algorithm was based on a more severe criterion than used in the previous run. At each iteration an additional number of points become localized until an asymptotic value is reached. In figure 8.4-1* a random-dot stereogram is shown, and figures 8.4-2 and 8.4-3 show two successive localizations by AUTOMAP-1. The black areas are undetermined; and, when viewing the stereogram of figure 8.4-1*, similar border areas appear ambiguous. This does not mean that AUTOMAP-1 is a model for human stereopsis, but the ambiguities that prevail at the boundaries of areas at different depth (that have no disparity since one eye's view is hidden) cannot be resolved without familiarity cues. As long as the input is devoid of the usual familiarity cues, it is not surprising that the human visual system cannot

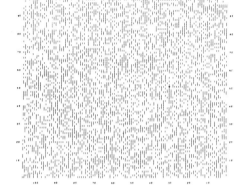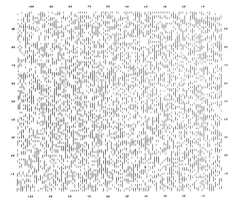

8.4-1*   Random-dot stereogram used as input for AUTOMAP-1.

8.4-2 and 8.4-3   Two successive localizations by AUTOMAP-1 of the input stereogram of figure 8.4-1*. Criterion of a cluster of points having the same binocular disparity becomes less severe after each iteration.

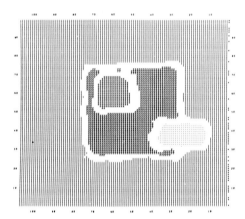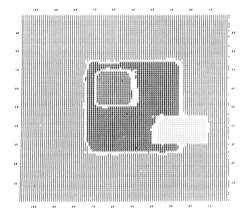

8.4-2   8.4-3

do better than a computer algorithm. Even for random-dot stereograms when one of the images is expanded or tilted, the visual system can still obtain stereopsis in a certain range, while AUTOMAP-1 fails under the slightest misalignment.

This is the reason why the spring-coupled dipole model was introduced in § 6.5 to replace the difference field model. In principle, one could computer simulate the spring-coupled dipole model which would be less critical of misalignments and show greater similarity to human stereopsis than does AUTOMAP-1. However, the cooperative magnetic dipole model is much simpler to realize as an analog mechanical device than as a computer algorithm. It remains to be seen, whether a computer simulation of the spring-coupled dipole model of adequate resolution is economically feasible with present computers. Nevertheless, the tedious task of human stereoscopic fusion and point-by-point subjective plotting of thousands of overlapping aerial photographs warrants large efforts at automation.

8.4-4* Stereoscopic close-up image of a 7.6 cm square lunar surface as photographed during the Apollo-11 mission. (Courtesy NASA.)

Of course, in situations where familiar complex objects are provided, stereopsis often becomes a secondary depth cue. It is well known that the pseudoscopic image (stereogram with interchanged left and right views) of a human face resists all efforts to appear as concave. As a matter of fact, one could define familiarity with an object by its resistance to change from convex to concave (or vice versa) under pseudoscopic viewing. The many monocular familiarity and depth cues will help resolve ambiguities in contour-map plotting as well, particularly when the terrain is photographed from low altitudes. At higher altitudes familiarity cues are less important, and in usual plotting tasks automatic algorithms without semantic considerations can be of considerable value.

The main message of this section is that human binocular depth perception can separate objects without having to recognize them. Thus, binocular disparity cues alone often can be as valuable as all the complex, monocular, familiarity, and depth cues altogether. This is particularly true for texture rich but unfamiliar environments. Such a case is shown in figure 8.4-4* that portrays a stereoscopic close-up image of the moon as taken during the Apollo-11 mission. Does this mean, that if only a monocular view of the environment is provided one cannot localize and separate objects without familiarity cues? This problem will be discussed next.

8.5 Three-Dimensional Reconstruction from Two-Dimensional Brightness Distribution

Is it possible to find the three-dimensional shape of an opaque object from its reflected two-dimensional brightness distribution? This problem was first brought to my attention by the late J. L. Kelly at Bell Labs, who assumed that this problem had a solution at least for Lambert surfaces (having a cosine reflectivity function).

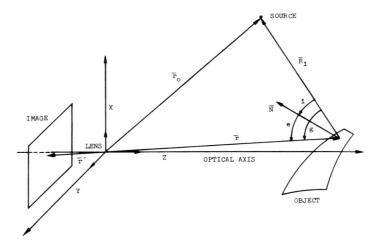

8.5-1 Illustration of relations between parameters in a task of three-dimensional reconstruction from two-dimensional brightness distributions. Source and reflectivity function of object is known. (From Horn 1969.)

Rindfleisch (1966) solved this "shape and shading" problem for the special problem of lunar topography. The general problem has been studied by Horn (1969) at M.I.T. under some simplifying assumptions. The location and intensity distribution of the light-source are known, and it is assumed that it is incoherent and time-invariant. The object (whose shape is unknown) must not be translucent, its reflectivity function has to be symmetrical with rotation in the tangent plane, the reflectivity function with continuous derivatives must be known, the surface of the object is assumed to be reasonably smooth, and so on. The image is assumed to be noise free and distortion free, and the optics of the imaging system are of known properties.

Let $b(x',y')$ be the intensity measured at the point x',y' in the image plane. Now, if the incident angle i, emission angle e and the total angle g determine the reflectivity function Φ, and $A(r)$ is the intensity function at the point r in figure 8.5-1, then

$$A(r)\Phi(i,e,g) - b(x',y') = 0 . \qquad (8.5.1)$$

Horn showed, by expanding the terms in this equation into differentials, that a first order nonlinear differential equation is obtained which can be transformed into a system of five ordinary differential equations. The conditions under which such a system of ordinary differential equations can be solved goes beyond the scope of this book. Here it suffices to say that a solution can be obtained if some boundary conditions along some initial curve on the surface of the object are known and this curve is nowhere parallel to certain curves (the characteristics of the differential equations). In complex cases the solution requires tedious computations that in practice can be solved only with computers. I included this example to show that between the case of given binocular disparity and given monocular familiarity cues a situation exists for which the semantic information is rather modest yet yields depth localization. The a priori knowledge of the light source and the reflectivity function of the object are less severe requirements than is familiarity with the object class itself.

In Horn's case, the reflectivity function of all surfaces have to be known, furthermore these surfaces must be textureless or their texture must be precisely specified. Because of this, the requirements are just the opposite of automatic depth localization by stereopsis, for which textureless objects should be avoided. Therefore the combination of both algorithms might greatly extend the usefulness of the two programs. Where one fails the other works optimally and vice versa.

8.6 How to Set Up Cyclopean Research Facilities

One purpose of this book is to make available a large number of cyclopean stimuli in the form of random-dot stereograms and cinematograms. In spite of my efforts to provide a gamut of slides ranging from tests of stereopsis to optical illusions and complex three-dimensional surfaces which can be used in human and animal experiments, such ready-made patterns have their obvious limitations. The reader who acquires the skills of cyclopean thinking may want to design cyclopean stimuli which are fitting for certain research problems not even thought of in this book. Furthermore, the clumsiness of the motion-picture format prevents its publication in a book. Because of these problems, this section is devoted to the question of how one could actively pursue cyclopean research.

The problem can be divided into two, depending on whether the researcher wants to use fixed stimulus material or desires to generate the stimuli in real time. While the first stimulus class is within the reach of any university department or small laboratory that can afford the smallest digital computer (e.g., ten to twenty thousand dollar range), the latter stimulus class (as of now) is only available for a lucky few (costing about one-third million dollars). What is even more, fixed stimuli can be generated by batch processing at any computation center, while real-time computation requires full access to a large computer. Unfortunately, visual patterns contain such large amounts of information that even on our present third-generation computers, time sharing is out of the question. Because of this great difference between the two stimulus classes, this section discusses only fixed-picture generation. Those few who try to generate detailed motion pictures in real time are at the frontier of present computer technology, and the description of such a facility (several of which are under construction in this country including one in my own department at Bell Laboratories) goes beyond the scope of this book.

There are two kinds of fixed-stimulus material one may want to generate in cyclopean format. The first kind portrays simple figures, such as rectangles (and polygons in general), circles, and surfaces in three-dimensions which can be described by explicit mathematical equations. For such cases no pictorial input facility is necessary, since the desired cyclopean forms can be easily generated by programming. In figure 8.6-1 a typical FORTRAN program is shown that generates a random-dot stereogram of 100 × 100 dots portraying a center disk of 50 dots diameter. The program consists of three parts (or subroutines). The first is a pseudorandom number generator that fills the 100 × 100 array with randomly chosen 0's and 1's. As most of the readers know a finite digital automaton can never produce random sequences. However, it is possible to select algorithms that multiply two properly chosen numbers, truncate

```
C
C
C
      DIMENSION LEFT(100,100),IRIGHT(100,100)
C
C FILL THE LEFT VIEW, USING A RANDOM NUMBER GENERATION ALGORITHM
C
      MAGIC1=0343277244615
      MAGIC2=MAGIC1
      MASK=0377777777777
      NVALUS=2
      NBITS=2**(35-NVALUS)
      DO 10 J=1,100
      DO 10 I=1,100
      MAGIC2=AND(MAGIC1*MAGIC2,MASK)
10    LEFT(I,J)=MAGIC2/NBITS
C
C CREATE RIGHT VIEW
C
C BACKGROUND IS COPIED FROM LEFT VIEW
C
      DO 12 J=1,100
      DO 12 I=1,100
12    IRIGHT(I,J)=LEFT(I,J)
C
C A CIRCULAR AREA IS FILLED RANDOMLY AND PLACED ONE ELEMENT TO THE RIGHT
C         OF CENTER IN THE LEFT VIEW, AND ONE ELEMENT TO THE LEFT OF
C         CENTER IN THE RIGHT VIEW TO PRODUCE A TOTAL DISPARITY OF 2
C         ELEMENTS BETWEEN THE TWO VIEWS
C
      DO 14 J=1,100
      DO 14 I=1,100
      IF((I-50)**2+(J-50)**2.GT.625)GO TO 14
      MAGIC2=AND(MAGIC1*MAGIC2,MASK)
      LEFT(I+1,J)=MAGIC2/NBITS
      IRIGHT(I-1,J)=MAGIC2/NBITS
4     CONTINUE

OUTPUT ARRAY ON MICROFILM PRINTER

      CALL CAMERA(LEFT)
      CALL CAMERA(IRIGHT)
      STOP
      END
```

8.6-1   Example of a FORTRAN
program for producing a random-dot
stereo pair.

their product and multiply it with a fixed number. The obtained sequence of products will repeat only after a cycle of several billions. Of course, only some very carefully chosen numbers will yield such long cycles. The program in figure 8.6-1 gives such a preferred pair of numbers called MAGIC1 and MAGIC2. In most applications one can avoid generating tens of thousands or millions of numbers by time consuming multiplications (which can take one minute on the fastest computers). It is usually adequate to store one row of the image (by computing 100–1,000 random numbers or type it in from a table of random numbers). The same random-number sequence can be used to shift the random number sequence in the second, third, or n-th row by a specified amount and copy this shifted sequence in the appropriate row. If the shift between adjacent rows is more than a few picture elements, the human visual system will not recognize that all rows are identical but out of phase and will perceive the image as a random texture.

The second subroutine defines the shape of the cyclopean disk ($x^2 + y^2 < 25^2$) in FORTRAN notation. The third subroutine generates a stereoscopic image pair copying all the numbers of the random-dot array into two memory arrays $LEFT(I,J)$ and $RIGHT(I,J)$ with the exception that in one array the dots that fell inside the disk are shifted horizontally by $SHIFT = 4$ units in one direction. How these two arrays are outputted is really the main subject matter of this section.

Computer centers that possess a microfilm plotter, such as the General Dynamics

(Stromberg-Carlson) 4020 or 4060 or a similar high-precision device, will have available several subroutines that will convert a series of numbers into black and white dots or alphanumeric characters. However, the usual laboratory has no access to such a facility. Those researchers who have either a small computer of their own or access to batch processing of their computer center can still produce high-quality video information. This can be accomplished by buying or renting a standard facsimile receiver of about 200 lines/inch resolution (available in most police stations in the United States for fingerprint identification and in many newspaper editing offices for picture transmission) which costs about six thousand dollars and includes an automatic paper print developer. Such a device can portray an 8×8 inch picture ($1,600 \times 1,600$ dots) in excess of 10 brightness levels within 10 minutes, including development time. Such a stereogram composed of $1,000 \times 1,000$ dots of 10 brightness levels is shown in figure 5.4-1*.

The second way of portraying the computer output is by the use of a high-resolution microfilm printer mentioned above. Such a picture of a complex surface (similar to fig. 5.4-1*) is shown in figure 5.4-2* in which 10^6 points are computed and only 10% of them are portrayed. Usually the large computers of a computation center do the computation together with the picture generation. However, as in the case of ambiguous stereograms portraying complex surfaces, generation of $1,000 \times 1,000$ dot arrays might require 24 hours computer time (on line). Such excessive demands are not only economically unreasonable but beyond the capabilities of a computation center. In such cases, provided the small or medium sized computer in one's possession has a digital tape unit and the computation center owns a microfilm printer, one can compute on the small computer and use the large computation center for portraying the images.

These examples all referred to cyclopean image generation of mathematically defined forms. If the figures cannot be mathematically defined, then it would be very difficult to define a complex drawing by giving its coordinates point by point. The easiest way to convert pictures into numbers is to use a facsimile transmitter. Such a device is even cheaper to buy or rent than is a facsimile receiver (since the automatic photoprocessing is not needed). One merely wraps the desired photograph or drawing around the rotating drum and in about 8 minutes $1,600 \times 1,600$ picture points are converted into digital numbers. The speed of scanning is slow enough that the analog-digital converters can handle the input data flow in real time. However, since at least 5 bits of brightness information are transmitted at each sample point and the resolution is high, the information rate is quite high.

Most of the classical stimuli which were converted into cyclopean stimuli were processed with such a facsimile transmitter which was connected to a DDP-24 medium-sized computer. The scanned input had two or three brightness levels, where white was converted to textured areas with zero disparity, black was selected to represent areas with some nasal disparity, while gray was converted into a third disparity value. Of course, any brightness levels could be used, but three depth levels seem adequate for most purposes. The size of the computer memory does not have to be much more than the number of samples contained in two lines. While one line is filled with the facsimile samples, the other line is read onto tape. For sufficiently fast computers the facsimile output can be digitalized and read onto tape directly.

Because of the very nature of stereograms, all the disparity shifts occur line by line, and therefore all computation can proceed for a single line in the left and right images at a given time. This means, at least in theory, that a small computer with a facsimile transmitter and receiver station could produce random-dot stereograms without the need for storing the entire images on tape or disk. It would suffice to process the image line by line. For instance, one could store a string of random numbers (first line) and generate a left image by recycling this string with different starting points. This image, after photoprocessing, could then be scanned as the input by the facsimile transmitter system. The computer would just delay a single line (of a string of random numbers) and perform the necessary horizontal shifts according to the desired disparity information to produce a right image. One problem is to be able to place the left image on the facsimile drum without skew. If the density is kept low—that is, two or three lines contain identical data strings—then the skew problem can be handled. Of course, such a tape- or disk-free system which uses the facsimile input-output as a sort of optical mass memory is not as flexible as a computer with tape or disk, but it might suffice for most purposes.

If one wants to be certain that all monocular cues have been removed from the stimulus, he has to use computers and accurate display devices, as discussed earlier. Nevertheless, after the introduction of computer-generated random-dot stereograms, many researchers tried simpler ways such as cutting out from sandpaper the desired form and overlaying it on another piece of sandpaper of the same texture and then shifting the cut out sheet by certain amounts. Unfortunately, it is very difficult to eliminate physical gaps, shadows, and other imperfections between the two sheets, and some monocular cues will usually result. More importantly, some of the straight edges of the cut-out form will cover the amorphic clusters of the random background in a regular fashion, thus giving rise to monocular contours. However, if the random pattern used contains only small granules of low density, then the overlapping figure of the same texture will very seldom cut a granule in half, and the monocular edges that can be traced through these incomplete granules might not be perceived. Good results can be obtained by using textures such as shown in figure 4.5-5* in which white (black) dots of 5% density are randomly scattered in a uniform black (white) field, but there are many other random designs available which will be suitable. It is relatively simple to make a sandpaper stereogram provided there are only a few depth planes. Before shifting the overlaid figures above the background, they are photographed, and after horizontal shifting they are photographed again. Of course, stereograms of complex surfaces cannot be manufactured by this simple method; they necessitate the use of computers. Also sandpaper-cinematograms are not easy to make. They require elaborate animation stands and are limited to the simplest scenes.

It is thus apparent that sandpaper correlograms are substitutes only for random-dot correlograms. Yet, in simple cases they provide a quick and inexpensive way to pretest certain ideas. They also permit the reader without computer facilities to participate in cyclopean experiments.

9.1 Problems of Semantics

Undoubtedly some of the readers who have struggled through the book to this chapter will be disappointed that some of the most intriguing problems of visual perception, form recognition, and the invariance of form under various transformations were not even mentioned. This was, of course, intentional, for it is my belief that without the technique of familiarity deprivation (using random textures instead of familiar shapes) most of the phenomena under study might have been elusively complex. The scientist's desire to stay away from enigmatic problems of semantics as long as possible was the driving force behind the linguistic school of Chomsky (1965), which was preoccupied until quite recently with problems of generative grammar. It appears that without semantics the linguist-grammarians were unable to cope with some relatively "simple" practical problems such as automatic language translation. Nevertheless, their approach helped to gain deep insights into linguistic structures. I can only hope that cyclopean perception will follow an analogous development.

In spite of the many reasons that justify the skipping of semantics, I share the disappointment of the reader. Such an approach reminds me of the intuitionist school of mathematicians who tried to clear mathematics from unresolvable paradoxes by removing from it the notion of infinity. The finite mathematics that was obtained was free from paradoxes and could still serve the applied sciences. Nevertheless, the essence of mathematics (mathematical induction, limit theorems, etc.) was lost in this purification process, and no creative mathematician was interested in this oversimplified subject matter. Similarly, visual perception without form recognition might be a useful discipline, yet a very restricted one. Of course, this avoidance of familiarity cues is not a goal in itself, but only a research strategy until problems of semantics are better understood. Nevertheless, let me illustrate with a few examples the interrelationship between findings of cyclopean perception and form recognition. This chapter will be more speculative than the rest of the book, and the reader who likes only facts is recommended to skip it, turning to chapter 10 instead.

The contributions of cyclopean perception to form perception are of two kinds. One incorporates actual findings that relate to form perception, while the other indicates how certain structures may be generalized in order to handle semantic problems. Examples of the first kind are numerous. For instance, the finding that in figure-ground perception the closer area always becomes the figure is of interest in form perception. Experiments with cyclopean optical illusions help evaluate the role of perspective cues and size constancy. Experiments with ambiguous stereograms portraying two (or more) surfaces of familiar shapes may permit the study of familiarity cues. One might expect that the more familiar surface will be more often perceived than the less familiar one if all other cues of stereopsis are identical.

The second kind of example indicates how some of the cooperative phenomena of stereopsis and movement perception may be generalized to cope with more complex phenomena. For instance, the spring-coupled magnetic dipole model illustrated how the various hysteresis phenomena of binocular depth perception can be explained. These hysteresis effects already belong to the perceptual constancies or invariances. After all, it takes a long time to obtain fusion of a complex array, but after fusion it is difficult to destroy the obtained state. Such hysteretic phenomena are most common for the complex perceptual constancies under transformations. For instance, if we view a photograph of a familar face, it is possible to rotate the photograph almost upside down without losing the percept of the face. However, if we initially show that photograph of the familiar face upside down, we may not recognize it and might have to rotate it by $90°$ or more to obtain recognition. Already the dipole model of stereopsis gives some insight why after fusion one of the dipole arrays can be rotated by $6°$ or more without destroying the interlocking between the dipoles; however, without prior fusion much less rotation can be tolerated in order to obtain interlocking. If instead of left and right arrays, one imagines that the two arrays correspond to the stimulus and its memorized version, respectively; and if one generalizes the arrays and couplings to hyperarrays of arrays coupled together at a higher hierarchical level, one can get some feel for the rotational constancy of perception. Later I will discuss such a generalized model.

In addition to the generalization of cooperative models describing binocular depth perception, there is another important relationship between binocular vision and form perception. My interest in binocular vision started with the insight that if in binocular vision most of the complex monocular familiarity cues are skipped, object separation is still possible. It is interesting to see how certain phenomena, which are easily grasped for binocular vision, become increasingly difficult to understand for monocular vision.

9.2 Binocular and Monocular Perceptual Invariances

We have seen that in order to explain perceptual invariance of stereopsis under dilation, the dipole arrays had to be suspended on a rubber sheet. Thus, for stereopsis one must generalize the metric of space from a rigid Euclidean one to a less rigid affine or topological space. That an affine space may suffice comes from other evidence, reviewed as follows.

That stereopsis cannot be explained in the framework of Euclidean metric is shown by the law of binocular depth constancy. A simple experiment is convincing. Consider two objects in space at different distances. As we change convergence from the first target to the second, the perceived distance of the first target does not seem to change. It was Schelling (1956) who exploited this observation in a mathematical theory, following Luneburg's proposal that only expressions which are invariant under eye and head movements are meaningful in binocular vision. He showed that a metric exists—that is, a distance $d(x,y)$ can be defined ($d(x,x) = 0$, $d(x,y) = d(y,x)$, and $d(x,y) \leq d(x,a) + d(a,y)$) which remains invariant under any affine transformation. His distance function is a special case of Cayley's metric that is invariant under projective transformations (see Bell 1945). This distance is non-Euclidean in an n-dimensional (in our case three-dimensional) space of negative constant curvature. Now, as illustrated in figure 9.2-1, if F is a point in space which is fixed, if R and L denote the focal points of the right and left eyes, and O marks the pivot of the neck, the tetrahedron $FRLO$ should be the subjectively fixed frame for binocular perception, independent of head and eye movements. Since an affine transformation changes any tetra-

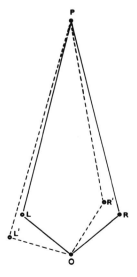

9.2-1 Illustration of the geometry underlying binocular depth constancy. See description in § 9.2.

hedron into any other one, Schelling's assumption states that binocular vision should be invariant to a class of affine transformations. Indeed, as we rotate our heads around O, the positions of the focal points of the eyes become R' and L', and the tetrahedron $FR'L'O$ undergoes an affine transformation without changing the perceived distance of $d(OF)$. An absolute binocular metric does not exist. It changes with the Euclidean distance between subject and fixation point. Yet the non-Euclidean curvature remains negative under all conditions. In the language of Luneburg's followers the metrics are always hyperbolic. It is interesting to note that Luneburg's followers often used biting boards and head-rests to immobilize the subject's head during their psychophysical experiments, when according to Schelling the essence of the hyperbolic metric of binocular vision is based on the distance-invariancy during the observer's unrestricted head and eye movements.

That this binocular depth-constancy can be obtained is the result of the physical constants of the positions of our eyes in the skull and the many physiological cues of vision and the vestibular system that inform us of our eye and head movements. I called the perceptual constancies that operate for stationary external objects during the subject's eye, head, and body movements, "perceptual constancies of the first kind" (Julesz 1969b). The problem of "perceptual constancies of the second kind" when the observer is at a standstill but the external objects change size, position, and shape is much more difficult. Even if we ignore topological shape changes and concentrate on positional, size, and shape changes of solid objects under projective geometrical transformations, the problem is still enigmatic. A possible, but unlikely solution was proposed by me (Julesz 1969b) using Cayley's method of constructing invariances under projective transformation (see review by Busemann and Kelly 1953). The only reason for including this idea is that all attempts at finding invariances must solve the problem of a known reference object, and Cayley's solution a century ago has great heuristic appeal.

We know that second-order surfaces (i.e., elliptical paraboloid, hyperbolic paraboloid, and ellipsoid) remain second-order surfaces under projective transformations. (Of course, an ellipsoid might become a paraboloid.) Furthermore, it is easy to verify that the logarithm of the cross-ratios among 4 points $\log[r(A,B,X,Y)]$ (out of which A,B are fixed and X,Y can vary) is a distance function. Now figure 9.2-2 shows Cayley's method of constructing a metric invariant under any projective transformation in two dimensions. If points X and Y are the points whose distance we want to define and C is any second-order curve (i.e., a circle or hyperbola) called

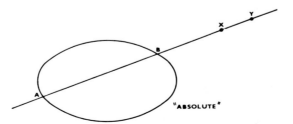

9.2-2 Derivation of Cayley's invariant metric under projective transformations.

Cayley's absolute such that the connecting line between X and Y will intersect C in two points A and B and, furthermore, if $d(X,Y) = k \log r(A,B,X,Y)$ where $r(A,B,X,Y) = (XA/YA)/(XB/YB)$, where XA is the Euclidean distance between points X and A, XB between X and B, and so on; then $d(X,Y)$ is a distance, invariant under any projective transformation. Thus, at least in theory, provided a second-order surface exists in the environment to serve as Cayley's absolute, it is possible to preserve the metric of any projection of solid objects in the environment. Perhaps the horizon or another large curve or surface serves as the absolute. (Some degenerate second-order surfaces, such as crossing planes, might serve as an absolute as well.) Of course, my proposal is very speculative and illustrates only one way a device could be constructed that would "perceive" invariantly the rigid environment under any projective transformation. This invariance is more general than what we are usually confronted with. In everyday life affine transformations and size dilations (which together are called "linear transformations") are adequate to cope with, and are only a subclass of the projective transformations. Indeed, very seldom, if ever, do coins project retinal images whose boundaries are not ellipses or circles, but are instead parabolas or hyperbolas.

That a reference object is necessary for monocular perceptual constancies of the second kind has been demonstrated by the trapezoid rooms of Ames (see Ittelson 1952). He placed familiar objects in distorted familiar environments such as trapezoid rooms or rhomboid-shaped windows. The observers would try to perceive these larger surfaces or curves as undistorted parallelepipeds or rectangles, and in this process the smaller objects would be seen distorted. For instance, a child appears taller than a grown-up man in Ames' room, provided the observer views the scene monocularly. (Binocularly, owing to disparity cues, the trapezoid shape of the room becomes apparent, and the test objects retain their natural size or form.) It is a large step from second-order reference surfaces to reference surfaces of familiar shapes, since for the latter, problems of form recognition enter. Nevertheless, it is again an interesting demonstration of how the metric of perceptual space is influenced by a few reference objects.

9.3 Klein's Erlanger Program and Perception

Another approach is by Hoffman (1964), who takes Hubel and Wiesel's findings and applies the method of continuous groups by Lie (1893). In a way he poses Klein's Erlanger program for psychology. The mathematician Klein in the last century tried to construct geometries whose objects remained invariant under various transformations. Similarly, Hoffman defined a certain "psychology" as the list of all transformations under which the percepts of the organism remain invariant. Thus, human adult psychology is characterized by translational, dilational, and rotational invariances for certain tasks. Conversely, for other tasks such as reading, the rotational invariancy will not hold. Therefore, the psychology of a reading person is different from the psychology of an observer when viewing simple familiar objects during rotations. In this section I review only Hoffman's first attempt (Hoffman 1964) based on Hubel and Wiesel's results, while his latest models (Hoffman 1966, 1968) on Lie algebras and their postulated cortical manifestations I leave to the interested reader, since they are speculations without convincing neurophysiological evidence.

The mathematical model of Hubel and Wiesel's findings in the cat's and monkey's cortex is as follows: The finding that each cortical column contains neural units whose receptive fields have edge or slitlike shapes of the same orientation is equivalent to a set of isoclines shown in figure 9.3-1. To a given retinal location (x,y) there are thousands of neural units in the cortex responding only to a line of a certain orientation dy/dx. Thus the cortex maps the input stimulus $f(x,y)$ into a set of isoclines. In other words, the "brain" converts the retinal input into a differential equation

$$y' = \frac{dy}{dx} = f(x,y) \text{ or, more generally, } F(x,y,y') = 0. \tag{9.3.1}$$

The task of the "mind" is to derive a solution of this differential equation under certain boundary conditions. The solution of this first-order differential equation is quite general without further constraints. It was Lie at the end of the last century (1893) who showed a method of how to solve equation (9.3.1) such that the solution would be invariant under several continuous transformation groups. The above-mentioned projective transformation of translations, dilations, rotations, forms a continuous parameter group since the product of, say, two rotations (applied subsequently) gives a new rotation (being the sum of the two rotations), and there exists an inverse operation which nullifies the result of a previous rotation. Ince (1926) gives a clear interpretation of how to solve differential equations by Lie's method such that in addition to 9.3.1 the partial differential equation

$$\xi(x,y)\frac{\partial F}{\partial x} + \eta(x,y)\frac{\partial F}{\partial y} = 0 \tag{9.3.2}$$

has to be satisfied as well. Here $\xi(\partial/\partial x) + \eta(\partial/\partial y)$ are the differential Lie operators. For instance, $\xi = x$, $\eta = 0$ is the operator for translation in the x direction; $\xi = x$, $\eta = y$ is the operator for dilation, while $\xi = y$, $\eta = -x$ is the operator for rotation. It can be shown that the solution of the partial differential equation (9.3.2) is equivalent to solving two ordinary differential equations called the "characteristic equations." It can be proved that for the affine transformations and dilation the solutions of the differential equations give two orthogonal families of curves. Hoffman argues that these explain the orthogonal afterimages of Hunter (1915) and MacKay (1961). For instance when showing radial lines (which is the eigenfunction for dilation, since this is the only pattern that does not change because of expansion-contraction) a subsequent noise field organizes into concentric circles (the eigenfunction for rotation). The importance of Hoffman's observation is in its heuristic value. It explains one possible reason for extracting the slopes of the input stimulus in the cortex. Only after such data reduction can the various constraints for invariances be satisfied. In this heuristic model no detailed description is given of the neuronal structure of the differential Lie-operators and such interpretations seem to me premature.

The reader, of course, could object that a first-order linear differential equation

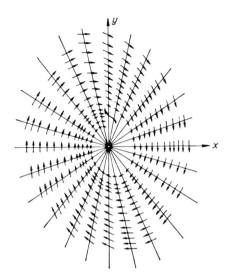

9.3-1   Illustration of a field of isoclines.

grossly oversimplifies the highly complex and nonlinear feature extractors of Hubel and Wiesel. This objection could be met by higher-order and nonlinear differential equations which could be solved by computer approximations.

Let us conclude with the relationship between these invariances preserving differential operators and optical illusions. We recall the hypothesis by Tausch and others that optical illusions might be perceptual correcting mechanisms preserving size and shape. Since the various "psychologies" can be defined as the collection of all perceptual invariances of an organism and several of these invariances are acquired by maturation and learning, the observation that the optical illusions also change with age and experience gives some support to Tausch's hypothesis. Yet, some optical illusions increase while others decrease with age, and it remains to be seen whether the postulated corresponding size and shape constancies vary similarly with age. Wohlwill (1960) in an excellent survey on the development of perception reviews several papers on the Müller-Lyer illusion which agree that it diminishes with age, yet size constancy in general increases with age. This is another serious criticism against the perspective theory of optical illusions.

9.4 Cooperative Phenomena and Perceptual Constancies

In the previous section we reviewed attempts to define psychologies as geometries invariant under a set of transformations. In spite of their heuristic appeal such programs can only serve as first approximations. In geometrical transformations, the underlying structure is the *group*. Each of the transformations has an inverse, and the product of two transformations is again a transformation. Unfortunately for psychology the transformations do not always have a group structure. For instance, take rotation of a simple object, say, a pencil. Under any amount of rotation the pencil is perceived as such and the rotation has a group structure. However, if we present a rotated picture of a certain human face, at some angle of rotation we would confuse its identity with an unrotated picture representing the same face or a different face. Within a certain degree of rotation there will be no error; surpassing this limit there will be an increasing uncertainty of identity of the picture. Hence in this case rotation has no group structure. Furthermore, there is a large hysteresis, depending on whether an identified figure is rotated from normal position, or whether an unidentified upside-down figure is rotated into normal position.

This and other evidence may suggest that perceptual invariances are closer to cooperative phenomena than to transformational groups. We could even try to generalize the spring-coupled dipole model to cope with form recognition, at least qualitatively. Imagine that instead of the two-dimensional brightness distributions on the retinae we have an *n*-dimensional space of attributes or features. One array of dipoles refers to the memory image, and depending on whether a given attribute is present in the stimulus or not, we align the dipoles with their north or south poles turned to a given direction. When perceiving a physical stimulus, the same feature extractors will act on a second dipole array and produce a similar dipole polarity distribution. If the stimulus pattern is simple, only a few of the features are activated, and the memory array and stimulus array can attract each other from a great distance, but with weak force. If the stimulus pattern is complex, the many north and south poles cancel each other at great distances and must be brought close to each other so that the corresponding dipoles can interlock. But after interlocking—which corresponds to recognition—great attraction exists between the two arrays, and it is quite difficult to lose the percept of a recognized shape. In a concrete example, the stored image of a simple object such as a pencil and a physically presented pencil in another orientation share only a few features, but the corresponding dipoles will attract each other even if the two arrays are rotated. On the other hand, a complex stimulus, such as a human face, requires a better alignment of orientation between the stored image and the concrete image for its recognition.

Let us imagine the workings of this model in greater detail. The two n-dimensional coupled dipole arrays (that represent the features of the input stimulus and its memorized version, respectively) are brought into alignment by merging with each other. In a certain position of the two arrays, that is obtained either by chance or by some learned strategy, a certain subspace (hypersurface) of coupled dipoles may become interlocked. After such an interlocked domain has been formed, the two n-dimensional arrays are further shifted in n-dimensions relative to each other. Because of the coupling the interlocked domain will remain locked for considerable amounts of shift. Thus one can explore a relatively large space in the neighborhood of the interlocked domain by searching for other hypersurfaces of dipoles that may align and become interlocked. The more such interlocked domains are formed the stronger the forces of interlocking become and the more they will resist the forces that tend to unlock them because of shifting. Thus wider and wider spaces can be explored around the interlocked domains.

If the first interlocked domain has no similar domains in its vicinity, the exploratory shifting will increase causing the first domain to unlock, and the shifting will continue until another interlocked domain forms. If this new domain contains several other domains in its vicinity that will interlock, the search ends. For stereopsis the interlocked dipoles constituted surfaces that are close to each other in three-space. It was impossible to fuse simultaneously two or more surfaces that had larger differences in their disparities than Panum's limit. If the surfaces had similar disparity values but were distant in the x-y plane, it was difficult to shift attention to all of them simultaneously. In n-space it is even more important to search for hypersurfaces of interlocked dipoles that are not too distant from each other. The coupling between dipoles may correspond to features that are associated with each other and the more distant or less coupled certain features are the more irrelevant their contribution must be to a recognition task.

Unfortunately, the details of the model are not worked out. The metric of the attribute (feature) space and the coupling between adjacent dipoles is not known. The problem of adjacency and the establishment of independent and coupled features is hard to quantify. Nevertheless, some great discoveries in the last decade on multidimensional nonparametric scaling indicate a way to establish the dimensions and metric of the perceptual space. We turn our attention next to results obtained by multidimensional scaling and clustering algorithms.

9.5 Toward a Psychophysics of Form

Throughout this book the cluster of proximate dots of similar values plays an important role. When similar values referred to brightness, the clusters of proximate dots of similar brightness values proved crucial in texture discrimination and symmetry perception. When the notion of similarity was generalized from brightness values to disparity values, then the clusters of proximate dots of similar disparity values were shown to be the basis of stereopsis. If the rigid Euclidean distance metric on which proximity is based is also generalized to a more flexible affine space, an even more realistic model of stereopsis can be formulated.

Customarily in nature after a powerful solution has been developed—in this case the search for clusters—the same solution is then applied to more elaborate situations after a step toward generalization has been found. Indeed, we have recently witnessed a birth of psychophysics of higher mental activities in which clusters play an important role (see Julesz 1968c for a review). Here let us briefly review the mathematical ideas that can be used in the study of form perception.

Before we turn to mathematics, let us note that there is something arbitrary about the "features" underlying form recognition. These features seem to reflect many atavistic influences acquired at various evolutionary stages of an organism. The original purpose of extracting a certain feature is long forgotten, yet it is embedded in the structure of the nervous system in a later evolving species. This ad hoc nature of form recognition makes its study very difficult. Teleological arguments are even more dangerous in visual perception than in other disciplines. Consider the beautifully organized visual system of the Limulus which shows scant behavioral evidence of ever using it. It is not surprising that students who are accustomed to the inherent consistency and simplicity of physics often feel disappointed when confronted with the complexity and arbitrariness of living structures.

With the advent of fast digital computation, nonmetric multidimensional scaling techniques (factor analysis) and hierarchical cluster-seeking algorithms became powerful new research tools. Particularly interesting was the development of multidimensional scaling by Shepard (1962) and Kruskal (1964). Although much work along similar lines had gone on under the name "numerical taxonomy" (Sokal and Sneath 1963), it was an ingenious insight by Shepard (1962) that started nonmetric multidimensional scaling. Shepard wanted to preserve the hierarchy (order) of magnitude of similarity judgments of perceived stimulus pairs when they were placed as geometrical points in an abstract space of a given dimensionality. Shepard's insight was as follows: N stimulus points can always be placed in an $(N-1)$-dimensional space with the hierarchical order of distances between point pairs being identical to the judged similarities. As the dimensions of the abstract space are reduced, the hierarchical order will be violated for a few points. The departure from the original ordering is measured by an expression called the "stress." This stress measure increases with a decrease in the number of dimensions. There is usually a minimum number of dimensions where the stress is still low, and further reduction causes a jump in the stress. The desired solution is this lowest dimensional abstract space, in which the maximum number of pairs can be placed such that their geometrical distances (according to some metric) preserve the order of judged similarities between them.

In addition to solving this minimax problem by varying the number of dimensions, the metric can be varied as well. For instance, when using the Minkowskian metric $ds = (\,|\,dx\,|^p + |\,dy\,|^p)^{1/p}$, can be varied. As an example, figure 9.5-1 shows the results of applying multidimensional scaling to a color-naming experiment of spectral colors performed by Boynton and Gordon (1965). Here a spectral color was presented (with constant brightness and saturation), and subjects were asked to name the colors using only four names: red, yellow, green, and blue. Thus, violet would sometimes be called blue, sometimes red; a blue color would always be called blue. From the distribution of responses the similarity matrix between colors could be computed and the minimax multi-dimensional scaling solution gave an optimum three-dimensional solution and a good two-dimensional solution. Figure 9.5-1 gives the two-dimensional solution for 23 stimulus points. Point 1 corresponds to 440 nanometer and successive points progress in 10 nm steps terminating with 660 nm for point 23. For details the reader is referred to Shepard and Carroll (1966). The most important finding is that the percept of a stimulus with one degree of freedom (only wavelength was varied) requires a three-dimensional perceptual space. It should be noted that here the three axes of the coordinate system have no "meaning" by themselves. It is *not* the case that the stimulus had three degrees of freedom and the obtained three dimensions of the perceptual space reflected this inherent structure. One might argue that color perception in general is applied to stimuli with three degrees of freedom such as brightness, saturation, and hue; and, therefore, the perceptual space for color is three-dimensional regardless of the fact that only one physical parameter has been varied. But this argument is not valid. An inspection of figure 9.5-1 reveals that the locus of the perceived hues is a twisted line in two dimensions (or three dimensions) where each dimension is related *only* to hue changes.

Besides color, there are many elementary percepts of one-dimensionally varying stimuli that must be embedded in a higher than one-dimensional perceptual space. Similarly, stimuli with two degrees of freedom might yield abstract perceptual spaces that are not two-dimensional. This mapping of dimensions of reality to dimensions of the mind is a new approach to the mind-body problem—a sort of "perceptual psychophysics." It is an intriguing thought that even though we do not know what a percept (such as a red or blue color sensation) by itself is, we can determine its dimensionality and metric!

The problem with all these methods is the vast number of measurements necessary to determine the similarities among the objects. For $n = 100$ objects, $n(n-1)/2 = 4{,}950$ object pairs have to be measured and rank-ordered with respect to their psychological similarity (confusability). Such numbers tend to defy even the most motivated researchers. Interestingly, there seems to be one way out of this dilemma. Humans are exceedingly fast and accurate in sorting objects into clumps. For instance, when the experimenter specifies the number of clusters the objects should be sorted into, the subjects usually are able to perform the task very quickly. If the majority of subjects arrive at the same clustering, the data presentation is already solved. If the

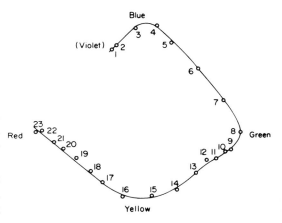

9.5-1 Multidimensional scaling solution in two dimensions of a color-naming experiment by Boynton. (From Shepard and Carroll 1966.)

subjects' sortings differ, similarity measures can be defined on the basis of the sortings. Each pair of objects is assigned weight one, if the pair is assigned to the same cluster by a given subject, and weight zero otherwise, and then these weights for each object pair are averaged over subjects. This procedure assigns a similarity value between zero and one to each pair, and the multidimensional scaling can be carried out.

With this technique both simple and complex problems of visual perception have been attacked. Applying multidimensional scaling to discrimination of textures composed of random 2×2 arrays, it was found that the most important factor was brightness and the next most important was orientation (Julesz 1964a). Studies by Brown and Owen (1967) of the multidimensional scaling of random polygonal figures of Attneave and Arnoult (1956) using a geometrical criterion of similarity reveal a dozen complex cluster parameters of compactness, jaggedness, skewness, rotation, and so on. Each was highly correlated with low-order moments of interior angles, side lengths, perimeter, and area measures. Which of these parameters is actually used by humans is hard to determine primarily due to shifts in human attention. These shifts and the large number of parameters greatly hamper the development of the psychophysics of form. Nevertheless, work by Thomas (1967) indicates that human discrimination of random polygons can be represented in a two-dimensional space whose axes are complexity and symmetry.

These examples served only as illustrations. Similar techniques revealed complex multidimensional data varying from Morse code perception to interpreting facial expressions. The "objects" under study might be visual patterns presented to subjects. Psychological measurements are obtained from the similarities (or confusions) among these objects. The problem is discovering whether there is any inherent structure in the psychological data themselves. Besides nonmetric multidimensional scaling there are clustering schemes that try to arrange objects into optimally homogeneous groups; such schemes were developed by Ward (1963) and refined by Johnson (1967). Multidimensional scaling and clustering algorithms can be combined (Shepard 1962).

These examples seem to indicate the usefulness of these refined statistical methods in revealing the main features underlying form perception. However, there are several problems with these techniques. One of the severest problems is our inability to interpret some of the main dimensions of a solution. In some cases, such as in Morse code perception, some dimensions have a clear interpretation (e.g., number of dots and dashes; ratio of dots and dashes); other dimensions, however, defy interpretation. In addition, problems of shifts in attention during data collection can introduce unwanted noise into the solution. For instance, one can base his judgments of faces on a few features such as ovalness or complexion. But when a bearded face is shown for the first time a new perceptual dimension becomes salient, one which is irrelevant to all the previous cases. This clearly would prevent finding a tight solution with minimum dimensionality.

This search for a minimum-dimensional space in which the departure from the monotony of ordered similarities has a minimum is itself an arbitrary criterion. But all of the sciences abound in such arbitrary "beliefs." Whether this "creed" is the acceptance of a "least-energy" principle in physics or the Darwinian doctrine of natural selection, no scientist outside mathematics can escape succumbing to such meta-scientific assumptions. Having witnessed the overthrow of some of the most sacred dogmas of physics, such as the "symmetry of parity," we must realize that in science the rules of the game are ever-changing. So, until better assumptions can be believed, multidimensional scaling has a useful role to play in perceptual studies.

The reason for including this section is to show how the notions of clusters can be extended to cope with problems as complex as form. For texture discrimination the two factors underlying grouping—proximity and similarity—had a straightforward meaning. Proximity meant adjacent points in Euclidean space, and similarity referred to points having almost identical brightness values. For stereopsis, the proximate points were those that were elastically coupled, and similarity referred to points with almost-identical disparity values. The clusters underlying form perception or higher mental functions are points that are grouped in a perceptual space with many dimensions. However, the metric of these spaces depends on very complex generalized notions of proximity and similarity. We cannot guess yet whether these generalized proximity and similarity parameters will turn out to be ad hoc, different for each perceptual task (as suggested in the opening paragraphs of this section) or will exhibit some common characteristics. Each possibility has important consequences for the basic question of what the nature of visual perception is.

10 Cyclopean Perception in Perspective

10.1 Other Psychoanatomical Techniques

Chapter 9 was more speculative and general than the rest of the book, so a "coda"—a return to the basic theme of the book—is now appropriate in order to end this treatise on the right note. The basic theme has been the tracing of the visual system by cyclopean techniques. However, as noted throughout the book, several other classical techniques can be used to tell retinal from cerebral processes. As long as the black box is not forced open any technique that can localize mental processes with respect to others can be regarded as belonging to psychoanatomy.

It is not always clear whether a given technique should be regarded as purely psychological or, rather, should be considered physiological. The "borderline" techniques, when carefully applied, do not penetrate walls of the black box, yet manipulate the inside of the box without following the usual course through the sensory transducers. For instance, the production of visual phosphenes by direct electrical stimulation of the eyeball or by very intense electromagnetic fields is a sort of cyclopean stimulation, though the site of penetration probably stays in the retina. Pressure blinding of the retina can be used to determine whether an afterimage or aftereffect is retinal or not. Since the applied pressure affects the retinal blood supply, this technique is even less psychological than the production of visual phosphenes by electromagnetic energy that penetrates to some depth into the central nervous system.

Psychotomimetic drugs, such as LSD or mescaline, can produce perceptual changes; and if their site of action were established, they could help to localize perceptual processes. However, at present the action of these chemicals is not known, and the direction of inference is typically the reverse: psychological and cyclopean techniques are used to ascertain the active sites of the chemicals (Julesz 1970c). It is also questionable whether experiments during chemically induced mental states can be regarded as purely psychological methods. How chemical stimulation will contribute to the understanding of visual perception remains to be seen. In spite of this, some observations during drug induced states are of interest. Szara (1957) conducted experiments on himself with the injection of DMT. He described eidetic phenomena, optical illusions and hallucinations of brilliantly colored oriental motifs. The phenomenon of dysmegalopsia (also called dysmetropsia)—the disturbance of the visual appreciation of the size of objects—often occurs with certain drugs. Macropsia (increase in object size) has often been described, particularly in alcoholic delirium, and the interested reader should consult the paper by Holmstedt and Lindgren (1967). Micropsia (the decrease in object size) occurs less often than macropsia, but can occur with either alcohol, cocaine, chloral hydrate, or cannabis intoxication (see also Holmstedt and Lindgren 1967). It is interesting that these experiences are described by people who are fully oriented in time and space, and that only living persons and animals but not dead objects, change size. This might indicate that a highly central zooming mechanism is affected by the chemicals.

Another borderline field is the measurement of electroretinograms (ERG), cortically evoked potentials (EP), and electroencephalograms (EEG) during sensory stimulation. These techniques are more psychological than the previous ones, since the measurements are passive and could be regarded as psychoanatomical. The only problem with these techniques at present, particularly with EP and EEG, is that it is not well established whether these potentials have any relation at all to sensory or perceptual events. The confusing state of this field is well documented in a bulletin of an NRP work session chaired by MacKay (1969). The majority of the reports show no relationship between perceptual state changes and EP changes. The few reports in the last few years that find such correlations are not unequivocal. For instance, Cobb, et al. (1967) studied EP during binocular rivalry. A subject pressed a key as the dominant or the suppressed state of one eye alternated and EPs corresponding to the dominant and suppressed states were significantly different in a component that had a latency of about 250 msec. However, Donchin and Cohen (1967) could not repeat this finding when the flashes were presented at random (and not after key pressing when dominance switched). As another equivocal example, Lehmann and Fender (1967) measured EP elicited by flashing a uniform field to one eye while the target structure to the other eye varied. They found monotonically decreasing EPs with increased stimulus complexity (spatial frequency) above which the EPs do not decrease. But this asymptote is reached much before the limit of visual acuity.

Riggs (1966, 1969) and Riggs and Sternheim (1969) reported experiments using colored stripes of two different hues, in which they varied the hue difference and compared the psychophysical function for visibility of the stripes with simultaneously recorded ERG and EP data. The ERG records were in good agreement with the psychophysical data and were reasonably well-described in terms of a simple, additive trichromatic theory in which there is no trace of a yellow response system or of opposed responses to complementary hues. However, the EP data was not easy to interpret.

The failure of EP recordings to correlate with complex perceptual phenomena is probably the result of the primitive state of the art. Although averaging techniques increased the signal-to-noise ratio by orders of magnitude, the electrodes customarily used pick up the activities of billions of uncorrelated neurons. In principle, it should be possible, by the simultaneous recording outputs of an array of electrodes, to resolve several neuron assemblies with different spatial and temporal properties; these assemblies might be more related to perception than the present indiscriminate sums of neurons.

Lehmann, et al. (1969) have recently reported several attempts to measure EP simultaneously by multiple electrodes. They used 48 scalp electrodes and averaged each of these channels individually. Field maps of EP could be observed. The fields slowly migrated, suggesting the invasion of different cortical areas by the response processes.

In spite of these controversies a recent development may be of interest. Regan and Spekreijse (1970) found differences between EPs when horizontal or vertical disparity changes were cyclically introduced in random-dot stereograms. There is a possibility, though, that these differences resulted from eye movements.

I received at the last moment a communication by Campbell and Maffei (1970) which I regard as a triumph for EP techniques. They altered in phase gratings of various spatial frequencies at 8 Hz and measured the EP (from the occiput of man) as a function of contrast. They found that a linear relation exists between the log of contrast and the amplitude of the EP; and extrapolation to zero amplitude predicts the psychophysical threshold. The slope of the EP-log (contrast) line was found to be independent of either the spatial frequency or the area of the stimulus grating. This slope could be markedly increased by using either two different spatial frequencies or two different orientations, presented simultaneously. Selectivity to orientation and spatial frequency was found to be high and a second grating 15° less tilted or having spatial frequency within an octave had no influence on the response to the original grating. These findings give further evidence for spatial frequency and orientation sensitive units in man. The findings suggest that EP depends on the number of channels (selectively sensitive to spatial frequency or orientation) and does not depend on the area of stimulation. But more importantly, these results herald an era, when psychophysicists may switch from threshold measurements to the more realistic suprathreshold observations.

So far, we have reviewed in this section only the few techniques of perceptual process localization that are not purely psychological. Earlier we have mentioned several purely psychological localization techniques. Interocular transfer experiments have been one of the most popular among such techniques. Usually a comparison is made between the outcomes of a monocular, a dichoptic (or dichopic) and a binocular control experiment. For illustration a typical experiment in this class carried out by Gilinsky and Doherty (1969) will be quoted. A masking grating (with given contrast, spatial frequency and orientation) was presented to the subjects and after this a test grating of various orientations was shown. In the monocular case both the masking and test grating were presented to the right eye. In the dichoptic case the masking grating and the test grating were presented separately to each eye. For the binocular control case identical masking gratings were binocularly presented to both eyes, and the test grating to the right eye. Both the monocular and dichoptic conditions resulted in a strong masking effect, which peaked at a test target orientation. The significant masking found with dichoptic viewing is evidence that interocular transfer of orientational effect does occur. The finding is in agreement with the notion of binocularly driven edge detectors in the visual cortex of the cat and monkey.

Results of interocular transfer experiments, however, are not as unequivocal as cyclopean findings. During the testing of the nonadapted eye the previously adapted eye is deprived of patterned stimulation; this can cause binocular rivalry and thus modify results in unpredictable ways. Furthermore, as Day (1958) pointed out, conclusions based on interocular transfer alone can be erroneous. He refers to the interocular transfer of negative afterimages (noted as early as Newton and reported several times since), which originally was regarded as evidence for their cerebral origin. It took considerable debate and research (including the "idea" to close both eyes and still see afterimages) to prove that, on the contrary, afterimages are of retinal origin. Day went a step further and criticized the assumption that the interocular transfer of visual aftereffects shows their central origin. Terwilliger (1963) voiced a similar opinion, and Scott (1964) gave another. Scott pointed out that negative afterimages are not really "transferred" since the afterimages can be clearly seen with both eyes covered. Scott made the further distinction that afterimages which have been seen with one eye may produce an image superimposed upon the signals that fall on corresponding areas on the other retina, but do not affect what is seen by that eye. The only conclusive evidence for interocular transfer is obtained under retinal pressure blinding. For instance, the outcome of the Gilinsky and Doherty experiment could have been differently explained if they had not known that the retina is devoid of edge detectors. Instead of believing that binocular edge detectors in the cortex were adapted by the monocular grating, one could have believed that edge detectors in the retina were stimulated by the masking grating. The result of this adaptation could have resulted in increased noise that superimposed on the test grating at the site of binocular combination, causing an increase in the visibility threshold.

This conjecture is most unlikely, yet without knowing the cortical location of the edge detectors, the interocular transfer of orientational aftereffects could have either a retinal or cerebral interpretation. In this respect, strong cyclopean stimuli are less dependent on neuroanatomy. The only a priori knowledge required is that the first site of binocular combination of the two visual pathways is in the cortex in order to tell retinal and cortical processes apart. Of course, with more precise knowledge of the site of global stereopsis one could determine whether a process occurs before, say, Area 18 or after; but, regrettably, binocular methods can only locate processes with respect to this cyclopean retina.

We have seen how random-dot correlograms, at least in principle, can portray information at cyclopean retinae other than the one corresponding to stereopsis. Are there other techniques of psychology that can trace the information flow beyond the site of binocular combination? Psychological studies on humans who have suffered brain injuries from penetrating missile wounds do not qualify for inclusion in this section (though such findings were cited in the book), because the observed perceptual deficits cannot be related to their cause without an exact post-mortem determination of all lesions. The black box has to be opened.

In searching for other psychological techniques, reaction time (RT) experiments come to mind. In our age of radar and sonar one might expect to measure the length of neural pathways by RT. This paradigm cannot work, unfortunately, in such a simplified form. The great variation of neuron diameters causes propagation times to vary by orders of magnitude; furthermore, the time delay at each synaptic junction is variable. Thus, localization by differences in RT alone cannot tell retinal processes apart from cortical ones even if the neuroanatomy of the CNS were roughly known. However, Sternberg (1967) introduced a new paradigm based on RT measurements. He has used his paradigm in studies of recognition and recall, and was able to establish the serial structure of retrieval. He also showed that for recognition based on short memorized lists, the serial retrieval process is exhaustive; that is, the search through the memorized list does not terminate when the correct item has been found, but only after the entire list is mentally scanned. Sternberg (1967) has used this paradigm to determine how much processing of the stimulus is performed before its representation is compared to memory. By analogy the paradigm can be used to decide whether size-zooming, rotation, or other form-preserving transformations are performed on the stimulus itself or on the stored images. A third alternative could be suggested, that the transformation is performed on the stimulus repeatedly, once before each comparison, but this seems implausible.

An application of Sternberg's method of localizing the site of a stimulus transformation to the process of size-zooming was suggested by Julesz and Sternberg who undertook a preliminary study. Imagine that the subject has memorized N symbols of a given size. During the test period a symbol of the same or different size is presented for a brief flash, and the subject has to determine whether (regardless of change in size) it is a new symbol or belongs to the memorized set. He responds by pressing a key. RT is a monotonically increasing function of the number of stored symbols. If a test symbol is larger or smaller than the stored representation of it, the RT will be somewhat longer. If size constancy (i.e., compensation for the size change by zooming) were achieved on the stimulus itself then the RT as a function of the stored number of items would increase by a constant, thus the RT function would be parallel to the control RT function (where the test stimulus has the same size as its stored version). On the other hand, if the zooming were performed on the stored symbols one at a time, then one by one all the stored symbols would have to be zoomed and compared for each test symbol. Thus the RT as a function of number of memorized symbols would increase more steeply than in the previous case.

Even this paradigm has some theoretical and practical difficulties. Variation in stimulus size can be regarded as a sort of distortion or added noise. The increase of RT produced by size variation can be simply the result of longer decision time that is necessary for comparing a noisy stimulus representation with a memorized symbol, as inferred by Sternberg (1967). Were this the case, we would find that the slope of the RT curve increases with the number of memorized symbols. However, the same finding would lead to a very different conclusion were we to propose another model. This model would be to evoke the existence of a noise-cleaning operation as an inverse transform that restores perceptual invariance. This assumption is particularly plausible for very distorted symbols that can be still easily recognized (such as cursive letters or handwritten letters when the stored symbols are assumed to be block letters). In the case of a noise cleaner, as Sternberg (1967) showed, one can decide from the RT function whether the noise cleaner acts once and for all on the stimulus or individually on each memorized symbol. Since we do not know which of the two alternatives holds (even if the inverse transform hypothesis appears more likely), every process localization by this method is open for criticism. Because of this theoretical difficulty, the practical problems with this paradigm should be mentioned only briefly. Humans are so good at recognizing slightly rotated, expanded or otherwise slightly distorted symbols that the RT increase can be expected to be marginal, and it is difficult to determine whether the new RT function rises more steeply than before. One has to introduce large rotations or distortions in order to obtain reliable data, while still trying to mention a reasonable error rate. Perhaps a good way to slow down perceptual performance would be to use as test symbols cyclopean forms portrayed by random-dot stereograms. Such a cyclopean text is portrayed by the random-dot stereogram of figure 10.1-1*.

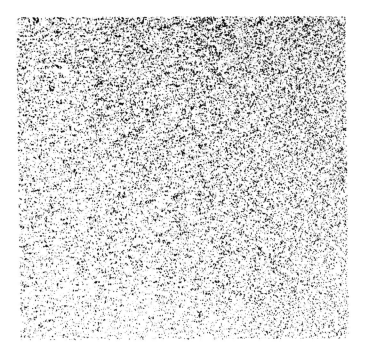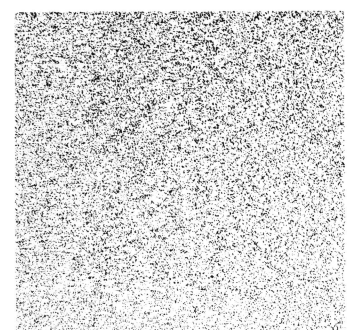

10.1-1* Text portrayed by random-dot stereogram.

Sternberg's work, particularly its recent development (Sternberg 1969), is of great interest but goes beyond the scope of this book. The fact that the researcher can reveal certain structures in a cascaded sequence of processors by RT measurements alone suggests a psychoanatomical procedure. However, Sternberg's technique cannot establish the order of these processors. Because of this it is not a true psychoanatomical procedure.

In passing, the perceptual adaptation experiments should be mentioned in this context. Whether the visual field is presented upside down, or a text is inverted about several symmetry axes, or color fringes and other distortions are seen through prisms, after some adaptation time the stimulus appears to be normal. Which of these phenomena are based on proprioceptive cues and which are strictly visual rearrangements are questions beyond the scope of this book. However, the complexity of the rearrangement and the long adaptation period makes their cerebral origin most likely. One input is the visual stimulus falling on the retinae, the other is the proprioceptive information of the subject's position and movement with respect to his environment. The final percept is the combination of these two messages into a third kind of information. Usually the two inputs are not independent of each other. In all the prism-adaptation experiments, strict and relatively simple correlations exist between stimulus changes and changes in subject's position. Nevertheless, this correlation is a cyclopean message. As adaptation proceeds, this cyclopean information is increasingly extracted and utilized.

I wanted to conclude this book with this chapter in order to place cyclopean methodology in proper historic perspective. As we have seen, psychoanatomical thinking is as old as experimental psychology itself. The quest of tracing the information flow in the central nervous system by purely psychological means is essential in physiological psychology. It is possible to trace the information flow in the visual system (to some extent) without opening the black-box. (A computer-programmer could also establish the information flow among various subroutines without knowing any detail about the hardware.) Nevertheless, without knowing at least approximately the site where the two optic pathways meet in man, much of the binocular findings might have been less important. Similarly, without the epoch-making discoveries of neurophysiology in the last decade, much of the cyclopean results would have been less interesting. Whether cyclopean stimulation by random-dot stereograms skips four, five, or six synaptic levels is not immaterial. But to determine the exact location of a certain cyclopean retina is beyond the realm of psychoanatomy. Cyclopean techniques can only determine these sites with respect to others. Absolute localization necessitates sophisticated techniques of anatomy and physiology.

In retrospect, I am not sure whether it was a wise decision on my part to extend cyclopean methods beyond strongly cyclopean stimulation. Had I restricted the book to strongly cyclopean stimuli, the premature problem of relating neurophysiological findings to psychological phenomena could have been avoided. Indeed, one does not have to be familiar with the latest neurophysiological findings to accept that the cyclopean retina stimulated by random-dot stereograms cannot reside prior to the site where the two optical pathways combine. On the other hand, one must be absolutely sure that no edge, movement, corner, or other feature extractors exist in the human peripheral system in order to devise weakly cyclopean stimuli. This requires a constant surveillance of the neurophysiological literature. Had I reviewed only results that were obtained by random-dot stereograms and cinematograms, much of the lengthy reviews of neurophysiology could have been skipped, and the strange mixture of physiological and psychological terms could have been avoided. Furthermore, the cyclopean retinae of global stereopsis and movement perception are highly cortical and of excellent spatial resolution and thus can be used for localizing a great variety of perceptual phenomena.

Yet, the inclusion of weakly cyclopean stimuli broadens the scope of the book. It gives access to several other cyclopean retinae in addition to that of global stereopsis and movement perception. Weakly cyclopean stimulation might also bring cyclopean perception closer to other branches of psychology.

In turn neurophysiologists should find psychological findings helpful in several ways. Psychoanatomists can tell the neuroanatomists where to seek or not seek for certain perceptual phenomena. For instance, since vernier acuity perception occurs before global stereopsis, the neurophysiological evidence of simple, complex, and hypercomplex units operating before global stereopsis should be able to account for this superresolution. Similarly, since symmetry perception probably occurs after stereopsis, some symmetrical connections should be found after the site of stereopsis. Furthermore, neurophysiologists can use cyclopean stimulation during their recordings. Cyclopean stimulation portrays the information closer to the recording level and thus simplifies the data.

10.2 Unsolved Problems of Cyclopean Research

I hope the reader who has struggled through the book has learned the basic skills of visual counterpoint and can in principle "compose" random-dot stereograms and cinematograms for most perceptual phenomena he wants to localize. I hope also that the reader has understood the essence of cyclopean methodology: the operational skipping of early processing stages. "Skipping" is not meant literally. When, say, an optical illusory figure is portrayed on the cyclopean retina (of stereopsis or movement perception) the black and white dots of the correlograms undergo processing by retinal receptors, bipolars, ganglions, LGN cells, and neural units in Area 17 of the cortex. However, it is only one or two stages deeper—where global stereopsis or movement is probably processed—where the illusory figure first emerges. This cyclopean figure is portrayed by depth or motion gradients that can be easily detected with good spatial resolution. Therefore, it is possible to portray on the cyclopean retinae of stereopsis or movement perception forms almost as detailed as those used in classical studies.

The book reviewed a few dozen cyclopean experiments (most of them reported here for the first time) that were carried out in order to localize the origin of some of the most famous perceptual phenomena. There are hundreds of less known perceptual phenomena (some of them discovered only very recently) which have not yet been tried under cyclopean conditions. Only time and expense interfere with studying all these effects. For instance, it is conceptually clear how to test spiral-aftereffects with cyclopean stimuli. However, the production of random-dot stereogram movies that portray a spiral in depth (with variable speeds of rotation and desired dynamic noise characteristics) is an elaborate undertaking. I am confident that real-time computer movies will speed up the study of complex spatiotemporal perceptual phenomena.

Can cyclopean methodology be extended beyond this? Would it be possible to have access to cyclopean retinae that lie *after* the ones we already have? As of now, the deepest cyclopean retina is stimulated by random-dot stereograms and cinematograms. Whether the cyclopean retina of global stereopsis or that of global movement is located at a more central site is a minor problem and can be overcome by portraying the phenomenon under study both by random-dot stereograms and cinematograms. We have seen that other cyclopean retinae exist, such as the one of eidetic imagery or the McCollough effect. Unfortunately, these cyclopean retinae are *before* the one of stereopsis, and because of this they are of limited usefulness.

It is not enough to search for new cyclopean retinae at deeper and deeper sites in the central nervous system. We have another requirement. These new cyclopean retinae should have adequate spatial resolution to enable us to portray complex shapes.

These two requirements of a highly central site and good spatial resolution seem to be incompatible. The more central a process is the less location-specific it appears to be. It might well be that global movement-perception or stereopsis are the highest levels in the visual system that still preserve the exact positions of their constituent local processes. If so, cyclopean stimulation can "only" bypass five or six synaptic levels. However, if deeper cyclopean retinae were to be discovered, the power of cyclopean methodology would significantly increase.

Even if it would turn out to be impossible to extend cyclopean stimulation beyond the site of stereopsis or movement perception in the visual system, some other modalities might be more promising. One extension of cyclopean perception could be achieved perhaps by using random-dot correlograms in the tactile domain. Since the tactile receptors on the skin surface are two-dimensionally organized, the skin surface is a better analog to the retina than is the cochlea. Because of this, many cyclopean techniques appear more promising for touch than for hearing. There is increasing activity to provide prosthesis for the blind by attaching sizeable arrays (in excess of 20×20) of tactile transducers to sensitive areas of the body surface (Bliss, et al. 1966, Bliss 1969, Bach-y-Rita, et al. 1969). It may not be far-fetched to present random-dot cinematograms to a tactile transducer array and test whether a cyclopean figure can be experienced by the subject. Perhaps even a bitactile analog of random-dot stereograms may have perceptual significance.

There is no need to introduce cyclopean techniques into cognition, since in the study of the highest mental processes they were often implicitly used. I started this book with the example of musical counterpoint. As another cyclopean-cognitive technique I referred to an experiment by Treisman (1964) in which delayed messages in two languages were presented to the attended and unattended ears, respectively. Bilingual subjects could detect the semantic identity of the messages even after considerable delays. Here the cyclops is a semantic processor that can be stimulated by two different languages, and a person who speaks only one language is blind to this stimulation. There are many similar cognitive examples and the only contribution of cyclopean perception to cognition might be to make researchers aware of these latent uses.

Nevertheless, these cognitive examples are only hypothetically cyclopean since no neurophysiological evidence has been found yet for cognitive feature extractors. This situation, however, may quickly change. Recently, Thompson, et al. (1970) found single neurons that can count in the association cortex of the cat. In a sequence of stimulus presentations—regardless of stimulus modality, intensity, and interstimulus interval—these cells characteristically discharge to a particular stimulus number. The investigators found in an adult cat five counting cells which code the numbers 2, 5, 6 (two cells), and 7, which is such a small sample that it is difficult to assess the total number of counting cells in the association cortex. However, the finding that the abstract notion of the number of items is neurophysiologically represented in single units is of great importance. When a subject is presented with a series of auditory clicks and another series of light flashes and has to decide whether they are of the same or different number, we have a really cyclopean cognitive process.

On the other hand, such an abstract task does not necessitate mental imagery, and therefore, is beyond the realm of cyclopean perception. Were global stereopsis and global movement perception to be processed in Area 19, then random-dot correlograms would be probably the highest cyclopean stimuli that still give rise to detailed imagery. If, however, these global processes were residing in Area 18 in man, then one could still search for a cyclopean stimulus that could stimulate a cyclopean retina in Area 19, and try to repeat all the previous experiments using these new stimulus classes.

The extension of cyclopean perception could come from quite unexpected directions. Let me illustrate such a development by a recent experiment. It is known that asymmetries in the processing of the visual field on either side of the midline are related to hemispheric specialization in man. In subhuman species, the two hemispheres appear to act symmetrically and each hemisphere serves the contralateral half field. In man, asymmetrical representation of cognitive functions, notably language, complicates the situation. While usually the left hemisphere contains the cognitive centers in man, the other hemisphere contains centers of spatial orientation. Kinsbourne (1970) was able to overload the speech hemisphere (by asking the subjects to keep in mind six one-syllable words) and show that the detection of a gap on the right side of a square was significantly better than the detection of the gap on the left side. Without cognitive loading there was no difference in the left and right gap-detection performance. Thus, it is possible to obtain, at least temporarily, a "split brain" subject by purely psychological means.[1] Such a tracing of the information flow within hemispheres is cyclopean even if it differs from the stimulus techniques that were discussed throughout the book.

Recent evidence of spectrum analysis in the visual system suggests a new cyclopean technique. In § 3.4 we briefly reviewed a demonstration of Blakemore and Campbell (1969b) in which a test drawing of a human figure was partly covered by raindrops (made up of segments of a vertical grating). When first an adapting grating was viewed for about a minute (having the same frequency as the raindrop grating), the raindrops disappeared in the test grating, and the human figure appeared clearly without any perturbation. The results of Julesz and Stromeyer (1970), who used filtered one-dimensional noise for masking, permit an even more powerful technique than the one mentioned above. This masking experiment verified the findings of Campbell and his co-workers that the visual system employs periodicity analyzers with a ± 2 octave-wide critical band. This permits the use of filtered noise that masks certain spectral components of an image but leaves others intact.

Let me illustrate this basic idea with an example, to be seen in figure 10.2-1. In figure 10.2-1*A* the word LEFT has horizontal line segments at equal spacings, corresponding to a basic spatial frequency (f_0) along the vertical axis. If a horizontal sinusoidal grating is added to the word with a frequency of Nf_0 (where N is an integer and

---

1. The same paradigm was used earlier by linguists in order to decide whether the decoder of speech is an auditory or linguistic mechanism. Several investigators have found that words presented to the right ear are better heard than those presented to the left, but the ear advantage is reversed for musical and other nonspeech sounds (Broadbent and Gregory 1963; Kimura 1961, 1964, and 1967). These results have been interpreted to mean that speech sounds are processed primarily by the linguistic mechanism in the left hemisphere and the nonspeech sounds by the auditory mechanism in the right.

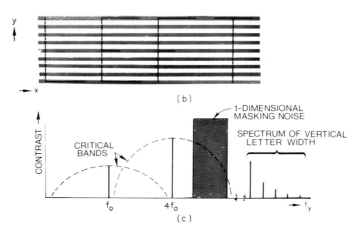

(a)

(b)

(c)

10.2-1 Illustration of a cyclopean technique based on critical bands in vision. (*A*) The original stimulus. (*B*) A sinusoidal perturbation grid added to the original stimulus renders it unrecognizable. (*C*) When one-dimensional filtered noise is added to the same critical band into which the perturbation grid falls, the original stimulus reappears. This perceived stimulus might differ somewhat from the original since a few lower harmonics of f_0 (not illustrated) will also be masked.

$N \geqslant 4$) in such a phase and adequate contrast that the horizontal line segments of the letters become hidden, the text becomes unreadable (fig. 10.2-1*B*).

We now add to this hidden stimulus filtered one-dimensional noise (composed of horizontal bars whose amplitudes are Gaussian) as shown in figure 10.2-1*C*. This noise falls into the critical band of the perturbation grating but spares the majority of critical bands around the stimulus components. This operation will mask the perturbation grid, rendering the word visible again. Thus, at the site where spectrum analysis takes place, one can unmask a masked stimulus. This is a new cyclopean retina. Whether it is before or after that of stereopsis remains to be seen.

The idea that in visual perception the sum of all spectral components are acting together and that one cannot ignore by will any of these channels is exploited in many displays. For instance, figure 10.2-2, generated by L. D. Harmon, is a coarsely quantized picture of a face. The information is contained in the low frequencies; yet, due to quantization, high frequency noise is generated, rendering the picture unrecognizable. Only if this high-frequency component is filtered out, by viewing the picture from great distance, or by squinting the eyes, or by moving the image back and forth, does the hidden image of Abraham Lincoln appear. For details, see Harmon (1970). The crucial difference between this demonstration and the filtered masking noise technique is that in the first case the perturbing frequencies have to be removed physically from the picture and are not contained on the retinal level, whereas in the second case the filtered noise is added to the stimulus and masks the unwanted spectrum components at a neural level where spectrum analysis does occur!

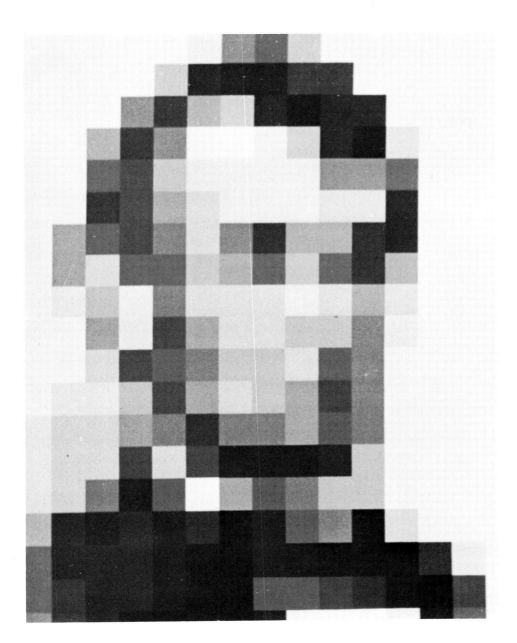

10.2-2 A coarsely sampled and quantized picture. One must blur the image (by viewing it from a great distance, or by squinting) in order to recognize the hidden face. (Courtesy Leon D. Harmon, Bell Laboratories.)

The discovery of a "cyclopean retina of critical bands" came too late to be used in these psychoanatomical investigations.[2] Although this stimulus class is only weakly cyclopean, many of the phenomena in this book could be retested with this technique. Were this cyclopean retina after stereopsis, the findings obtained would be of great interest. But even if this cyclopean retina were before stereopsis, one would possess a useful technique that could be used in place of the eidetic-cyclopean or the inverse-cyclopean stimulation.

10.3 The Role of Cyclopean Perception in Psychology

An epilogue ends a book on a personal note outside its main framework. This section is intended to end it within. These final paragraphs are devoted to the question of how cyclopean perception fits into psychology.

There is no consensus among experimental psychologists even as to what their discipline is about. Some believe that psychology should be restricted to the shaping of animal and human behavior. Others adhere to the notion that psychology should strive to fathom the entire depth of human abilities. There are those who want to expand human perception and consciousness. Still others are convinced that psychology is saturated with enigmatic mental phenomena and what is needed is their quantification and clarification by the use of models.

Cyclopean perception is broad enough to satisfy several of these schools of researchers. First of all, by demonstrating several new, hitherto unobserved phenomena, cyclopean perception may appeal to those who are mainly interested in finding the limits of the human perceptual experience. The fact that some of the stimuli that elicit these new percepts do not exist in real-life situations may be of interest to the "mind-expanders." But most importantly, the basic aspect of cyclopean perception —the tracing of the information flow in the visual system—may be of use to the model-builders. There are many other features of cyclopean perception that may attract the specialists. For instance, some of the clinical aspects of cyclopean perception may appeal to those who think that psychology should help medicine. Those who reject introspective reports and anecdotal accounts may be fascinated by the objective, unfakeable nature of cyclopean techniques that permit us to clarify the structure of some elusive perceptual phenomena. There are psychologists and neurophysiologists who may regard the earliest finding of cyclopean perception as the most important: the demonstration that binocular localization of an object does not require its recognition, thus is a simpler process than originally thought.

My opinion is that what psychology needs the most are bold, new methodological advances. I do not mean by "new advances" unprecedented "breakthroughs." The history of psychology teaches us that almost all radical innovations had some precursors. The "unconscious" was known much before Freud; and animal trainers, reportedly, were quite successful millennia before Skinnerian methods. However, a

---

2. Most likely the masked stimulus has to be restricted to one-dimensional structures, which limits the generality of this technique. Furthermore, the masking noise will interfere with some of the stimulus components, thus "blurring" the cyclopean percept.

methodological innovation can be introduced on such an advanced level of abstraction, precision, and generality that it is difficult to trace its humble origins.

Cyclopean perception is no exception. Its basic stimulus: the random-dot stereogram—generated by a computer and purified of all monocular depth and familiarity cues—is a new entity. However, the underlying principle has been exploited by nature (in a less pure way) since time immemorial. Animals that developed camouflage can successfully hide from predators that have no stereopsis, but often fall prey to those who are endowed with stereopsis. This natural camouflage is quite imperfect, since without a computer it is impossible to analyze the many minute details of the environment and compensate for them and also for the omnipresent shadows and perspective cues that produce monocularly visible features.

Even if natural camouflage were to be ideal, the basic idea of cyclopean perception differs from that of camouflage. The purpose of camouflage is to hide animals (or objects) from creatures that have no stereopsis. In cyclopean perception the monocularly camouflaged stimulus is presented to observers that possess good stereopsis. We want to hide the stimulus only from the peripheral system, but it is our intention to present certain desired stimulus features as a vivid percept at a central site.

As cyclopean techniques become further generalized to encompass the stimulation of any central location (that is accessible by psychological means without stimulating earlier sites with the desired forms) the resemblance to camouflage is lost. What we have gained is a novel methodology. Thus, cyclopean perception provides us with a clear history of how an idea originates, develops, and metamorphoses into newer forms. Perhaps, in some indirect ways, this emphasis on methodology—so characteristic of cyclopean perception—will have a deeper impact on psychology than the findings themselves which were obtained by its use.

The older and more accessible a field is the more difficult it becomes to make significant discoveries that are not brought about by methodological innovations. This is particularly true of visual perception. Throughout the history of mankind billions of people have shared the same sort of perceptual experiences when observing objects and events of everyday life. Even in the laboratory it is difficult to observe novel phenomena when using customary stimuli. In the last few years the many newly discovered perceptual phenomena that turned out to be rediscoveries of old findings attest to this fact. If we were to up-date the works of Helmholtz, Hering, Mach, Panum, or Exner by introducing modern neurophysiological terminology—but otherwise keeping the psychological material intact—these classics would often strike us as contemporary texts. Thus, it is very difficult to have something really new under the psychological sun!

I admit that quantification of old findings can bring us new insights and neurophysiological interpretation of old discoveries can lead us into profitable research. Nevertheless, in order to confront ourselves with unexperienced phenomena that may clarify psychological processes at work we must invent really novel stimuli. This quest for meaningful stimuli in space and time that can elicit in us unexplored, new percepts is shared by both the artist and psychologist. I hope that this new artistic-creative stream in psychology will not dry up for a long time to come.

In conclusion, I hope that the reader has become familiar with random-dot stereo-grams and cyclopean techniques in general and can use them for his own research problems. On the other hand, I cannot assure the reader that cyclopean perception constitutes a real paradigm. The word paradigm is defined by Kuhn (1962) as "achievements that share two essential characteristics. The achievement is sufficiently unprecedented to attract an enduring group of adherents away from competing modes of scientific activity. Simultaneously, it is sufficiently open-ended to leave all sorts of problems for the redefined group of practitioners to resolve." Only researchers through their results will be able to prove whether or not cyclopean perception is a paradigm.

This book can be read on two levels. One is the usual intellectual level. This is the level at which most scientific books convey their message. The reader accepts the reported findings as true and makes his judgment on the internal logic and generality of the material. Some books, including this one, however, can be enjoyed on a second level. For those who have stereopsis it is possible to share the excitement and direct appeal of the perceptual experience. This active participation by the reader is an important asset in the study of perception, and every effort has been made to help the reader to achieve stereoscopic fusion of the stereograms.

Almost every stereogram in this book is printed twice: first, as a black and white pair side-by-side at the proper place in the text; it is also repeated (provided that its figure number has an asterisk) in red-green anaglyph format following this appendix. The black-and-white stereograms can be fused by crossing your eyes or using a wedge (prism) in front of one eye (or one for each eye). Since almost all stereograms are portrayed by two brightness levels, most of them can be easily Xeroxed or photographed (using high-contrast film), and the copy can be viewed in a stereoscope or projected by polarizing techniques. The anaglyphs are only an ersatz solution, and for scientific research the original black and white stereograms should be used.

The anaglyphs in the book are of high quality, and the color of the pigments is matched to the color of the filters in the viewers (anaglyphoscopes). Unfortunately, at present the available green filters are not perfect, and do not reject the green channel completely. However, the red filter is quite satisfactory. Because of this, the image through the red filter is sharp, whereas the image through the green filter has considerable "crosstalk" from the green channel.

The anaglyphs were intentionally printed such that the left and right images are horizontally shifted in the nasal direction (if red is in front of the right eye) by one or two picture elements. This is done to hide the cyclopean forms in the anaglyphs when viewed without the viewers. Because of the shift the background is also in depth, above the plane of the printed page, but otherwise the perceptual experience is identical to that which can be obtained when the original stereogram is fused. For some anaglyphs this slight horizontal shift is in the temporal direction and causes the background to be perceived above the printed page.

With all their limitations, the anaglyphs in the book suffice and give the flavor of the perceptual demonstrations. They provide a quick and easy way to obtain stereoscopic fusion. The reader can exchange the viewers (have the green filter over the right eye) and all the perceived depth relations will reverse. This cannot be done with the black and white stereograms unless they are copied and the left and right images are interchanged. The anaglyphs are also less sensitive to head rotation than when polarizing techniques are used.

Nevertheless, there will be a few readers who have no stereopsis or are stereopsis-deficient. For them I can give only this advice. First, they should accept the reported findings and read the book only on the intellectual level. They can also show the stereograms to someone they trust (and who has good stereopsis) in order to verify the reports. However, if the reader with no stereopsis would still like to participate in cyclopean phenomena, he can view the stereograms as cinematograms. With the

Appendix: Some Comments on the Viewing of the Anaglyphs and Other Practical Matters

proper apparatus he can present the left and right images to one eye in quick temporal alternation. A 16-mm computer-generated movie, in the process of being made, portrays several of the cyclopean phenomena by dynamic noise of different speeds. The interested reader may write to the University of Chicago Press to find out how the movie may be obtained.

Anaglyphs

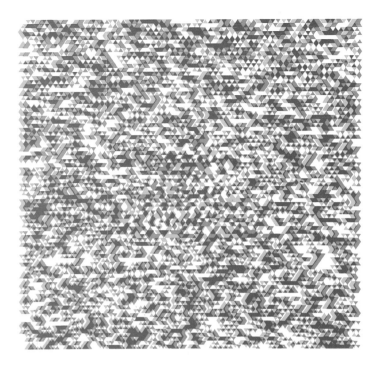

1.0-1* Random-dot stereogram composed of two arrays of black and white randomly selected cells of triangular shape. When monocularly viewed, the arrays appear as formless random textures. However, when stereoscopically fused, a large triangle is perceived above a background in vivid depth.

2.4-1* Strongly cyclopean stimulus, portraying the global information as a random-dot stereogram. When the images are monocularly inspected they appear uniformly random, yet when stereoscopically fused a center square is seen above the surround in vivid depth.

When the anaglyph of this stereogram is viewed with the red-green viewer, a center square is seen in front of the surround (or behind—depending whether the right or left eye sees through the red filter).

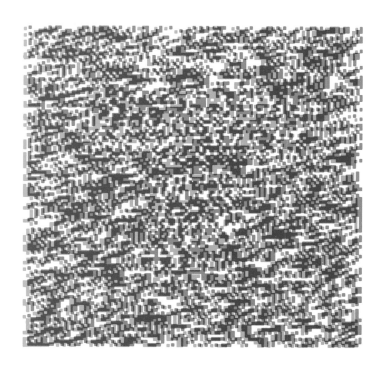

2.4-4* Random-dot stereogram portraying a T-shaped area in depth: (*A*) portrayed by 100 × 100 black and white square-shaped cells;

(*B*) portrayed by 100 × 100 dots.

2.4-5* Random-dot stereogram in which a monocularly visible global organization of a text (the word YES) is scrambled when stereoscopically viewed. If the reader has a strongly dominant left eye, he should reverse the viewer, that is, view the anaglyphs with the green filter over the right eye. (From Julesz 1967a.)

2.6-2* The cyclopean Müller-Lyer illusion.

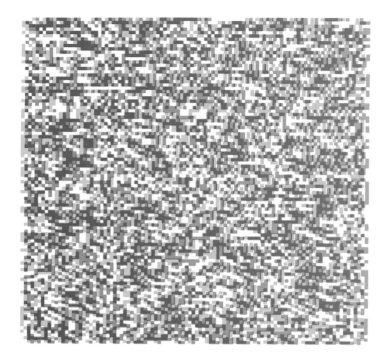

2.7-2* The cyclopean Hermann-Hering grid. Unlike the classical case, no inhibitory effects are experienced at the intersections.

2.7-3* The inverse cyclopean Hermann-Hering grid.

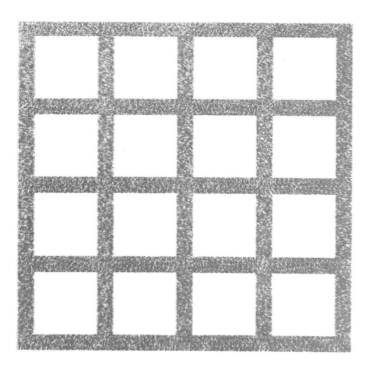

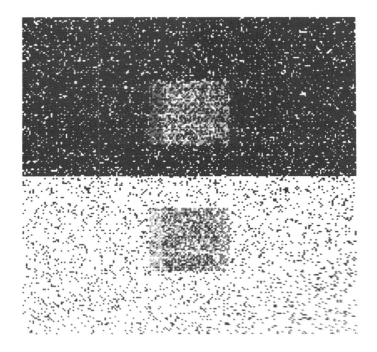

2.7-4* Simultaneous contrast phenomena under cyclopean stimulation.

2.8-3* Strongly cyclopean version of figure 2.8-1 in the form of a random-dot stereogram. Only the area closest to the viewer can be perceived as the figure. For figure-ground reversal one has to reverse the red-green viewer as well.

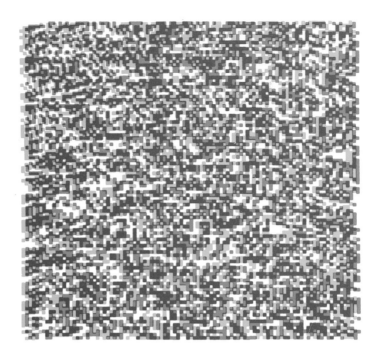

2.8-5* Random-dot stereogram portraying figure 2.8-4 cyclopeanly in depth. (From Julesz and Payne 1968.)

2.8-7* Classical stereogram in which one of the images is 15% expanded. Stereopsis is difficult to obtain.

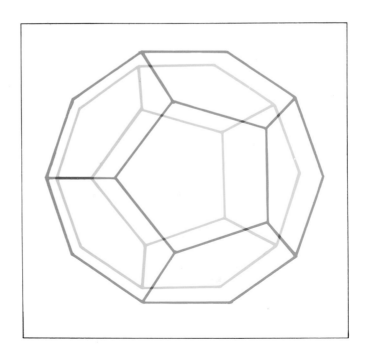

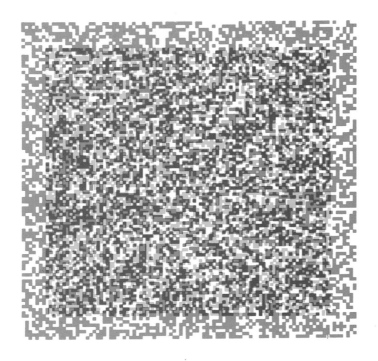

2.8-8* Random-dot stereogram in which one of the images is 15% expanded. Stereopsis can be easily obtained.

2.10-2* The cyclopean Benussi (Koffka)-ring phenomenon.

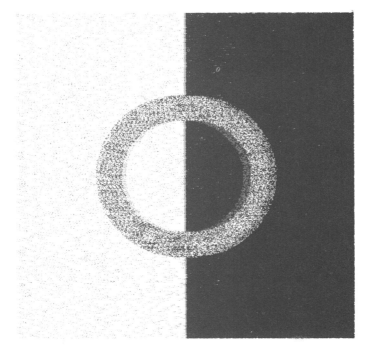

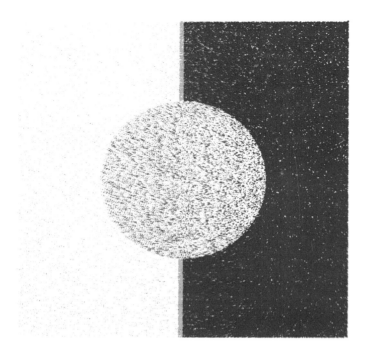

2.10-3* Similar to figure 2.10-4* except that a cyclopean edge in depth cuts the disk into two halves.

2.10-4* The cyclopean modified version of figure 2.10-1 using a gray disk and a cyclopean cut of the two halves of the disk in depth.

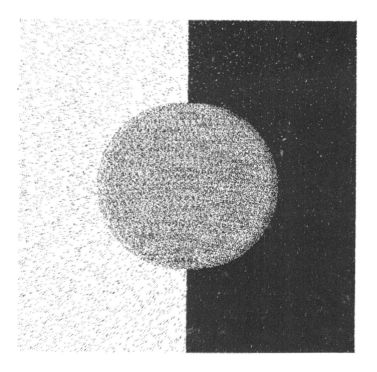

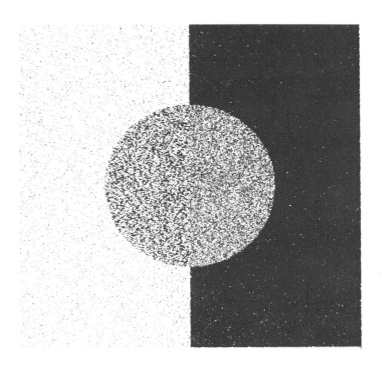

2.10-5* Similar to figure 2.10-4* except the cyclopean cut is obtained by dynamic and static stroboscopic movement. This is not a stereogram. For its viewing see text.

3.6-1* Random-line stereogram. The minute breaks in the vertical lines are correlated in the two images, and a center square is seen in depth. (From Julesz and Spivack 1967.)

3.6-3* Random-line stereogram. Similar to figure 3.6-1* except that lines are horizontal. (From Julesz and Spivack 1967.)

3.9-1* Random-dot stereogram, derived from figure 2.4-1* except that in the left image the diagonal connectivity is broken. Eighty-four percent of the dots remain identical, and stereopsis is easy to obtain.

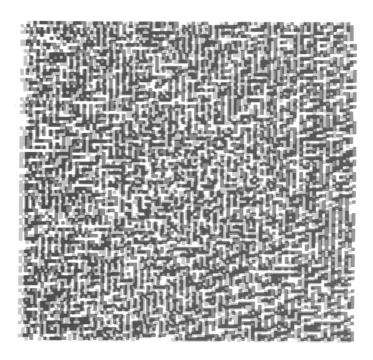

3.9-2* Random-dot stereogram in which, similar to figure 3.9-1*, 16% of the dots are complemented but at randomly chosen positions.

3.9-4* Random-line stereogram of correlated horizontal and diagonal line segments. When stereoscopically viewed, a hovering center square is seen.

3.9-5* Random-line stereogram portraying a Kaufman effect. Orientation portrays a diamond in front, and brightness (line width) portrays a square behind the surround.

3.9-6* Random-line stereogram consisting of diagonal line segments that are orthogonal to each other in the two eyes' views except for correlated black areas having ⅛ probability. In spite of binocular rivalry, stereopsis is easily obtained.

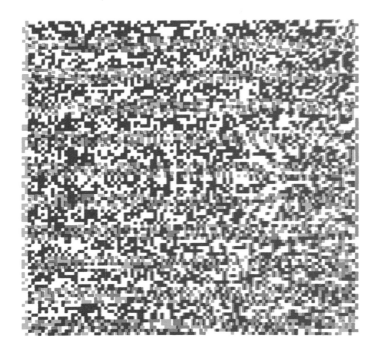

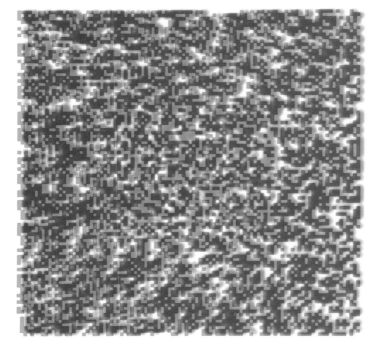

3.10-1* Inverse cyclopean stimulus. The monocularly apparent one-fold symmetry is suppressed in the binocular view. If the reader has a strongly dominant left eye, he should reverse the viewer, that is, view the anaglyphs with the green filter over the right eye. (From Julesz 1967a.)

3.10-3* Random-dot stereogram, similar to figure 2.4-1* except that the left image is strongly blurred.

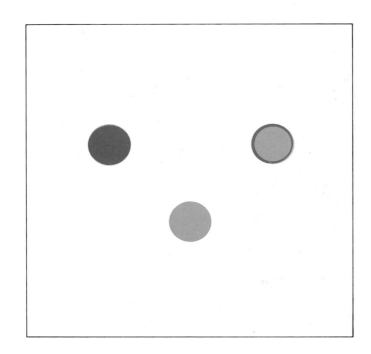

3.10-4* Experiment by Levelt, showing how the richness of contour determines binocular combination.

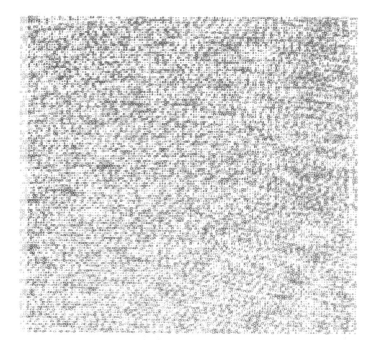

3.10-5* Random-dot stereogram portraying a center square. The eight brightness levels are approximated by eight characters of the microfilm printer.

3.10-9* Stereogram in which the left image is identical to the left image of figure 3.10-5*, whereas the right image is the inverse Fourier transform of figure 3.10-7 (low frequencies missing).

3.10-10* Stereogram in which the left image is identical to the left image of figure 3.10-5*, whereas the right image is the inverse Fourier transform of figure 3.10-8 (high frequencies missing).

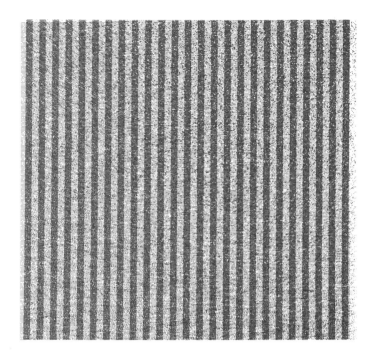

3.10-11* Random-dot stereogram that demonstrates how a monocular grating can resist binocular scrambling.

4.5-2* Random-line stereogram containing line segments of ± 45° orientation.

4.5-3* Hyperbolic paraboloid with torus portrayed by a random-dot stereogram.

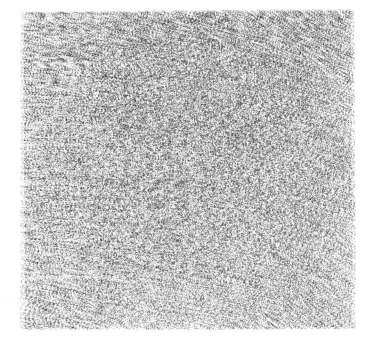

4.5-4* Spiral surface portrayed by a random-dot stereogram.

4.5-5* Random-dot stereogram with 5% black dots. The square in depth is vividly seen as forming a uniformly textured sheet.

4.5-7* Random-dot stereogram showing a 10 $\times$ 10 square of two and eight brightness levels and eight picture-element disparity. Only the square having eight brightness levels yields stereopsis.

5.3-1* Random-line stereogram, such as used by Marlowe; corresponding line segments in the left and right fields differ by ± 7.5°. Global stereopsis can easily be obtained.

5.3-2* Similar to figure 5.3-1* except that the corresponding line segments in the two fields differ by ± 15°. Global stereopsis can still be obtained.

338

5.4-1* Same as figure 5.4-2* except a facsimile system is used portraying each of the 10^6 dots in four brightness levels.

5.4-2* Truncated hyperbolic paraboloid of $1,000 \times 1,000$ resolution, but only 10% of the computed dots are portrayed, using two brightness levels.

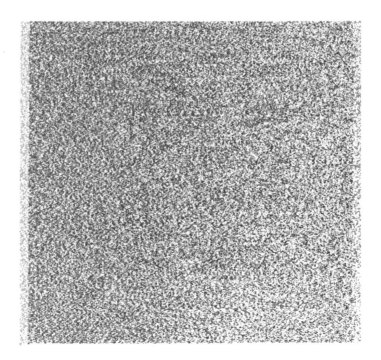

5.4-3* Random-dot stereogram of 33 × 33 resolution. The center square is only 15 × 15, yet stereopsis can be easily obtained.

5.5-1* Simple outlined stereogram showing that a positive and negative corresponding image yield fusion. (After Helmholtz 1909.)

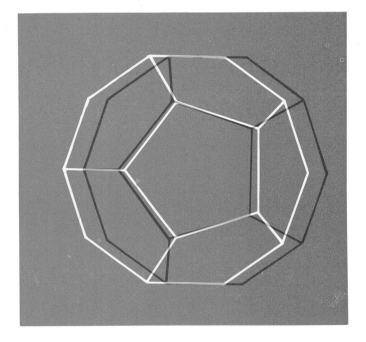

5.5-2* Complex (random-dot) stereogram showing that in this case
a positive and negative image pair
cannot be fused.

5.5-3* Positive and negative image of
the outlined version of figure 2.4-1*.
Again fusion is impossible. (From
Julesz 1964b.)

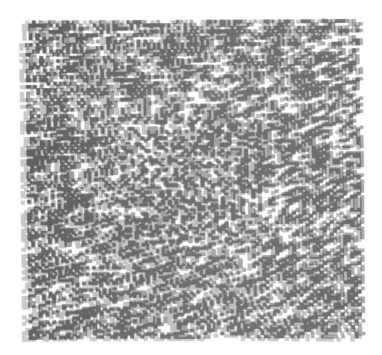

5.5-4* Random-dot stereogram similar to figure 2.4-1* except the black cells in the left image are expanded by 20% $(f_L : Bf_R)$. (From Julesz 1963a.)

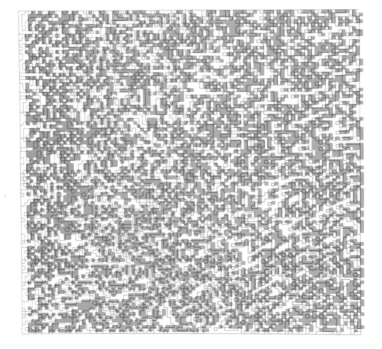

5.5-5* Random-dot stereogram showing $f_L : Cf_R$. Stereopsis is medium. (From Julesz 1963a.)

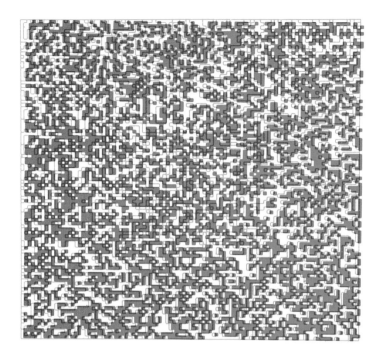

5.5-6* Random-dot stereogram showing $f_L : CBf_R$. Stereopsis cannot be obtained. (From Julesz 1963a.)

5.6-5* A cyclopean Necker-cube. This can be seen in perspective depth and reversed like the classical one.

5.7-1* Random-dot stereogram containing three transparent depth planes.

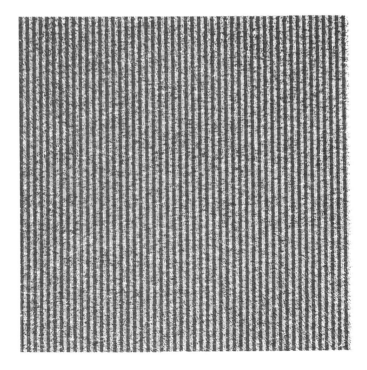

5.7-2* Inverse cyclopean stimulus of a vertical grating. Monocular bars and grids resist binocular scrambling.

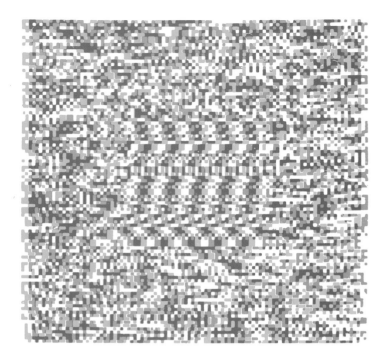

6.2-2* Stereogram of ambiguously perceivable center square flanked with unambiguous areas in front of and behind the surround.

6.2-5* Random-dot stereogram portraying an unambiguous ascending pyramidal staircase in front of the plane of the printed page. (From Julesz and Johnson 1968b.)

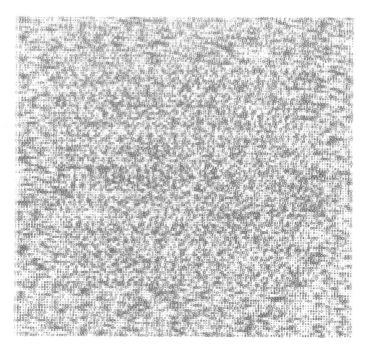

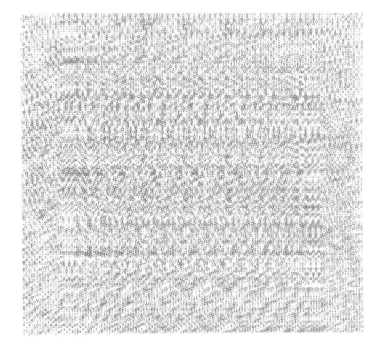

6.2-7* Random-dot stereogram portraying an ambiguous pyramidal staircase either ascending or descending at the will of the observer. (From Julesz and Johnson 1968b.)

6.2-8* Ambiguous stereogram with unambiguous margins in the upper and lower areas. Surface *A* is a horizontal plane in front of the printed page, while surface *B* is a wedge behind the printed page. (From Julesz and Johnson 1968b.)

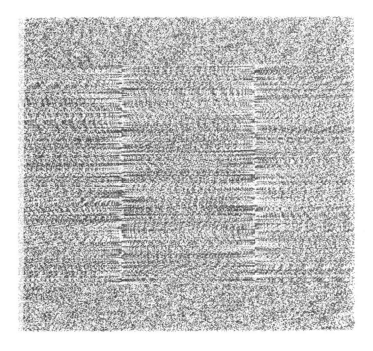

6.2-9* Ambiguous stereogram with unambiguous margins in the upper and lower areas. Two slanted planes (above the plane of the printed page) that intersect each other are portrayed. Figure 6.2-10 shows the various percepts which can be obtained. (From Julesz and Johnson 1968b.)

6.2-11* Ambiguous stereogram with unambiguous margins in the upper and lower areas. It portrays a cosine function and a cosine function of lesser amplitude (height) and half periodicity. Both surfaces appear in front of the printed plane. (From Julesz and Johnson 1968b.)

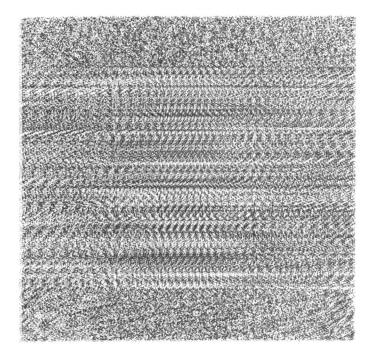

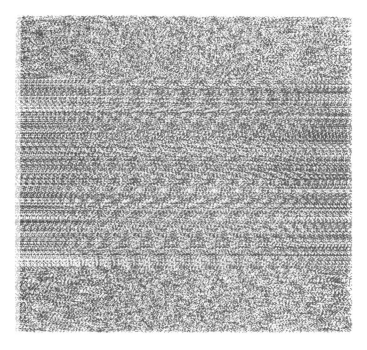

6.2-12* Ambiguous stereogram with two unambiguous rectangles in the upper and lower areas. It portrays three surfaces. A cosine function in front of the plane of the printed page, the same function behind the plane, and the printed plane itself (a plane with zero disparity). (From Julesz and Johnson 1968b.)

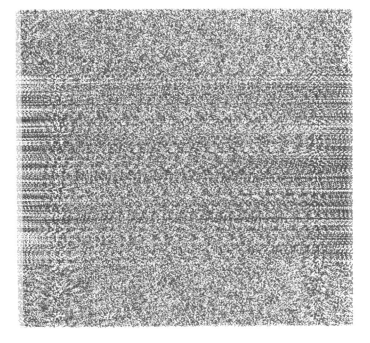

6.2-13* Ambiguous stereogram with wide unambiguous margins. It portrays a cosine function in front of the printed page and a wedge behind the plane. (From Julesz and Johnson 1968b.)

6.3-2* Random-dot stereogram portraying two intersecting transparent ellipsoid surfaces.

6.4-1* Random-dot stereogram containing 3 $\times$ 3 squares with 1, 2, 4, 6, and 8 picture element disparities.

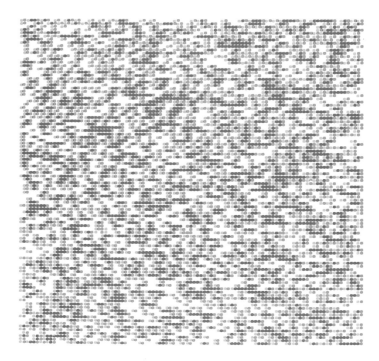

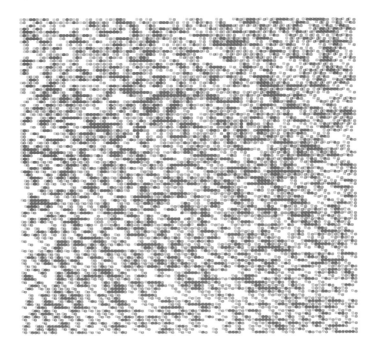

6.4-2* Random-dot stereogram containing 5×5 squares with 1, 2, 4, 6, and 8 picture element disparities.

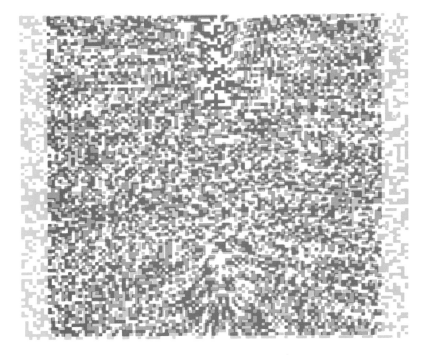

6.5-3* Random-dot stereogram in which one image is expanded in the horizontal direction by 15%. The planes of the surround and center square appear tilted.

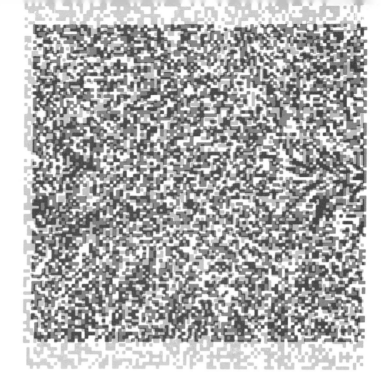

6.5-4* Random-dot stereogram in which one image is expanded in the vertical direction by 15%. Tilt of planes is opposite of figure 6.5-3*.

6.5-5* Ambiguous stereogram that contains a center square either in front of or behind the surround.

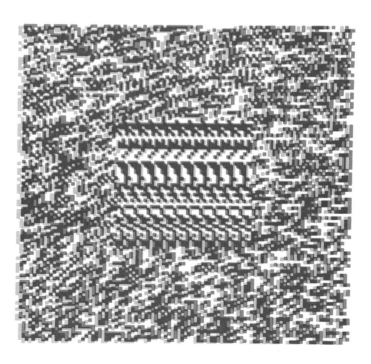

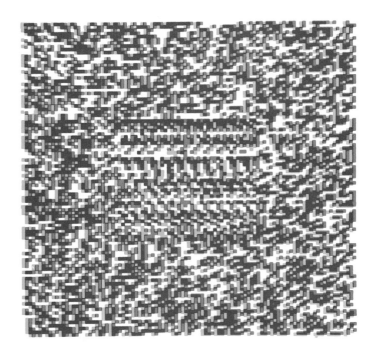

6.5-6* Ambiguous stereogram, similar to figure 6.5-5* except that one of the depth organizations is biased by 10%.

6.6-1* Complex random-dot stereogram portraying a recessed ellipsoid.

7.2-3* The cyclopean Ponzo illusion. Effect is similar to the classical case.

7.2-4* Modified cyclopean Ponzo illusion. The test and inducing parts of the figure are at different depths. Illusory effect greatly diminished.

7.2-6* The cyclopean vertical-horizontal illusion. Illusory effect similar to the classical case.

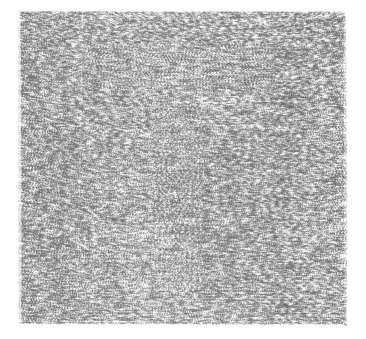

7.2-7* Modified cyclopean vertical-horizontal illusion. Vertical bar is closer than the horizontal one.

7.2-8* Modified cyclopean vertical-horizontal illusion. Vertical bar is farther away than the horizontal one.

7.2-10* The cyclopean Ebbinghaus illusion. Illusory effect similar to classical one.

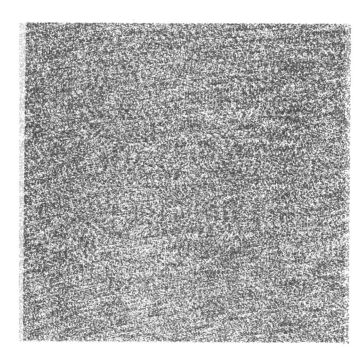

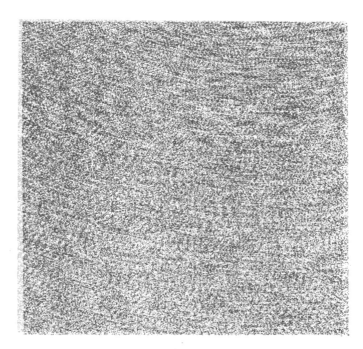

7.2-11* Modified cyclopean Ebbinghaus illusion. Test and inducing parts of figure at different depths. Illusory effect greatly reduced.

7.2-13* The cyclopean Poggendorff illusion. Illusory effect similar to classical case.

7.2-14* Modified cyclopean Poggendorff illusion. Test figures are behind
inducing figure. Illusory effect
disappears.

7.2-15* Müller-Lyer illusion portrayed by a random-dot cinematogram.
For viewing see text.

7.2-16* Modified cyclopean Müller-Lyer illusion. Test figure closer than the inducing arrowheads. Illusory effect greatly reduced.

7.2-18* The cyclopean Zöllner illusion.

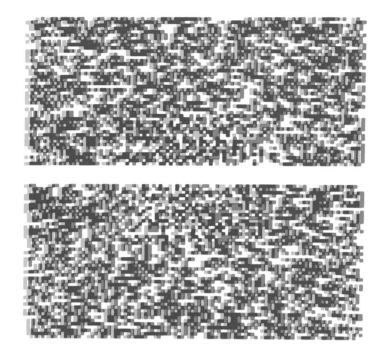

7.7-1* Anomalous contour after Shipley. Closure of cyclopean square across white gap can be experienced.

7.7-2* Similar to figure 7.7-1*, however, the diamond-shaped cyclopean figure does not give rise to a corner-shaped anomalous contour.

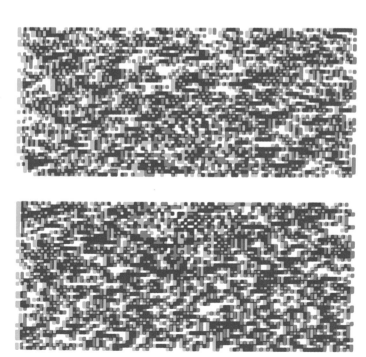

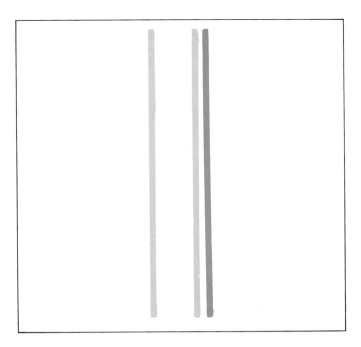

7.8-1* Panum's limiting case. When stereoscopically fused the right-hand line is seen in front of the left-hand one.

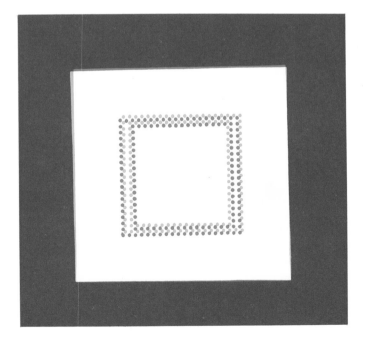

7.8-2* Classical case of stereopsis after Lawson and Gulick for which each dot has a well-defined disparity. The white surfaces within the inner and outer square are perceived at the depth of the printed page. (From Lawson and Gulick 1967.)

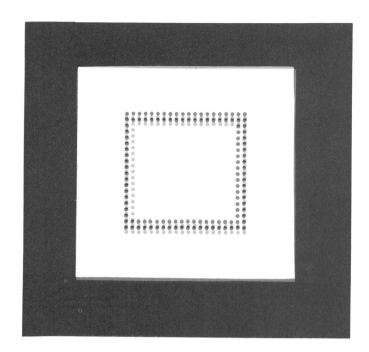

7.8-3* Anomalous contour by Lawson and Gulick. Fusion gives rise to the inner square surface in front of the printed page and anomalous contours (without disparity). (From Lawson and Gulick 1967.)

7.9-4* Adaptation stereogram that demonstrates the Blakemore and Julesz (1971) three-dimensional aftereffect. After fixating at the center mark during fusion for about 1 min, the test stereogram of figure 7.9-5* should be viewed.

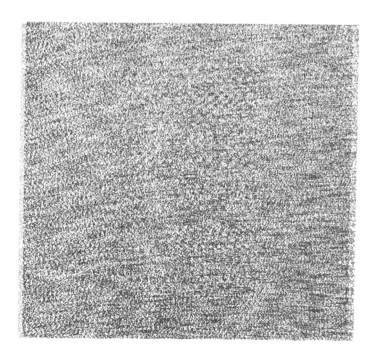

7.9-5* Test stereogram that demonstrates the Blakemore and Julesz (1971) three-dimensional aftereffect. After adaptation to figure 7.9-4*, when fixating at the center mark in depth, the upper and lower squares appear briefly at different depths. (From Julesz and Blakemore 1971.)

8.1-2* Test figures for determining stereopsis deficiency. The stereograms have diminishing binocular correlation: *A*, 100%;

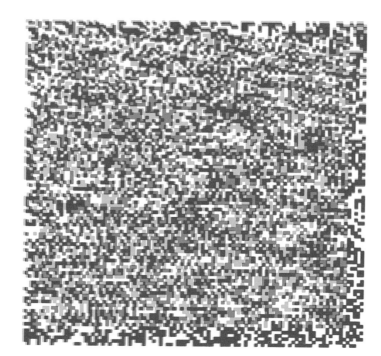

B, 90%;

C, 80%;

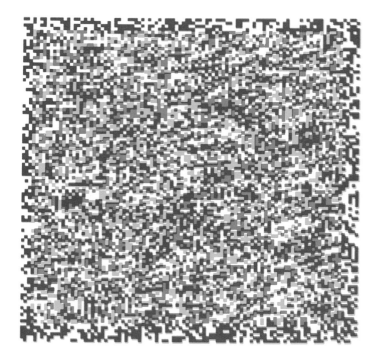

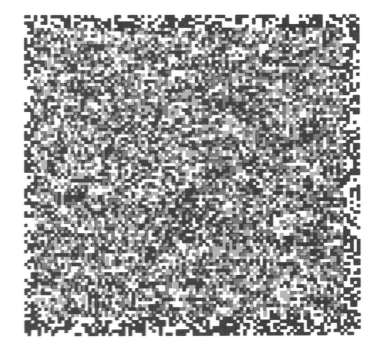

D, 70%;

E, 60%;

F, 50%;

and *G*, 40%.

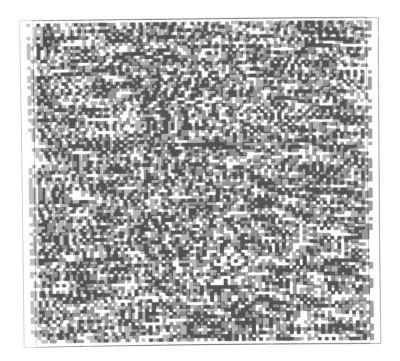

8.3-3* Random-dot stereogram portraying a cyclopean vertical bar in depth.

8.3-4* Random-dot stereogram portraying a cyclopean horizontal bar in depth.

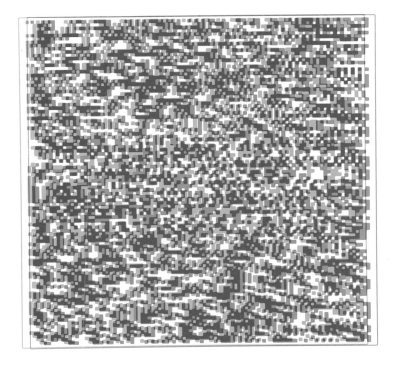

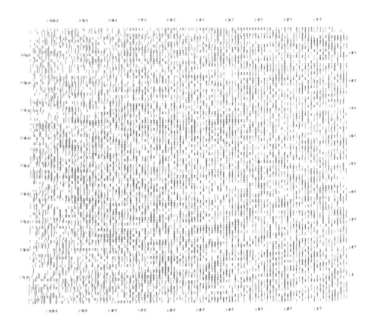

8.4-1* Random-dot stereogram used as input for AUTOMAP-1.

8.4-4* Stereoscopic close-up image of a 7.6 cm square lunar surface as photographed during the Apollo-11 mission. (Courtesy NASA.)

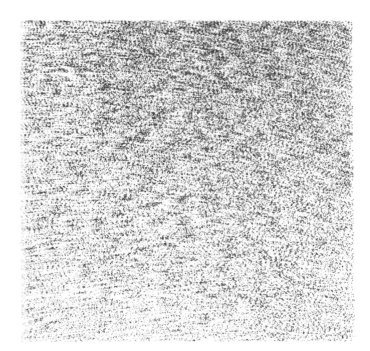

10.1-1* Text portrayed by random-dot stereogram.

Fig. 2.4-2. Reprinted from John Homans, *A textbook of surgery*, 6th ed., 1945. Courtesy of Charles C. Thomas, Publisher, Springfield, Illinois.

Fig. 2.4-5*. Reprinted, by permission, from Julesz, Suppression of monocular symmetry during binocular fusion without rivalry, *Bell System Tech. J.* 46 (1967). Copyright © 1967, The American Telephone and Telegraph Co.

Fig. 2.8-1. Reprinted, by permission, from Wyburn, Pickford, and Hirst, *Human senses and perception* (Toronto: University of Toronto Press; Edinburgh and London: Oliver & Boyd, 1964).

Figs. 2.8-5* and 2.8-6. Reprinted, by permission, from Julesz and Payne, Differences between monocular and binocular stroboscopic movement perception, *Vision Res.* 8 (Pergamon Press, 1968).

Figs. 3.1-1 and 3.1-2. Reprinted, by permission, from Julesz, Visual pattern discrimination, *IRE Transactions on Information Theory* IT-8 (1962).

Figs. 3.2-4, 3.2-5, and 3.2-6. Reprinted, by permission, from Julesz, Suppression of monocular symmetry during binocular fusion without rivalry, *Bell System Tech. J.* 46 (1967). Copyright © 1967, The American Telephone and Telegraph Co.

Fig. 3.4-1. Reprinted, by permission, from Attneave, Some information aspects of visual perception, *Psychology Review* 61 (1954).

Fig. 3.4-2. Reprinted, by permission, from Green and Courtis, Information theory and figure perception: The metaphor that failed, *Acta Psychologica* 25 (1966).

Figs. 3.6-1*, 3.6-2, and 3.6-3*. Reprinted, by permission, from Julesz and Spivack, Stereopsis based on vernier acuity cues alone, *Science* 157 (1967): 563–65. Copyright 1967 by the American Association for the Advancement of Science.

Fig. 3.10-1. Reprinted, by permission, from Julesz and Payne, Differences between monocular and binocular stroboscopic movement perception, *Vision Res.* 8 (Pergamon Press, 1968).

Fig. 4.3-1. Reprinted, by permission, from L. Löfgren, Connection possibilities and distinguishable stimulus patterns in cellular layers, in *Some Principles of Preorganization in Self-Organizing Systems,* T.R. 2, Contract Nonr 1834 (21), Biological Computer Laboratory, University of Illinois, Urbana (1960).

Figs. 4.4-1, 4.4-2, 4.4-3, 4.4-4. Reprinted, by permission, from Julesz, Visual pattern discrimination, *IRE Transactions on Information Theory* IT-8 (1962).

Figs. 4.4-5 and 4.4-6. Reprinted, by permission, from Frisch and Julesz, Figure-ground perception and random geometry, *Perception and Psychophysics* 1 (1966).

Fig. 4.4-7. Reprinted from *Perceptrons* by Minsky and Papert by permission of The M. I. T. Press, Cambridge, Massachusetts. Copyright © 1969 by The Massachusetts Institute of Technology.

Fig. 4.5-1. Reprinted, by permission, from Julesz, Experiment in perception, *Psychology Today Magazine* 2 (July 1968). CRM, Inc.

Figs. 4.6-1, 4.6-2, 4.6-3, and 4.6-4. Reprinted from Julian E. Hochberg, *Perception,* © 1964. By permission of the author and Prentice-Hall, Inc., Englewood Cliffs, New Jersey.

Figs. 5.2-1, 5.2-2, and 5.2-3. Reprinted, by permission, from Blakemore, Binocular depth discrimination and the nasotemporal division, *J. Physiol.* 205 (1969).

Fig. 5.5-3*. Reprinted, by permission, from Julesz, Binocular depth perception without familiarity cues, *Science* 145 (1964): 356–62. Copyright 1964 by the American Association for the Advancement of Science.

Figs. 5.5-4*, 5.5-5*, and 5.5-6*. Reprinted, by permission, from Julesz, Stereopsis and binocular rivalry of contours, *J. Opt. Soc. Am.* 53 (1963).

Figs. 5.6-1, 5.6-2, 5.6-3. Reprinted from *Perceptrons* by Minsky and Papert by permission of The M. I. T. Press, Cambridge, Massachusetts. Copyright © 1969 by The Massachusetts Institute of Technology.

Figs. 5.8-1, 5.8-2, 5.8-3, 5.8-4 and 5.8-5. Reprinted, by permission, from Fender and

Julesz, Extension of Panum's fusional area in binocularly stabilized vision, *J. Opt. Soc. Am.* 57 (1957).

Fig. 5.9-1. Reprinted, by permission, from Holst, Aktive Leistungen der menschlichen Gesichtswahrnehmung, *Studium Generale* 10 (1957).

Figs. 5.9-2, 5.9-3, and 5.9-4. Reprinted, by permission, from Fender and Julesz, Extension of Panum's fusional area in binocularly stabilized vision, *J. Opt. Soc. Am.* 57 (1967).

Figs. 6.2-3, 6.2-4, 6.2-5*, 6.2-6, 6.2-7*, 6.2-8*, 6.2-9*, 6.2-10, 6.2-11* 6.2-12*, and 6.2-13*. Reprinted, by permission, from Julesz and Johnson, Mental holography: Stereograms portraying ambiguously perceivable surfaces, *Bell System Tech. J.* 49 (1968). Copyright, 1968, The American Telephone and Telegraph Co.

Figs. 7.4-3 and 7.4-4. Courtesy John Merritt, Harvard University.

Fig. 7.5-1. Reprinted, by permission, from Julesz and Payne, Differences between monocular and binocular stroboscopic movement perception, *Vision Res.* 8 (Pergamon Press, 1968).

Figs. 7.6-1. Reprinted, by permission, from Julesz and White, Short term visual memory and the Pulfrich Phenomenon, *Nature* 222, no. 5194 (1969).

Figs. 7.8-2* and 7.8-3*. Reprinted, by permission, from Lawson and Gulick, Stereopsis and anomalous contour, *Vision Research* 7 (Pergamon Press, 1967).

Fig. 7.9-5*. Reprinted, by permission, from Julesz and Blakemore, Stereoscopic depth after-effect produced without monocular cues, *Science* 171 (1971): 286–88. Copyright 1971 by the American Association for the Advancement of Science.

Fig. 8.4-4. Courtesy National Aeronautics and Space Administration, Washington, D. C.

Fig. 8.5-1. Courtesy B.K.P. Horn (M.I.T., Cambridge, Mass.)

Fig. 9.5-1. Reprinted, by permission, from Shepard and Carroll, Parametric representation of nonlinear data structures, in *Multivariate Analysis* (New York: Academic Press, 1966).

Fig. 10.2-2. Courtesy Leon D. Harmon, Bell Laboratories. Copyright Bell Telephone Laboratories, Inc.

NOTE: The dates of the references usually agree with the publication dates of the journals or books in which they first appeared. An exception to this is made with preprints (accepted for publication but not yet published)—in this case, the date refers to the date of the manuscript. Another exception is made with papers delivered at symposia whose proceedings were published after years of delay—in this case, the date of the symposium is used.

Abplanalp, P. 1968. An experimental neuroanatomical study of the visual system. Ph.D. dissertation, M.I.T.

Addams, R. 1834. An account of a peculiar optical phaenomenon seen after looking at a moving body. *Philosophical Mag.* 5 (3d s.) 373–74. Reprinted in W. N. Dember, ed., *Visual perception: The nineteenth century.* New York: Wiley, 1964, pp. 81–83.

Andersen, E. E., and Weymouth, F. W. 1923. Visual perception and the retinal mosaic: 1. Retinal mean local sign—an explanation of the fineness of binocular perception of distance. *Am. J. Physiol.* 64: 561–94.

Anstis, S. M. 1970. Phi movement as a subtraction process. *Vision Res.* 10: 1411–30.

Anstis, S. M., and Atkinson, J. 1967. Distortions in moving figures viewed through a stationary slit. *Am. J. Psychol.* 80: 572–85.

Aschenbrenner, C. M. 1954. Problems in getting information into and out of air photographs. *Photogramm. Eng.* 20 (3): 398–401.

Attneave, F. 1954. Some informational aspects of visual perception. *Psychol. Rev.* 61: 183–93.

Attneave, F., and Arnoult, M. D. 1956. The quantitative study of shape and pattern perception. *Psychol. Bull.* 53: 452–71.

Bach-y-Rita, P.; Collins, C. C.; Saunders, F. A.; White, B. W.; and Scadden, L. 1969. Vision substitution by tactile image projection. *Science* 221: 963–64.

Barlow, H. B.; Blakemore, C.; and Pettigrew, J. D. 1967. The neural mechanism of binocular depth discrimination. *J. Physiol.* 193: 327–42.

Barlow, H. B., and Brindley, G. S. 1963. Inter-ocular transfer of movement after-effects during pressure blinding of the stimulated eye. *Nature* 200: 1347.

Barlow, H. B., and Hill, R. M. 1963. Selective sensitivity to direction of movement in ganglion cells of the rabbit retina. *Science* 139: 412–14.

Bassett, M. F., and Warne, C. J. 1919. On the lapse of verbal meaning with repetition. *Am. J. Psychol.* 30: 415–18.

Baumgartner, G. 1960. Indirekte Grössenbestimmung der rezeptiven Felder der Retina beim Menschen mittels der hermannschen Gittertäuschung. *Pfluegers Archiv.* 272: 1–21.

Beck, J. 1967. Perceptual grouping produced by line figures. *Perception and Psychophysics* 2: 491–95.

Beeler, G. W. 1965. Stochastic processes in the human eye movement control system. Doctoral dissertation, Calif. Inst. Tech.

Békésy, G. von. 1960. *Experiments in hearing.* New York: McGraw-Hill.

Bell, A.; Forster, R. G.; Finnegan, F. J.; Katz, M.; Lasusa, J. I.; and Mayzner, M. S. 1969. Sequential blanking and visual form perception. *Psychon. Sci.* 15 (2): 103–4.

Bell, E. T. 1945. *The development of mathematics.* New York: McGraw-Hill, pp. 331, 350–53.

Bender, M. B.; Feldman, M.; and Sobin, A. J. 1968. Palinopsia. *Brain* 91: 321–38.

Bender, M. B., and Teuber, H. L. 1947. Spatial organization of visual perception following injury to the brain. *Archs. Neurol. Psychiat. Chicago* 58: 721–39.

———. 1948. Spatial organization of visual perception following injury to the brain. *Archs. Neurol. Psychiat. Chicago* 59: 39–62.

Benham, C. E. 1894. Artificial spectrum top. *Nature* 51: 113–14.

References

Berry, R. N. 1948. Quantitative relations among vernier, real depth and stereoscopic depth acuities. *J. Exp. Psychol.* 38: 708–21.

Bidwell, S. 1897. On subjective colour phenomena attending sudden changes of illumination. *Proc. Roy. Soc.* 50: 368–77.

Bishop, P. O. 1963. The neurology of visual direction. Talk presented at M.I.T. psychology dept. colloquium, Cambridge, Mass., October. (See Richards 1968.)

————. 1969a. Neurophysiology of binocular single vision and stereopsis. In *Handbook of sensory physiology,* ed. R. Jung, vol. 7. Springer, in press.

————. 1969b. Cortical beginning of visual form and binocular depth discrimination. In *Neurosciences Res. Study Program Boulder Sympos. 1969 August,* ed. F. O. Schmitt. New York: Rockefeller Univ. Press, 1970, pp. 471–85.

Bishop, P.O.; Burke, W.; and Davis, R. 1959. Activation of single lateral geniculate cells by stimulation of either optic nerve. *Science* 130: 506–7.

Bizzi, E. 1966a. Charges in orthodromic and antidromic response of optic tract during the movements of sleep. *J. Neurophysiol.* 29: 861–70.

————. 1966b. Discharge patterns of single geniculate neurons during the rapid eye movements of sleep. *J. Neurophysiol.* 29: 1087–95.

Blakemore, C. 1968. Binocular interaction in animals and man. Ph.D. dissertation, University of California, Berkeley.

————. 1969. Binocular depth discrimination and the nasotemporal division. *J. Physiol.* 205: 471–97.

————. 1970. A new kind of stereoscopic vision. *Vision Res.* 10: 1181–99.

Blakemore, C., and Campbell, F. W. 1969a. On the existence of neurones in the human visual system selectively sensitive to the orientation and size of retinal images. *J. Physiol.* 203: 237–60.

————. 1969b. Adaptation to spatial stimuli. *J. Physiol.* 200: 11–13.

Blakemore, C.; Carpenter, R. H.; and Georgeson, M. A. 1970. Lateral inhibition between orientation detectors in the human visual system. *Nature* 228: 37–39.

Blakemore, C., and Cooper, G. F. 1970. Development of the brain depends on visual environment. *Science* 228: 477–78.

Blakemore, C., and Julesz, B. 1971. Stereoscopic depth aftereffect produced without monocular cues. *Science* 171: 286–88.

Blakemore, C.; Nachmias, J.; and Sutton, P. 1970. The perceived spatial frequency shift: Evidence for frequency-selective neurones in the human brain. *J. Physiol.* 210: 727–50.

Blakemore, C., and Sutton, P. 1969. Size adaptation: A new aftereffect. *Science* 166: 245–47.

Bliss, J. C. 1969. Optical-to-tactile image conversion. *Final Report, Stanford Univ.*

Bliss, J. C.; Crane, H. D.; and Link, S. W. 1966. Effect of display movement on tactile perception. *Perception and Psychophysics* 1: 195–202.

Blum, G. S. 1960. *A model of the mind.* New York: Wiley.

Bosche, C. 1967. Computer-generated random-dot images. In *Design and planning: 2. computers in design and communication,* ed. M. Krampen and P. Seitz. New York: Hastings House, pp. 87–91.

Bough, E. W. 1970. Stereoscopic vision in the macaque monkey: A behavioural demonstration. *Nature* 225: 42–44.

Bower, T. G. R. 1968. Morphogenetic problems in space perception. In *Proc. of the assoc. for research in nervous and mental diseases,* ed. D. Hamburg and K. Pribram. Stanford, Calif.: Stanford Univ. Press.

Boynton, R. M., and Gordon, J. 1965. Bezold-Brücke hue shift measured by color-naming technique. *J. Opt. Soc. Am.* 55: 78–86.

Braunstein, M. L. 1966. Sensitivity of the observer to transformations of the visual field. *J. Exp. Psychol.* 72: 683–89.

Breese, B. B. 1899. Binocular rivalry: On inhibition. *Psychol. Monogr.* 3 (1): 18–21, 44–

48, 59–60. Reprinted in *Visual perception: The nineteenth century,* ed. W. N. Dember. New York: Wiley, 1964, pp. 207–15.

Brindley, G. S. 1960. *Physiology of the retina and the visual pathway.* London: Edward Arnold.

———. 1962. Two new properties of foveal afterimages and a photochemical hypothesis to explain them. *J. Physiol.* 164: 168–79.

Brindley, G. S., and Lewin, W. S. 1968. The sensations produced by electrical stimulation of the visual cortex. *J. Physiol.* 196: 479–93.

Brindley, G. S., and Merton, P. A. 1960. The absence of position sense in the human eye. *J. Physiol.* 153: 127–30.

Broadbent, D. E. 1958. *Perception and Communication.* London: Pergamon.

Broadbent, D. E., and Gregory, M. 1963. Accuracy of recognition for speech presented to right and left ears. *Quart. J. Exp. Psychol.* 65: 103–5.

Brown, D. R., and Owen, D. H. 1967. The metrics of visual form. *Psychol. Bull.* 68: 243–59.

Busemann, H., and Kelly, P. J. 1953. *Projective geometry and projective metrics.* New York: Academic Press.

Campbell, F. W., and Kulikowski, J. J. 1966. Orientational selectivity of the human visual system. *J. Physiol.* 187: 437–45.

Campbell, F. W., and Maffei, L. 1970. Electrophysiological evidence for the existence of orientation and size detectors in the human visual system. *J. Physiol.* 207: 635–52.

Campbell, F. W., and Robson, J. G. 1968. Application of Fourier analysis to the visibility of gratings. *J. Physiol.* 197: 551–66.

Campenhausen, C. von. 1968. Über den Ursprungsort von musterinduzierten Flickerfarben im visuellen System des Menschen. *Z. Vergl. Physiol.* 61: 355–60.

Canaday, R. H. 1962. The description of overlapping figures. M.S. thesis, M.I.T.

Carmon, A., and Bechtoldt, H. P. 1969. Dominance of the right cerebral hemisphere for stereopsis. *Neuropsychologia* 7: 29–39.

Cherry, E. C. 1953. Some experiments on the recognition of speech with one and with two ears. *J. Acoust. Soc. Am.* 25: 975–79.

Chomsky, N. 1965. *Aspects of the theory of syntax.* Cambridge, Mass.: M.I.T. Press.

Choudhury, B. P.; Whitteridge, D.; and Wilson, M. E. 1965. The function of the callosal connections in the visual cortex. *Quart. J. Exp. Physiol.* 50: 214–19.

Chow, K. L. 1969. Personal communication, to be published in the *Handbook for Sensory Physiology,* ed. R. Jung. Springer, in press.

Chow, K. L.; Lindsley, D. F.; and Gollender, M. 1968. Modification of response patterns of lateral geniculate neurons after paired stimulation of contralateral and ipsilateral eyes. *J. Neurophysiol.* 31: 729–39.

Cobb, W. A.; Morton, H. B.; and Ettlinger, G. 1967. Cerebral potentials evoked by pattern reversal and their suppression in visual rivalry. *Nature* 216: 1123–25.

Cohen, H. H.; Bill, J. C.; and Gilinsky, A. S. 1968. Simultaneous brightness contrast: variations of Koffka's ring. *Proc. 76th ann. conv. APA,* pp. 99–100.

Cooley, J. W., and Tukey, J. W. 1965. An algorithm for the machine calculation of complex Fourier series. *Math. Computation* 19: 297–301.

Cowan, J. D. 1965. The problem of organismic reliability. In *Progress in brain research,* ed. N. Wiener and J. P. Schade, vol. 17. Amsterdam: Elsevier.

Cragg, B. G., and Temperley, H. N. V. 1954. The organization of neurones: A cooperative analogy. *Electroencephalog. Clin. Neurophysiol.* 6: 85–92.

Craik, K. J. W. 1940. Origin of visual afterimages. *Nature* 145: 512.

Cramer, E. M., and Huggins, W. H. 1958. Creation of pitch through binaural interaction. *J. Acoust. Soc. Am.* 30: 801–2.

Cutler, C. C. 1956. Stereoscopic presentation of radio detection signals, as an aid to recognition. *Airforce Rep. No. 24269/K,* Mar. 30.

Day, R. H. 1958. On interocular transfer and the central origin of visual after-effects. *Am. J. Psychol.* 71: 784–89.

De Valois, R. 1960. Color vision mechanisms in the monkey. *J. Gen. Physiol.* 43 (suppl. pt. 2): 115–28.

Diamond, A. L. 1958. Simultaneous brightness contrast and the Pulfrich phenomenon. *J. Opt. Soc. Am.* 48: 887–90.

Diamond, I. T., and Hall, W. C. 1969. Evolution of neocortex. *Science* 164: 251–62.

Disbrow, J. 1964. Studies in stereoscopy. M.S. thesis, M.I.T.

Ditchburn, R. W., and Ginsborg, B. L. 1952. Vision with stabilized retinal image. *Nature* 170: 36–7.

Dodwell, P. C., and Engel, G. R. 1963. The theory of binocular fusion. *Nature* 198: 39–40, 73–74.

Donchin, E., and Cohen, L. 1967. Averaged evoked potentials and intramodality selective attention. *Electroencephalog Clin. Neurophysiol.* 19: 325–35.

Dorff, J. E.; Mirsky, A. F.; and Mishkin, M. 1965. Effects of unilateral temporal lobe removals in man on tachistoscopic recognition in the left and right visual fields. *Neuropsychologia* 3: 39–51.

Dove, H. W. 1841. Über Stereoskopie. *Ann. Phys. series* 2 110: 494–98.

Efron, R. 1963. Stereoscopic vision: Effect of binocular temporal summation. *Brit. J. Ophthalmol.* 41: 709–30.

Egeth, H. E. 1966. Parallel versus serial processes in multidimensional stimulus discrimination. *Perception and Psychophysics* 1: 245–52.

Emmert, E. 1881. Grössenverhältnisse der Nachbilder. *Klin. Monatsbl. Augenheilk* 19: 443–50.

Engel, G. R. 1966. A test of the existence of monocular stereoscopic depth perception. *Perceptual and Motor Skills* 23: 235–38.

Enroth-Cugel, C., and Robson, J. G. 1966. The contrast sensitivity of retinal ganglion cells of the cat. *J. Physiol.* 187: 517–52.

Erlebacher, A., and Sekuler, R. 1969. An explanation of the Müller-Lyer illusion: The confusion theory examined. *J. Exp. Psychol.* 80: 462–67.

Erulkar, S. D., and Fillenz, M. 1960. Single unit activity in the lateral geniculate body of the cat. *J. Physiol.* 154: 206–18.

Evans, C. R. 1965. Some studies of pattern perception using a stabilized retinal image. *Brit. J. Psychol.* 56 (2 and 3): 121.

Evans, C. R., and Clegg, J. M. 1967. Binocular depth perception of Julesz patterns viewed as perfectly stabilized retinal images. *Nature* 215: 893–95.

Exner, S. 1875. Über das Sehen von Bewegungen und die Theorie des zusammengesetzen Auges. *S. B. Akad. Wiss.* (Wien) 72: 156–90.

Fechner, G. T. 1838. Über eine Scheibe zur Erzeugung subjektiver Farben. *Ann. Phys. (Leipzig)* 45: 227–32.

Feldman, M., and Cohen, B. 1968. Electrical activity in the lateral geniculate body of the alert monkey associated with eye movements. *J. Neurophysiol.* 31: 455–66.

Fender, D. H., and Julesz, B. 1967a. Extension of Panum's fusional area in binocularly stabilized vision. *J. Opt. Soc. Am.* 57: 819–30.

———. 1967b. MTF-S of random-dot patterns in normal and stabilized vision. *J. Opt. Soc. Am.* 57: 581.

Fender, D. H., and Nye, P. W. 1961. An investigation of the mechanisms of eye movement control. *Kybernetik* 1: 81–88.

Fender, D. H., and Saint-Cyr, A. J. 1969. Information transfer between the oculomotor systems of the two eyes during tracking tasks. *J. Opt. Soc. Am.* 59: 512.

Fisher, G. H. 1968. An experimental and theoretical appraisal of the inappropriate size-depth theories of illusions. *Brit. J. Psychol.* 59: 373–83.

Fox, R., and Blake, R. R. 1970. Stereopsis in the cat. Talk at 10th Psychonomic Soc. Meeting, San Antonio, Texas, Nov. 5–7.

Freud, S. L. 1964. The physiological locus of the spiral aftereffect. *Am. J. Psychol.* 77: 422–28.

Frisch, H. L., and Julesz, B. 1966. Figure-ground perception and random geometry. *Perception and Psychophysics* 1: 389–98.

Gabor, D. 1948. A new microscopic principle. *Nature* 161: 777–78.

Garner, W. R. 1966. To perceive is to know. *Am. Psychologist* 21: 11–19.

Gazzaniga, M. S.; Bogen, J. E.; and Sperry, R. W. 1965. Observations on visual perception after disconnection of the cerebral hemispheres in man. *Brain* 88: 221–36.

Gelfan, S., and Carter, S. 1967. Muscle sense in man. *Exp. Neurology* 18: 469–73.

Gibson, J. J. 1933. Adaptation, after-effect and contrast in the perception of curved lines. *J. Exp. Psychol.* 16: 1–31.

———. 1950. *The perception of the visual world*. Boston: Houghton Mifflin.

———. 1957. Optical motions and transformations as stimuli for visual perception. *Psychol. Rev.* 64: 288–95.

Gilinsky, A. S., and Doherty, R. S. 1969. Interocular transfer of orientational effects. *Science* 164: 454–55.

Gogel, W. C. 1965. Size cues and the adjacency principle. *J. Exp. Psychol.* 70: 289–93.

Graham, R. E., and Kelly, J. L., Jr. 1958. A computer simulation chain for research on picture coding. *I.R.E.Wescon Conv. Rec. (Computer Applications)*, pp. 41–46.

Green, B. F., Jr. 1961. Figure coherence in the kinetic depth effect. *J. Exp. Psychol.* 62: 272–82.

Green, B. F.; Wolf, A. K.; and White, B. W. 1959. The detection of statistically defined patterns in a matrix of dots. *Am. J. Psychol.* 72: 503–20.

Green, R. T., and Courtis, M. C. 1966. Information theory and figure perception: The metaphor that failed. *Acta Psychologica* 25: 12–36.

Gregory, R. L. 1963. Distortion of visual space as inappropriate constancy scaling. *Nature* 199: 678–80.

Gregory, R. L.; Wallace, J. G.; and Campbell, F. W. 1959. Changes in the size and shape of visual afterimages observed in complete darkness during changes in position in space. *Quart. J. Exp. Psychol.* 11: 54–55.

Gross, C. G.; Bender, D. B.; and Rocha-Miranda, C. E. 1969. Visual receptive fields of neurons in inferotemporal cortex of the monkey. *Science* 166: 1303–5.

Grüsser, O.-J.; Grüsser-Cornehls, U.; Finkelstein, D.; Henn, V.; Patutschnik, M.; and Butenandt, E. 1967. A quantitative analysis of movement-detecting neurons in the frog's retina. *Pfluegers Arch. Ges. Physiol.* 293: 100–106.

Guillery, R. W. 1969. An abnormal retinogeniculate projection in Siamese cats. *Brain Res.* 14: 739–41.

Guttman, N., and Julesz, B. 1963. Lower limits of auditory periodicity analysis. *J. Acoust. Soc. Am.* 35: 610.

Guzman, A. 1968. Decomposition of a visual scene into three-dimensional bodies. *Proc. Fall Joint Computer Conf.*, Dec. 1968, AFIPS 33. Washington: Thompson Book, pp. 291–304.

Haber, R. N. 1966. Nature of the effect of set on perception. *Psychol. Rev.* 73: 335–51.

———. 1969. Eidetic images. *Sci. Am.* 64: 36–44.

Haber, R. N., and Haber, R. B. 1964. Eidetic imagery: 1. Frequency. *Percept. Mot. Skills* 19: 131–38.

Harmon, L. D. 1970. Some aspects of recognition of human faces, ed. J. O. Grüsser. 4th Kybernetic Kongress, Berlin, 1970. Heidelberg: Springer (in press).

Harmon, L. D., and Knowlton, K. C. 1969. Picture processing by computer. *Science* 164: 19–29.

Harris, C. S. 1970. Effect of viewing distance on a color aftereffect specific to spatial frequency. Talk at the 10th Annual Meeting of the Psychonomic Society, Nov. 5–7, 1970.

Harris, C. S., and Gibson, A. R. 1968. Is orientation-specific color adaptation in human vision due to edge detectors, afterimages, or "dipoles"? *Science* 162: 1506–7.

Hartline, H. K. 1949. Inhibition of activity of visual receptors by illuminating nearby retinal areas in the limulus eye. *Fed. Proc.* 8: 69.

Hebb, D. O. 1949. *The organization of behavior.* New York: Wiley.

Heinemann, E. G.; Tulving, E.; and Nachmias, J. 1969. The effect of oculomotor adjustments on apparent size. *Am. J. Psychol.* 72: 32–45.

Helmholtz, H. von. 1909. *Physiological optics.* The Opt. Soc. 1924 ed. republished by Dover, N.Y. (translated from the 3d German ed.).

Henn, V., and Grüsser, O.-J. 1969. The summation of excitation in the receptive fields of movement-sensitive neurons of the frog's retina. *Vision Res.* 9: 57–69.

Henry, G. H.; Bishop, P. O.; and Coombs, J. S. 1969. Inhibitory and sub-liminal excitatory receptive fields of simple units in cat stirate cortex. *Vision Res.* 9: 1289–96.

Hepler, N. 1968. Color: A motion-contingent aftereffect. *Science* 162: 376–77.

Hering, E. 1879. Der Raumsinn und die Bewegung des Auges. In *Handbuch der Physiologie,* ed. L. Hermann, trans. C. A. Radde, *Am. Acad. Optom.,* Baltimore, 1942.

―――. 1964. *Outlines of a theory of the light sense.* Cambridge, Mass.: Harvard Univ. Press. Translated from German original that appeared in separate sections (1905, 1907, 1911) and last section completed from Hering's notebooks posthumously by C. Hess (1920).

Hess, E. H. 1952. Subjective colors: Retinal vs. central origin. *Am. J. Psychol.* 65: 278–80.

―――. 1956. Space perception in the chick. *Sci. Am.* 195: 71–80.

Hesse, Hermann. 1943. *The glass bead game (Magister Ludi)* (Original German title: Das Glasperlenspiel.) English ed., New York: Holt, Rinehart and Winston, 1969.

Hirsch, H. V. B., and Spinelli, D. N. 1970. Visual experience modifies distribution of horizontally and vertically oriented receptive fields in cats. *Science* 168: 869–71.

Hirsh, I. J., and Fraisse, P. 1964. Simultanéité et succession de stimuli hétérogènes. *L'Année Physiologique* 64: 1–19.

Hochberg, J. 1963. Illusions and figural reversals without lines. Talk at the 4th meeting Psychonomic Soc., Bryn Mawr, Penn.

―――. 1964. *Perception.* New Jersey: Prentice-Hall.

―――. 1968. In the mind's eye. In *Contemporary theory and research in visual perception,* ed. R. N. Haber. New York: Holt, Rinehart and Winston, pp. 309–31.

―――. 1970. Elementary units in the visual perception of objects. In *Handbook of cognitive psychology,* ed. J. Mehler. New York: Prentice-Hall, in press.

Hoffman, W. C. 1964. Pattern recognition by the method of isoclines: A mathematical model for the visual integrative process. *Boeing Scientific Res. Labs. Math Note No. 351,* Seattle.

―――. 1966. The Lie algebra of visual perception. *J. Math. Psychol.* 3: 65–98.

―――. 1968. The neuron as a Lie group germ and Lie product. *Quart. J. Appl. Math.* 25: 423–40.

Holmstedt, B., and Lindgren, J. E. 1967. Chemical constituents and pharmacology of South American snuffs. *Psychoactive Drugs.* Workshop Series of Pharmacology, NIMH, No. 2. Washington, D.C.: U.S. Govt. Printing Office.

Holst, E. von. 1957. Aktive Leistungen der menschlichen Gesichtswahrnehmung. *Studium Generale* 10: 231–43.

Holst, E. von., and Mittelstaedt, H. 1950. Das Reafferenzprinzip. *Naturwissenschaften* 37: 464–76.

Holway, A. H., and Boring, E. G. 1941. Determinants of apparent visual size with distance variant. *Am. J. Psychol.* 54: 21–37.

Horn, B. K. P. 1969. Shape from shading. Manuscript (M.I.T. Cambridge, Mass.) to be published. Abstract in *Special Interest Group on Artificial Intelligence of the Assoc. for Computing Machinery,* August 1969, p. 16.

Horn, G., and Hill, R. M. 1969. Modifications of receptive fields of cells in the visual cortex occurring spontaneously and associated with body tilt. *Nature* 221: 186–88.

Hubel, D. H. 1960. Single unit activity in lateral geniculate body and optic tract of unrestrained cats. *J. Physiol.* 150: 91–104.

Hubel, D. H., and Wiesel, T. N. 1960. Receptive fields of optic nerve fibres in the spider monkey. *J. Physiol.* 154: 572–80.

———. 1961. Integrative action in the cat's lateral geniculate body. *J. Physiol.* 155: 385–98.

———. 1962. Receptive fields, binocular interaction and functional architecture in the cat's visual cortex. *J. Physiol.* 160: 106–54.

———. 1965a. Receptive fields and functional architecture in two non-striate visual areas (18 and 19) of the cat. *J. Neurophysiol.* 28: 229–89.

———. 1965b. Binocular interaction in striate cortex of kittens reared with artificial squint. *J. Neurophysiol.* 28: 1041–59.

———. 1967. Cortical and callosal connections concerned with the vertical meridian of visual fields in the cat. *J. Neurophysiol.* 30: 1561–73.

———. 1968. Receptive fields and functional architecture of monkey striate cortex. *J. Physiol.* 195: 215–43.

———. 1969. Anatomical demonstration of columns in the monkey striate cortex. *Nature* 221: 747–50.

———. 1970. Stereoscopic vision in macaque monkey. *Nature* 225: 41–42.

Humphrey, N. K., and Morgan, M. J. 1965. Constancy and the geometric illusions. *Nature* 206: 744–45.

Hunter, W. S. 1915. Retinal factors in visual after-movement. *Psychol. Rev.* 22: 479–89.

Imai, S., and Garner, W. R. 1965. Discriminability and preference for attributes in free and constrained classification. *J. Exp. Psychol.* 69: 596–608.

Ince, E. L. 1926. *Ordinary differential equations.* Republished by Dover, N. Y., 1956. Chapt. 4, pp. 93–113.

Irvine, S. R., and Ludvigh, E. 1936. Is ocular proprioceptive sense concerned in vision? *A.M.A. Arch. Opthalmol.* 15: 1037–49.

Ising, E. 1925. Contribution to the theory of ferromagnetism. *Z. Physik.* 31: 253–58.

Ittelson, W. H. 1952. *The Ames demonstrations in perception.* Princeton, N.J.: Princeton University Press.

———. 1960. *Visual space perception.* New York: Springer.

Ives, H. E. 1928. A camera for making parallax panoramagrams. *J. Opt. Soc. Am.* 17: 435–39.

Jaensch, E. R. 1930. *Eidetic imagery and typological methods of investigation,* 2nd ed., translated by O. Oeser New York: Harcourt, Brace and Co.

Johnson, S. C. 1967. Hierarchical clustering schemes. *Psychometrica* 32: 241–54.

378 **References**

Joshua, D. E., and Bishop, P. O. 1969. Binocular single vision and depth discrimination: Receptive field disparities for central and peripheral vision and binocular interaction on peripheral single units in cat striate cortex. *Exp. Brain Res.*, in press.

Julesz, B. 1959. Method of coding television signals based on edge detection. *Bell System Tech. J.* 38: 1001–20.

————. 1960a. Binocular depth perception of computer-generated patterns. *Bell System Tech. J.* 39: 1125–62.

————. 1960b. Binocular depth perception and pattern recognition. *Information Theory, Fourth London Symposium, 1960,* E.C. Cherry. London: Butterworth, 1961.

————. 1962a. Visual pattern discrimination. *IRE Transactions on Information Theory* IT-8: 84–92.

————. 1962b. Towards the automation of binocular depth perception (AUTOMAP-1). *Proceedings of the IFIPS Congress, Munich 1962,* ed. C. M. Popplewell. Amsterdam: North-Holland, 1963.

————. 1963a. Stereopsis and binocular rivalry of contours. *J. Opt. Soc. Am.* 53: 994–99.

————. 1963b. Effects of contour flicker or apparent motion on retinally stabilized images. *J. Opt. Soc. Am.* 53: 1336.

————. 1964a. Some recent studies in vision relevant to form perception. Symposium Nov. 1964 Boston, Mass. *Models for the perception of speech and visual form,* ed. W. Whaten-Dunn. Cambridge, Mass.: M.I.T. Press, 1967, pp. 136–54.

————. 1964b. Binocular depth perception without familiarity cues. *Science* 145: 356–62.

————. 1965a. Texture and visual perception. *Sci. Am.* 212: 38–48.

————. 1965b. Some neurophysiological problems of stereopsis. In P. W. Nye, ed., *Proc. of symposium on information processing in sight sensory systems,* Caltech., Pasadena, Calif., Nov. 1-3 1965, pp. 135–42.

————. 1965c. Visual perception and roentgenography. *Proc. of 1965 conference of teachers of radiology, Feb. 13, 1965,* Am. College of Radiology, Chicago.

————. 1966a. Binocular disappearance of monocular symmetry. *Science* 153: 657–58.

————. 1966b. Computers, patterns and depth perception. *Bell Labs. Record* 44: 261–66.

————. 1967a. Suppression of monocular symmetry during binocular fusion without rivalry. *Bell System Tech. J.* 46: 1203–21.

————. 1967b. Visual perception of repetition, rotation, mirror and centric symmetry in random textures. Talk at International Information Theory Symposium, San Remo, Italy.

————. 1968a. Binocular depth perception. In W. Reichardt, ed., *Proc. of the International School of Physics "Enrico Fermi," course XLIII.* Varenna, Italy, July 1968. "Processing of optical data by organisms and by machines." New York: Academic Press, 1969, pp. 589–605.

————. 1968b. Pattern discrimination. In W. Reichardt, ed., *Proc. of the International School of Physics "Enrico Fermi," course XLIII.* Varenna, Italy, July 1968. "Processing of optical data by organisms and by machines." New York: Academic Press, 1969, pp. 580–88.

————. 1968c. Cluster formation at various perceptual levels. In S. Watanabe, ed., *Proc. int. conf. on methodologies of pattern recognition.* Hawaii, January 1968. New York: Academic Press, pp. 297–315.

————. 1968d. Experiment in perception. *Psychol. Today* 2: 16–23.

————. 1969a. Foundations of cyclopean perception (abstract). *J. Opt. Soc. Am.* 59: 1544.

————. 1969b. Optical-constancy phenomena. In W. Reichardt, ed., *Proc. of the International School of Physics "Enrico Fermi," course XLIII.* Varenna, Italy, July 1968. "Processing of optical data by organisms and by machines." New York: Academic Press, 1969: 417–30.

————. 1970a. The separation of retinal and central processes in vision. In *Psychotomimetic Drugs,* ed. D. H. Efron. New York: Raven Press, pp. 183–91.

―――. 1970b. Effects of Fourier domain operations on stereopsis. Talk presented at the Gordon Res. Conf. on Biomathematics, Andover, N. H., Aug. 3–7, 1970.

Julesz, B., and Bosche, C. 1966. Studies on visual texture and binocular depth perception. A computer-generated movie series containing monocular and binocular movies. Bell Tel. Labs., Inc.

Julesz, B., and Guttman, N. 1963. Auditory memory. *J. Acoust. Soc. Am.* 35: 63.

―――. 1965. High-order statistics and short-term auditory memory. *Proc. 5th Int'l Congress on Acoustics.* Liege, Belgium.

Julesz, B., and Hesse, R. I. 1970. Inability to perceive the direction of rotation movement of line segments. *Nature* 225: 243–44.

Julesz, B., and Hirsh, I. J. 1968. Visual and auditory perception: An essay of comparison. In *Human communication, A unified view,* ed. E. E. David, Jr., and P. Denes. McGraw-Hill, to be published.

Julesz, B., and Johnson, S. C. 1968a. Stereograms portraying ambiguously perceivable surfaces. *Proc. Natl. Acad. Sci.* 61: 437–41.

―――. 1968b. Mental holography: Stereograms portraying ambiguously perceivable surfaces. *Bell System Tech. J.* 49: 2075–83.

Julesz, B., and Miller, J. E. 1962. Automatic stereoscopic presentation of functions of two variables. *Bell System Tech. J.* 41: 663–76.

Julesz, B., and Payne, R. A. 1968. Differences between monocular and binocular stroboscopic movement perception. *Vision Res.* 8: 433–44.

Julesz, B., and Pennington, K. S. 1965. Equidistributed information mapping: An analogy to holograms and memory. *J. Opt. Soc. Am.* 55: 604.

Julesz, B.; Slepian, D.; and Sondhi, M. M. 1969. Correction for astigmatism by lens rotation and image processing. *J. Opt. Soc. Am.* 59: 485.

Julesz, B., and Spivack, G. J. 1967. Stereopsis based on vernier acuity cues alone. *Science* 157: 563–65.

Julesz, B., and Stromeyer, C. F., III. 1970. Masking of spatial gratings by filtered one-dimensional visual noise. Talk at the 10th Annual Meeting of the Psychonomic Society, San Antonio, Texas, Nov. 5–7.

Julesz, B., and White, B. W. 1969. Short term visual memory and the Pulfrich Phenomenon. *Nature* 222: 639–41.

Kaufman, L. 1963. On the spread of suppression and binocular rivalry. *Vision Res.* 3: 401–15.

―――. 1964. On the nature of binocular disparity. *Am. J. Psychol.* 77: 393–402.

―――. 1965. Some new stereoscopic phenomena and their implications for the theory of stereopsis. *Am. J. Psychol.* 78: 1–20.

Kaufman, L., and Pitblado, C. 1965. Further observations on the nature of effective binocular disparities. *Am. J. Psychol.* 78: 379–91.

Kaufman, L., and Rock, I. 1962. The moon illusion. *Sci. Am.* 207: 120–30.

Kawamura, H., and Marchiafava, P. L. 1966. Modulation of transmission of optic nerve impulses in the alert cat. *Brain Res.* 2: 213–15.

Kimura, D. 1961. Cerebral dominance and perception of verbal stimuli. *Canadian J. Psychol.* 15: 166–74.

―――. 1963. Right temporal lobe damage. Perception of unfamiliar stimuli after damage. *Arch. Neurol.* 18: 264–71.

―――. 1964. Left-right differences in the perception of melodies. *Quart. J. Exp. Psychol.* 16: 335–58.

―――. 1967. Functional asymmetry of the brain in dichotic listening. *Cortex* 3: 163–78.

Kinsbourne, M. 1970. The cerebral basis of lateral asymmetries in attention. *Acta Psychol.* 33: 193–201.

Kinston, W. J.; Vadas, M. A.; and Bishop, P. O. 1970. Multiple projection of the visual field onto the medial portion of the dorsal lateral geniculate nucleus and the adjacent nuclei of the cat. *J. Comp. Neurol.,* in press.

Klix, F. 1962. *Elementaranalysen zur Psychophysik der Raumwahrnehmung.* Berlin: Deutscher Verlag der Wissenschaften, VEB.

Knowlton, K. C. 1965. Computer-produced movies. *Science* 150: 1116–20.

Köhler, W., and Emery, D. A. 1947. Figural after-effects in the third dimension of visual space. *Am. J. Psychol.* 40: 159–201.

Köhler, W., and Wallach, H. 1944. Figural after-effects: An investigation of visual processes. *Proc. Am. Phil. Soc.* 88: 269–357.

Koestler, A. 1964. *The act of creation.* New York: MacMillan.

Kohler, I. 1964. The formation and transformation of the visual world. *Psychol. Issues* 3: 28–46, 116–33.

Kohlrausch, A. 1925. Der Verlauf der Netzhautströme und der Gesichtsempfindungen nach Momentbelichtung. *Pfluegers Arch. Ges. Physiol.* 209: 607–10.

Kolers, P. A. 1963. Some differences between real and apparent visual movement. *Vision Res.* 3: 191–206.

Korte, W. 1915. Kinematoskopische Untersuchungen. *Z. Psychol.* 72: 193–296.

Krauskopf, J., and Srebro, R. 1965. Spectral sensitivity of color mechanisms: Derivation from fluctuation of color appearance near threshold. *Science* 150: 1477–79.

Kroh, O. 1922. Subjektive Anschauungsbilder bei Jugendlichen: Eine psychologisch-pädagogische Untersuchung. Göttingen: Vandenhoeck and Ruprecht.

Kruskal, J. B. 1964. Multidimensional scaling by optimizing goodness of fit to a nonmetric hypothesis. *Psychometrica* 29: 1–27, 115–29.

Kuhn, T. S. 1962. *The structure of scientific revolutions.* Chicago: Univ. Chicago Press.

Külpe, O. 1904. Versuche über Abstraktion. *Berlin Intern. Congr. Exp. Psychol.,* pp. 56–68.

Kuffler, S. W. 1953. Discharge patterns and functional organization of mammalian retina. *J. Neurophysiol.* 16: 37–68.

Lambert, W. E., and Jakobovits, L. A. 1960. Verbal satiation and changes in the intensity of meaning. *J. Exp. Psychol.* 60: 376–83.

Land, E. H. 1959. Experiments in color vision. *Sci. Am.* 200: 84–94, 96, 99.

———. 1964. The retinex. *Am. Scientist* 52 (2): 247–64.

Lau, E. 1922. Versuche über das stereoskopische Sehen. *Psychol. Forsch.* 2: 1–4.

Lawson, R. B., and Gulick, W. L. 1967. Stereopsis and anomalous contour. *Vision Res.* 7: 271–97.

Lawson, R. B., and Mount, D. C. 1967. Minimum condition for stereopsis and anomalous contour. *Science* 158: 804–06.

Ledley, R. S. 1964. High-speed automatic analysis of biomedical pictures. *Science* 146: 216–23.

Lee, D. N. 1969. Theory of the stereoscopic shadow-caster: An instrument for the study of binocular kinetic space perception. *Vision Res.* 9: 145–56.

———. 1971. Binocular stereopsis without spatial disparity. *Perception and Psychophysics* 9: 216–21.

Leeper, R. 1935. A study of a neglected portion of the field of learning: The development of sensory organization. *J. Genet. Psychol.* 46: 41–73.

Lehmann, D., and Fender, D. H. 1968. Component analysis of human averaged evoked potentials: Dichoptic stimuli using different target structure. *Electroencephalog. Clin. Neurophysiol.* 24: 542–53.

Lehmann, D.; Madey, J. M.; Koukkou, M.; and Fender, D. H. 1969. Mapping of visually evoked EEG responses on the human scalp. *Invest. Ophthalmol.* 8: 651.

Leibowitz, H., and Moore, D. 1966. Role of changes in accommodation and convergence in the perception of size. *J. Opt. Soc. Am.* 56: 1120–23.

Leicester, J. 1968. Projection of the visual vertical meridian to cerebral cortex of the cat. *J. Neurophysiol.* 31: 371–82.

Leith, E. N., and Upatnieks, J. 1962. Reconstructed wavefronts and communication theory. *J. Opt. Soc. Am.* 52: 1123–30.

Lettvin, J. Y.; Maturana, H. R.; McCulloch, W. S.; and Pitts, W. H. 1959. What the frog's eye tells the frog's brain. *Proc. Inst. Radio Engrs.* 47: 1940–51.

Lettvin, J. Y.; Maturana, H. R.; Pitts, W. H.; and McCulloch, W. S. 1961. Two remarks on the visual system of the frog. In *Sensory Communication,* ed. W. A. Rosenblith. Cambridge, Mass.: M.I.T. Press; New York: Wiley.

Levelt, W. J. M. 1965. *On binocular rivalry*. Soesteberg, The Netherlands: Institute of Perception.

Licklider, J. C. R. 1948. The influence of interaural phase relations upon the masking of speech by white noise. *J. Acoust. Soc. Am.* 20: 150–59.

Lie, S., and Scheffers, G. 1893. *Vorlesungen über continuerliche Gruppen*. Leipzig: Teuber.

Lippmann, G. 1908. Epreuves réversibles donnant la sensation du relief. *J. Phys.* 7 (no. 11, 4th s.): 821.

Lit, A. 1949. The magnitude of the Pulfrich stereo-phenomenon as a function of binocular differences of intensity at various levels of illumination. *Am. J. Psychol.* 62: 159–81.

Löfgren, L. 1960. Connection possibilities and distinguishable stimulus patterns in cellular layers. In *Some principles of preorganization in self-organizing systems*. Univ. Illinois Electr. Eng. Res. Lab. Tech. Rep. no. 2 for ONR No. 049–123: 43–60.

Luneburg, R. K. 1950. The metric of binocular visual space. *J. Opt. Soc. Am.* 40: 627–42.

Luria, A. R. 1968. *The mind of a mnemonist*. New York: Basic Books. Original Russian edition, 1965.

McBride, P., and Reed, J. B. 1952. The speed and accuracy of discriminating differences in number and texture density. *Human Eng. Tech. Rep.,* SDC-131-1-3.

McCollough, C. 1965. Color adaptation of edge-detectors in the human visual system. *Science* 149: 1115–16.

Mach, E. 1886. *The analysis of sensations and the relation of the physical to the psychical*. First German edition in 1886; republished by Dover, N. Y., 1959 using the 5th German edition revised and supplemented by S. Waterlow.

MacKay, D. M. 1957. Moving visual images produced by regular stationary patterns. *Nature* 180: 849–50, 1145–46.

———. 1961. Interactive processes in visual perception. In *Sensory communication,* ed. W. A. Rosenblith. Cambridge, Mass.: M.I.T. Press; New York: Wiley, pp. 339–55.

———. 1965. Visual noise as a tool of research. *J. Gen. Psychol.* 72: 181–97.

MacKay, D. M., ed. 1969. Evoked brain potentials as indicators of sensory information processing. *Neurosci. Res. Program,* vol. 7 no. 3.

McLaughlin, S. C. 1964. Visual perception in strabismus and amblyopia. *Pschol. Monogr.* 78 (12): 1–23.

MacNichol, E. F., and Svaetichin, G. 1958. Electric responses from isolated retinas of fishes. *Am. J. Ophthalmol.* 46: 26–40.

Marlowe, L. H. 1969. Orientation of contours and binocular depth perception. Ph.D. dissertation, Brown University.

Masland, H. R. 1969. Visual motion perception: Experimental modification. *Science* 165: 819–21.

Mathews, M. V.; Miller, J. E.; Moore, F. R.; Pierce, J. R.; and Risset, J. C. 1969. *The technology of computer music.* Cambridge, Mass.: M.I.T. Press.

Matin, L.; Pearce, D.; Matin, E.; and Kibler, G. 1966. Visual perception of direction in the dark: Roles of local sign, eye movements, and ocular proprioception. *Vision Res.* 6: 453–69.

Mayzner, M. S.; Tresselt, M. E.; and Helfer, M. S. 1967. A provisional model of visual information processing with sequential inputs. *Psychonomic Monogr. Suppl.* 2 (7, no. 23): 91–108.

Milner, B. 1964. Discussion of M. Mishkin's paper. *Acta Psychol.* 23: 308.

Minsky, M., and Papert, S. 1968. *Perceptrons.* Cambridge, Mass.: M.I.T. Press.

Moray, N. 1959. Attention in dichotic listening: Affective cues and the influence of instructions. *Quart. J. Exp. Psychol.* 11: 56–60.

Morrell, F. 1967. Electrical signs of sensory coding. In *The Neurosciences—A study program,* ed. G. C. Quarton, T. Nelnechuk, and F. O. Schmitt. New York: Rockefeller Univ. Press: 452–69.

Mountcastle, V.B. 1957. Modality and topographic properties of single neurons of cat's somatic sensory cortex. *J. Neurophysiol.* 20: 408–34.

Nachmias, J.; Sachs, M. B.; and Robson, J. G. 1969. Independent spatial-frequency channels in human vision. *J. Opt. Soc. Am.* 59: 1538.

Narasimhan, R., and Mayoh, B. H. 1963. The structure of a program for scanning bubble chamber negatives. Digital Computer Lab. File No. 507. Univ. of Illinois.

Necker, L. A. 1832. On an apparent change of position in a drawing of engraved figure of a crystal. *Philosophical Mag.* 1 (3d s.): 329–37. Republished in *Visual Perception.* W. N. Dember, ed., New York: Wiley, 1964. 78–80.

Neisser, U. 1967. *Cognitive psychology.* New York: Appleton-Century-Crofts.

Neuhaus, W. 1930. Experimentelle Untersuchungen der Scheinbewegung. *Arch. Ges. Psychol.* 75: 315–458.

Newell, F. G., and Montroll, E. W. 1953. On the theory of the Ising model of ferromagnetism. *Rev. Modern Phys.* 25: 353–89.

Novikoff, A. E. J. 1962. Integral geometry as a tool in pattern perception. In *Principles of self organization,* ed. H. von Foerster and G. W. Fopf. New York: Pergamon Press, p. 11.

O'Brien, V. 1958. Contour perception, illusion or reality. *J. Opt. Soc. Am.* 48: 112–19.

Ogle, K. N. 1952. Disparity limits of stereopsis. *Arch. Ophthal.* 48: 50–60.

———. 1959. Theory of stereoscopic vision. In *Psychology: A study of science,* ed. S. Koch. New York: McGraw-Hill, vol. 1, pp. 362–94.

———. 1962. The optical space sense. *The eye,* ed. H. Davson. New York: Academic Press, vol. 4, pp. 211–432.

Ohwaki, S. 1960. On the destruction of geometrical illusions in stereoscopic observation. *Tohoku Psychol. Folia* 29: 24–36.

Oppel, J. J. 1854. Über geometrisch-optische Täuschungen. *Jahresber., D. Physical. Ver. Frankfurt,* (1854-55) pp. 37–47, (1856-57) pp. 47–55, (1860-61) pp. 26–37.

Over, R. 1968. Explanations of geometrical illusions. *Psychol. Bull.* 70 (6): 545–62.

Pantle, A., and Sekuler, R. W. 1968a. Velocity-sensitive elements in human vision: Initial psychophysical evidence. *Vision Res.* 8: 445–50.

———. 1968b. Contrast response of human visual mechanisms sensitive to orientation and detection of motion. *Vision Res.* 9: 397–406.

————. 1968c. Size-detecting mechanisms in human vision. *Science* 162: 1146–48.

Panum, P. L. 1858. *Physiologische Untersuchungen über das Sehen mit zwei Augen.* Kiel: Schwers.

Papert, S. 1961. Centrally produced geometrical illusions. *Nature* 191: 733.

————. 1964. Stereoscopic synthesis as a technique for localizing visual mechanisms. *M.I.T. Quart. Progr. Rep.* 73: 239–43.

Parks, T. E. 1965. Post-retinal visual storage. *Am. J. Psychol.* 78: 145–47.

Penfield, W. 1952. Memory mechanisms. *Arch. Neurol. Psychiat.* 67: 178–91.

Pettigrew, J. D.; Nikara, T.; and Bishop, P. O. 1968. Binocular interaction on single units in cat striate cortex: Simultaneous stimulation by single moving slit with receptive fields in correspondence. *Exp. Brain Res.* 6: 391–410.

Pollack, I. 1968. Periodicity discrimination for auditory pulse trains. *J. Acoust. Soc. Am.* 43: 1113–19.

————. 1969. Depth of sequential auditory information processing. *J. Acoust. Soc. Am.* 46: 952–64.

Pollehn, H., and Roehrig, H. 1970. Effect of noise on the modulation transfer function of the visual channel. *J. Opt. Soc. Am.* 60: 842–48.

Polyak, S. L. 1941. *The retina.* Chicago: Univ. of Chicago Press.

————. 1957. *The vertebrate visual system.* Chicago: Univ. of Chicago Press.

Popov, N. A., and Popov, C. 1954. Contribution à l'étude des fonctions corticales chez l'homme par la méthode des réflexes conditionnés électrocorticaux: V. Deuxième système de signalisation. *C.R. Acad. Sci.* 238: 2118–20.

Powell, T. P. S., and Mountcastle, V. B. 1959. The cytoarchitecture of the postcentral gyrus of the monkey macaca mulatta. *Bull. John Hopk. Hosp.* 105: 108–31.

Pritchard, R. M. 1961. Stabilized images on the retina. *Sci. Am.* 204: 72–78.

Pulfrich, C. 1922. Die Stereoskopie im Dienste der isochromen und heterochromen Photometric. *Naturwissenschaften* 10: 533–64, 569–601, 714–22, 735–43, 751–61.

Ratliff, F. 1965. *Mach bands.* San Francisco: Holden-Day.

Regan, D., and Spekreijse, H. 1970. Electrophysiological correlate of binocular depth perception in man. *Nature* 225: 92–94.

Richards, W. 1967a. Apparent modifiability of receptive fields during accommodation and convergence and a model for size constancy. *Neurophysiologia* 5: 63–72.

————. 1967b. Size scaling and binocular rivalry. *J. Opt. Soc. Am.* 57: 576.

————. 1968. Spatial remapping in the primate visual system. *Kybernetik* 4: 146–56.

————. 1969. The influence of oculomotor systems on visual perception. *Report AFOSR69-1934TR*, M.I.T., Cambridge, Mass., July 1969.

————. 1970. Stereopsis and stereoblindness. *Exp. Brain Res.* 10: 380–88.

Richards, W., and Smith, R. A. 1969. Midbrain as a site for the motion after-effect. *Nature* 223: 533–34.

Riggs, L. A. 1969. Progress in the recording of human retinal and occipital potentials. *J. Opt. Soc. Am.* 59: 1558–66.

Riggs, L. A.; Johnson, E. P.; and Schick, A. M. L. 1966. Electrical responses of the human eye to changes in wavelength of the stimulating light. *J. Opt. Soc. Am.* 56: 1621–27.

Riggs, L. A.; Ratliff, F.; Cornsweet, J. C.; and Cornsweet, T. N. 1953. The disappearance of steadily fixated visual test objects. *J. Opt. Soc. Am.* 43: 495–501.

Riggs, L. A., and Sternheim, C. E. 1969. Human retinal and occipital potentials evoked by changes of the wavelength of the stimulating light. *J. Opt. Soc. Am.* 59: 635–40.

Rindfleisch, T. 1966. Photometric method for lunar topography. *Photogramm. Eng.* 32: 262–75.

Roberts, L. G. 1963. Machine perception of three-dimensional solids. In *Optical and Electrooptical Information Processing,* ed. J. T. Tippett, et al. Cambridge, Mass.: M.I.T. Press, pp. 159–97.

Rock, I., and Halper, F. 1969. Form perception without a retinal image. *Am. J. Psychol.* 82, 425–40.

Ronchi, L., and Bottai, G. 1964. Simultaneous contrast effects at the center of figures showing different degrees of symmetry. *Atti della Fondazione G. Ronchi* 19 (1): 84–100.

Ronchi, L., and Salvi, G. 1965. The neural organization of the central retina as revealed by an experiment on simultaneous contrast. *Pubblicazioni dell'Istituto Nazionale di Ottica* 20: 192–99.

Rose, J. R., and Mountcastle, V. B. 1959. Touch and kinesthesis. In *Handbook of physiology: A critical, comprehensive presentation of physiological knowledge,* eds. J. Field, H. W. Magoun, and V. E. Hall, vol. 1, pp. 409–15. Washington, D. C.: Am. Physiol. Soc.

Rosenblatt, M., and Slepian, D. 1962. Nth order Markov chains with any set of N variables independent. *J. Soc. Indust. Appl. Math.* 10: 537–49.

Rubin, E. 1915. Visuell wahrgenommene Figuren. Copenhagen: Gyldenalske Boghandel. Reprinted as "Figure and ground" in *Readings in perception,* ed. D. C. Beardslee and M. Wertheimer. Princeton, N.J.: Van Nostrand 1958, pp. 194–203.

Rudel, R. G., and Teuber, H. L. 1963. Decrement of visual and haptic Müller-Lyer illusion on repeated trials: A study of crossmodal transfer. *Quart. J. Exp. Psychol.* 15: 125–31.

Sanderson, K. J.; Darian-Smith, I.; and Bishop, P. O. 1969. Binocular corresponding receptive fields of single units in the cat dorsal lateral geniculate nucleus. *Vision Res.* 9: 1297–1303.

Schelling, H. von. 1956. Concept of distance in affine geometry and its applications in theory of vision. *J. Opt. Soc. Am.* 46: 309–15.

Schiller, P., and Wiener, M. 1962. Binocular and stereoscopic viewing of geometric illusions. *Percept. Mot. Skills* 15: 739–47.

Schouten, J. F. 1940. The perception of pitch. *Philips Tech. Rev.* 5: 286–94.

Schumann, F. 1904. Einige Beobachtungen über die Zusammenfassung von Gesichtseindrucken zu Einheiten. *Psychol. Stud.* 1: 1–32.

Scott, T. R. 1964. On the interpretation of "interocular transfer." *Percept. Mot. Skills* 18: 455–56.

Scott, T. R., and Wood, D. Z. 1966. Retinal anoxia and the locus of the after-effect of motion. *Am. J. Psychol.* 79: 435–42.

Sekuler, R. W., and Pantle, A. 1967. A model for after-effects of seen movement. *Vision Res.* 7: 427–439.

Selfridge, O. G. 1958. Pandemonium: A paradigm for learning. In *Mechanization of thought processes.* Proc. Symp. at Natl. Phys. Lab., Nov. 1958 (HMSQ, London), pp. 511–35.

Severence, E., and Washburn, M. F. 1907. The loss of associative power in words after long fixation. *Am. J. Psychol.* 18: 182–86.

Shepard, R. N. 1962. The analysis of proximities: Multidimensional scaling with an unknown distance function. I and II. *Psychometrica* 27: 125–40, 219–46.

———. 1964. Book review of E. Feigenbaum and J. Feldman, eds., *Computers and thought.* New York: McGraw-Hill, 1963. In *Behavioral Sci.:* 57–65.

———. 1968. Psychological representation of speech sounds. In *Human communication, a unified view,* ed. E. E. David, Jr. and P. Denes. New York: McGraw-Hill, to be published.

Shepard, R. N., and Carroll, J. D. 1966. Parametric representation of nonlinear data structures. In *Multivariate analysis,* ed. P. R. Krishnaiah. New York: Academic Press, pp. 561–92.

Shepard, R. N., and Metzler, J. 1971. Mental rotation of three-dimensional objects. *Science* 171: 701–3.

Sherrington, C. S. 1898. Further note on the sensory nerves of the eye muscles. *Proc. Roy. Soc.* 64: 120.

——. 1906. *Integrative action of the nervous system.* New Haven: Yale Univ. Press.

Shipley, T. 1965. Visual contours in homogeneous space. *Science* 150: 348–50.

Shlaer, S.; Smith, E. L.; and Chase, A. M. 1942. Visual acuity and illumination in different spectral regions. *J. Gen. Physiol.* 25: 553–69.

Skavenski, A. A. 1970. Mechanisms underlying control of eye position in the dark. Ph.D. dissertation, Univ. of Maryland.

Skavenski, A. A., and Steinman, R. M. 1970. Control of eye position in the dark. *Vision Res.* 10: 193–203.

Sokal, R. R., and Sneath, P. H. A. 1963. *Principles of numerical taxonomy.* San Francisco: Freeman.

Sperling, G. 1960. The information available in brief visual presentations. *Psychol. Monogr.,* vol. 74, no. 11.

——. 1965. Visual spatial localization during object motion, apparent object motion, and image motion produced by eye movements. *J. Opt. Soc. Am.* 55: 1576.

——. 1970. Binocular Vision: A physical and neural theory. *J. Am. Psychol.* 83: 461–534.

Sperry, R. W. 1964. Brain bisection and mechanisms of consciousness. In *Brain and conscious experience,* ed. J. C. Eccles. New York: Springer, 1966, pp. 298–313.

Springbett, B. M. 1961. Some stereoscopic phenomena and their implications. *Brit. J. Psychol.* 52: 105–9.

Stasiak, E. A. 1965. Monocular stereoscopic presentation of disparate images. *Percept. Mot. Skills* 21: 371–74.

Sterling, P., and Wickelgren, B. G. 1969. Visual receptive fields in the superior colliculus of the cat. *J. Neurophysiol.* 32: 1–15.

Sternberg, S. 1967. Two operations in character recognition: Some evidence from reaction-time measurements. *Perception and Psychophysics* 2: 43–53.

——. 1969. Memory-scanning: Mental processes revealed by reaction-time experiments. *Am. Scientist* 57: 421–57.

Stratton, G. M. 1917. The mnemonic feat of the "Shass Pollak." *Psychol. Rev.* 24: 244–47.

Stromeyer, C. F., III. 1969. Further studies of the McCollough effect. *Perception and Psychophysics* 6: 105–10.

——. 1970. Eidetikers. *Psychol. Today* 4: 76–80.

——. 1971. McCollough analogs of two-color projections. *Vision Res.,* in press.

Stromeyer, C. F., III., and Mansfield, R. J. W. 1970. Colored after-effects produced with moving edges. *Perception and Psychophysics* 7 (2): 108–14.

Stromeyer, C. F., III., and Psotka, J. 1970. The detailed texture of eidetic images. *Nature* 225: 346–49.

Sutherland, N. S. 1964. The learning of discrimination by animals. *Endeavour* 23: 148–52.

Szara, S. 1957. The comparison of the psychotic effect of tryptamine derivatives with the effects of mescaline and LSD-25 in self-experiments. In *Psychotic Drugs,* ed. S. Garattini and V. Ghetti. (Ocilarco), pp. 460–67.

Tausch, R. 1954. Optische Täuschungen als artifizielle Effekte der Gestaltungsprozesse von Grössen- und Formenkonstanz in der natürlichen Raumwahrnemung. *Psychol. Forsch.* 24: 299–348.

ten Doesschate, G. 1955. Results of an investigation of depth perception at a distance of 50 metres. *Ophthalmologica* 129: 56–57.

Terwilliger, R. F. 1963. Evidence for a relationship between figural aftereffects and afterimages. *Am. J. Psychol.* 76: 306–10.

Teuber, H. L. 1962. Effects of brain wounds implicating right or left hemisphere in man: Hemispheric differences and hemisphere interactions in vision, audition, and somesthesis. In *Interhemispheric relations and cerebral dominance*, ed. V. B. Mountcastle. Baltimore: J. Hopkins Press, pp. 131–58.

Teuber, H. L.; Battersby, W.; and Bender, M. B. 1960. *Visual field defects after penetrating missile wounds of the brain*. Cambridge, Mass.: Harvard Univ. Press.

Thiery, A. 1896. Über geometrisch-optische Täuschungen. *Phil. Stud.* 12: 67–126.

Thomas, H. 1967. Multidimensional analysis of similarities judgments to twenty visual forms. *Psychonomic Bull.* 1(2): 3.

Thompson, R. F.; Mayers, K. S.; Robertson, R. T.; and Patterson, C. J. 1970. Number coding in association cortex of the cat. *Science* 168: 271–73.

Thompson, S. P. 1880. Optical illusions of motion. *Brain* 3: 289–98. Reprinted in W. N. Dember, *Visual perception: The nineteenth century*. New York: Wiley, 1964, pp. 84–94.

Titchener, E. B. 1902. *Experimental Psychology*. New York: Macmillan.

Treisman, A. 1964. Monitoring and storage of irrelevant messages in selective attention. *J. Verbal Learning and Verbal Behavior* 3: 449–59.

Walls, G. L. 1942. *The vertebrate eye and its adaptive radiation*. Cranbrook Inst. of Sci. Bull. No. 19, Cranbrook, Mich.

Ward, J. H., Jr. 1963. Hierarchical grouping to optimize an objective function. *J. Am. Statistical Assoc.* 58: 236–44.

Wenzel, E. L. 1926. Neue Untersuchungen über zöllnersche anorthoskopische Zerrbilder. *Z. F. Psych.* 100: 289–324.

Werblin, F. S., and Dowling, J. E. 1969. Organization of the retina of the mudpuppy, Necturus maculosus: II. Intracellular recording. *J. Neurophysiol.* 32: 339–54.

Wertheimer, M. 1912. Experimentelle Studien über das Sehen von Bewegung. *Z. Psychol.* 61: 161–265. Translated in large part in T. Shipley, ed., *Classics in psychology*. New York: Philosophical Library, 1961.

West, R. A. 1968. Recognition of periodicity in repeated random patterns: A basis for short-term memory. Ph.D. dissertation, Stanford University.

Wheatstone, C. 1838. On some remarkable, and hitherto unobserved, phenomena of binocular vision. *Philosophical Trans., Royal Soc.* (London) 128: 371–94. Reprinted in W. Dember, ed., *Visual perception: The nineteenth century*. New York: Wiley, 1964, pp. 371–77, 386–87.

White, B. W. 1962. Stimulus-conditions affecting a recently discovered stereoscopic effect. *Am. J. Psychol.* 75: 411–20.

Wickelgren, B. G., and Sterling, P. 1969. Influence of visual cortex on receptive fields in the superior colliculus of the cat. *J. Neurophysiol.* 32: 16–23.

Wiesel, T. N., and Hubel, D. H. 1965a. Comparison of the effects of unilateral and bilateral eye closure on cortical unit responses in kittens. *J. Neurophysiol.* 28: 1029–40.

―――. 1965b. Extent of recovery from the effects of visual deprivation in kittens. *J. Neurophysiol.* 28: 1060–72.

―――. 1966. Spatial and chromatic interaction in the lateral geniculate body of the rhesus monkey. *J. Neurophysiol.* 29: 1115–56.

Witasek, S. 1899. Über die Natur der geometrisch-optischen Täuschungen. *Z. Psychol. Physiol. Sinn.* 19: 81–174.

Wohlgemuth, A. 1911. On the aftereffect of seen movement. *Brit. J. Psychol. Monogr.* 1: 1–117.

Wohlwill, J. F. 1960. Developmental studies of perception. *Psychol. Bull.* 57: 249–88.

Woodworth, R. S. 1938. *Experimental Psychology*. New York: Holt.

Yarbus, A. L. 1957. A new method of studying the activity of various parts of the retina. *Biofizika* 2: 165–67.

Zöllner, F. 1862. Über eine neue Art anortoskopischer Zerrbilder. *Ann. Phys. Pogg. Ann.* 117: 477–84.

Zweig, H. 1956. Autocorrelation and granularity: 1. Theory. *J. Opt. Soc. Am.* 46: 805–11.

Zwicker, E. 1964. Negative afterimage in hearing. *J. Acoust. Soc. Am.* 36: 2413–15.